Network Design Basics

for Cabling Professionals

Network Design Basics
for Cabling Professionals

McGraw-Hill

New York Chicago San Francisco Lisbon
London Madrid Mexico City Milan
New Delhi San Juan Seoul
Singapore Sydney Toronto

McGraw-Hill

A Division of The ***McGraw-Hill*** *Companies*

1 2 3 4 5 6 7 8 9 0 DOC/DOC 0 7 6 5 4 3 2

ISBN 0-07-139916-X

The sponsoring editor for this book was Marjorie Spencer and the production supervisor was Sherri Souffrance. It was set in New Century Schoolbook by Patricia Wallenburg.

Printed and bound by R.R. Donnelley & Sons Company.

This book was printed on recycled, acid-free paper containing a minimum of 50% recycled, de-inked fiber.

Contents

Preface

Acknowledgments

The BICSI officers and membership wish to thank the following members of the BICSI Technical Information and Methods Committee Panel 100 who contributed to the development of the *Network Design Reference Manual*, 4th edition, and provided important feedback through their reviews of this manual.

Chair:

Peter Olders, RCDD/LAN/OSP Specialist, *Terra Communications, Inc.*

Committee:

Sharon Ballas, *BICSI*
Cory Boon, RCDD/LAN Specialist, *techformatique, inc.*
Ray Craig, RCDD/LAN Specialist, *ComNet Communications, Inc. (CCI)*
Lamar Davis, RCDD/LAN Specialist
R.S. (Bob) Erickson, RCDD/LAN/OSP Specialist, *Jber Enterprises, LLC.*
Robert Faber, RCDD/LAN Specialist, *The Siemon Company*
Joan Hersh, *BICSI*
Nelda Hills, *BICSI*
Steve Kepekci, RCDD/LAN Specialist, *techformatique, inc.*
Dave Labuskes, RCDD/LAN Specialist, *RTKL Associates, Inc.*
Thomas Leis, PE, RCDD/LAN Specialist, *CTEEC*
Russ Oliver, RCDD/LAN Specialist, *CTC Communications*
Ron Shaver, RCDD/LAN/OSP Specialist, *BICSI*
Andrew Young, RCDD/LAN/OSP Specialist, *DESA Australia Pty Ltd.*

About BICSI

BICSI, a not-for-profit telecommunications association, was founded in 1974 to serve and support the professionals responsible for the design and distribution of telecommunications wiring in commercial and multi-family buildings.

BICSI has grown dramatically since those early days and is now recognized worldwide as an educational resource for the cabling infrastructure industry. Our membership spans the globe and our services cover the broad spectrum of voice, data, and video technologies. BICSI offers training, conferences, publications, and registration programs for cabling distribution designers, as well as commercial, and most recently, residential installers.

BICSI member benefits and opportunities

BICSI members receive substantial discounts on quality education—design courses, conferences, and manuals. Members also gain access to valuable telecommunications information with the *BICSI News, Region News, District News*, standards and regulatory updates, and BICSI's Web site including a Members Only section.

Membership offers ample opportunities for professional networking, and career development and advancement.

Members may pursue and obtain prestigious credentials—RCDD, RCDD/LAN Specialist, RCDD/OSP Specialist, Registered Commercial Installer and Technician, and Registered Residential Installer.

BICSI Network Design Courses, Publications, and RCDD/LAN Specialty Program

Today's high-performance communications cabling has been developed to enable connections between current and emerging high-speed networking devices. Therefore, anyone involved in the sales, purchasing, design, installation, or maintenance of connectivity solutions can benefit from an in-depth understanding of networking technologies.

As one of the largest professional telecommunications associations in the world, BICSI can provide you with the resources you need to advance in the area of network design. Our many data distribution design courses and publications serve as an excellent introduction or a detailed review of all the latest in network design. They also serve as a challenging career path for those pursuing the prestigious RCDD/LAN Specialty designation, recognized in many parts of the world.

BICSI's RCDD/LAN Specialty

BICSI's RCDD/LAN Specialty is available to current BICSI members who possess the Registered Communications Distribution Designer (RCDD®) designation. Industry wide, RCDDs are recognized as having superior design expertise. With that foundation, those achieving the LAN Specialty designation possess outstanding network design qualities that give them significant competitive recognition.

The RCDD/LAN Specialty registration program enables recipients to gain a broad, vendor-neutral view of networking. While other industry certifications focus on the mastery of specific hardware or software products, the LAN Specialty emphasizes the networking technologies relevant to all organizations, such as:

- Local area networking.
- Remote access.
- Internetworking.
- Cabling and wireless connectivity.
- Storage.
- Security.
- Management.
- Internet protocol.
- Network services and applications.

To summarize, BICSI's LAN Specialty program provides the tools needed to understand an organization's network infrastructure, thereby complementing an RCDD's knowledge of premises connectivity. The recipient of an RCDD/LAN Specialist designation can be counted on to provide expert guidance on suitable connectivity solutions, both for an organization's physical facilities as well as the networking environments within.

BICSI's data distribution design courses

BICSI has developed a comprehensive curriculum of network design courses, ranging from an introductory class, to a review course for the RCDD/LAN Specialty exam, to a LAN and internetworking update course, and more. These classes are ideal for those wanting to remain current in LAN and internetworking and actually "design" a network, as well as for those studying for the RCDD/LAN Specialty exam.

DA100: Introduction to LANs and Internetworks enables students to enhance their professional status through the study and understanding of LANs and internetworks. The fundamentals of data networking are explained using a practical, real-world approach. This three-day course is recommended for anyone requiring an overview of data networks and their influence on cabling system designs.

DA110: Designing LANs and Internetworks provides an in-depth understanding of the design rules, configuration options, and media selection criteria for today's most popular LAN and internetworking technologies. Students are taught networking design skills from a real-world perspective, with significant class time devoted to actual network design and its impact on the cabling infrastructure.

DA200: LAN Specialty Review features a fast-paced review of the BICSI *Network Design Reference Manual*, 4th edition, highlighting the critical areas students should know for the RCDD/LAN Specialty exam. Material reviewed includes cabling systems, Ethernet and token ring designs, internetworking, bridges and routers, and high-speed campus backbones. Students focus on design and installation strategies, the functions, components, and topologies of LANs, and assessment of LAN performance.

DA300: LAN and Internetworking Technologies Update offers an insightful look into present and future trends in LAN and internetworking technologies, as well as how these technologies affect the marketplace. Significant innovations and trends in LAN cabling infrastructure are explored, as well as the developing wireless market, ever-changing cabling standards and specifications, and the future of remote access to networks. Candidates for renewal of the RCDD/LAN Specialty designation, as well as telecommunications LAN designers should attend this course.

BICSI's network design publications

To adapt to the increasing demand for professionals qualified in LAN and internetworking technologies, BICSI developed several publications specific to this audience—the *Network Design Reference Manual* and the *LAN and Internetworking Applications Guide*.

The *Network Design Reference Manual (NDRM)*, 4th edition, 2001, brings readers the latest in modern network technology, where the Internet makes it possible to connect any device to any network, regardless of distance. Indexed and illustrated with more than 360 figures, tables, and examples, the two-volume manual contains approximately 900 pages. Complete with a glossary and bibliography, the NDRM's 15 chapters and three appendices describe all aspects of networking, including fundamentals and technologies.

In 2002, BICSI partnered with McGraw-Hill Professional to publish two reference books—*Network Design Basics* and *Networking Technologies*. *Network Design Basics* contains information from Volume 1 of the *NDRM*, while *Networking Technologies* contains information from Volume 2 of the *NDRM*.

The BICSI *LAN and Internetworking Applications Guide* provides a comprehensive overview and design guidelines for the integration of emerging applications into existing LANs. Topics cover multimedia, Groupware, and access applications, as well as advances in LAN and internetworking technologies. Each chapter uses a case format to evaluate and reconfigure an existing network and cabling system to accommodate the proposed application.

For Further Information and to Obtain Errata

For a complete packet of information on BICSI's wide range of products and services—including BICSI's network design courses, publications, and RCDD/LAN Specialty program—please contact:

BICSI World Headquarters
8610 Hidden River Parkway
Tampa, FL 33637-1000 USA
813-979-1991 or
800-242-7405 (USA/Canada toll free)
fax: 813-971-4311
e-mail: bicsi@bicsi.org
Web site: www.bicsi.org

BICSI strives to provide up-to-date and accurate information. If errors are found in this manual, corrections will be posted on the BICSI Web site. Visit www.bicsi.org and select Publications. Comments about the manual, including possible errors, may be e-mailed to publications@bicsi.org.

BICSI Policy for Numeric Representation of Units of Measure

International System of Units (SI)

BICSI technical manuals primarily follow the modern metric system, known as the International System of Units (SI). The SI is intended as a basis for worldwide standardization of measurement units. All units of measure in this manual are expressed in SI terms, followed by an equivalent empirical (U.S. customary) unit of measure in parentheses (see exceptions listed below).

Style guidelines

- In general, SI units of measure are converted to an empirical unit of measure and placed in parentheses. Exception: When the reference material from which the value is pulled is provided in empirical units only, the empirical unit is the benchmark.
- In general, soft (approximate) conversions are used in this manual. Soft conversions are considered reasonable and practicable; they are not precise equivalents. In some instances, precise equivalents (hard conversions) may be used when it is a:
  - Manufacturer requirement for a product.
  - Standard or code requirement.
  - Safety factor.
- For metric conversion practices, refer to ANSI/IEEE/ASTM SI 10-1997, *Standard for Use of the International System of Units (SI): The Modern Metric System.*
- Trade size is approximated for both SI and empirical purposes. Example: 103 mm (4 trade size).
- American wire gauge (AWG) and plywood are not assigned dual designation SI units. Dimensions shown in association with AWGs represent the equivalent solid conductor diameter. When used in association with flexible wires, AWG is used to represent stranded constructions whose cross-sectional area (circular mils) is approximately equivalent to the solid wire dimensions provided.
- In some instances (e.g., optical fiber media specifications), the physical dimensions and operating wavelengths are designated.
- When Celsius temperatures are used, an equivalent Fahrenheit temperature is placed in parentheses.

Chapter

1

Network Design Overview

Chapter 1 presents an introduction to personal computer-based networking, with an emphasis on office environments. This chapter is also a guide to understanding the organization of this manual.

Overview

General

This chapter presents an introduction to the networks that enable the exchange of data between users equipped with various types of personal computers (PCs). Office environments are emphasized, although it is possible to implement the same types of networks in other settings (e.g., homes or classrooms). This chapter can also be used as a guide to understanding the organization of this manual, which describes the concepts and types of products commonly considered when designing a network. The manual is structured as follows:

- Chapter 1: Network Design Overview
- Chapter 2: Networking Fundamentals
- Chapter 3: Network Components
- Chapter 4: Network Connections
- Chapter 5: Network Communications
- Chapter 6: Switching and Virtual LANs (VLANs)
- Chapter 7: Network Services and Applications
- Chapter 8: Network Security

- Chapter 9: Network Management
- Appendix A: Binary and Hexadecimal Numbering Fundamentals and Appendix B: Internet Protocol (IP) Addressing Fundamentals describe commonly used numbering and addressing formats for networks and devices.
- Appendix C: Network Design Examples contains examples of the network design process for various types of networks

Introduction to Network Design

General

A variety of networks are implemented in most organizations to provide access to available resources from multiple locations. Individuals responsible for network design can be expected to address requirements such as the following:

- A network is needed in an office to link existing stand-alone desktop and laptop PCs. Some of the laptops will also connect to the network from remote locations when personnel are at home or traveling.
- An existing office network must be linked to three other networks.
  - The first is on another floor in the same building
  - The second is in a nearby building on the same campus
  - The third is in a branch office in another city
- An existing or planned network must be made more secure and easier to manage.
- An existing network is upgraded to a newer or different technology or network operating system (NOS).
- New services must be added to an existing network. Under consideration are the following:
  - Access to the Internet
  - An organizational Web site with electronic commerce capabilities
  - Videoconferencing and multimedia training
  - Collaborative publishing and document management

The term network design is used in this manual to describe the planning that precedes the implementation of the following types of networks:

- Local area networks (LANs)
- Metropolitan area networks (MANs)
- Remote access
- Linked LANs

- Backbone networks
- Wide area networks (WANs)

Local Area Networks (LANs)

A LAN interconnects two or more PCs in a common location, such as an office (see Figure 1.1).

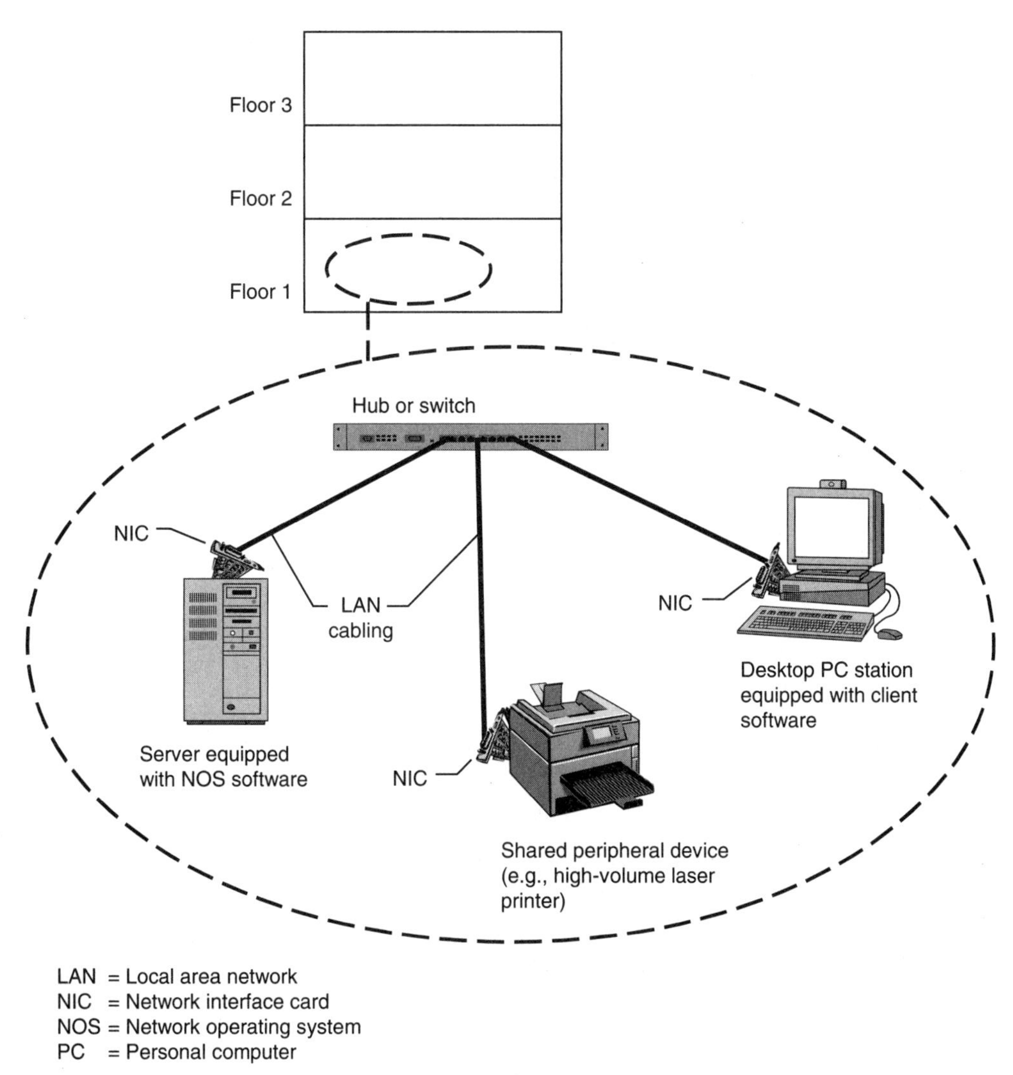

Figure 1.1 Example of a local area network.

NOTE: LAN and other network components are described in detail in Chapter 3: Network Components.

Remote access

Remote access connects a stand-alone PC to a LAN over an extended distance through any of the following:

- Direct connections using dial-up or leased telecommunications circuits (see Figure 1.2)

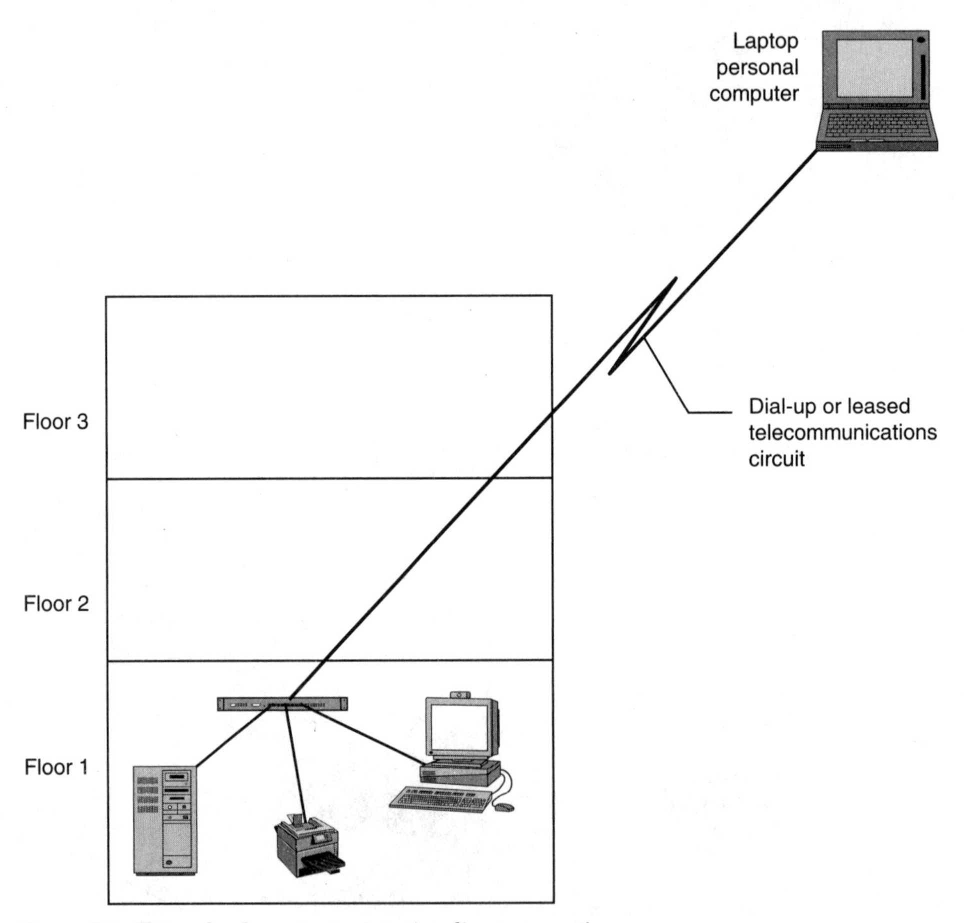

Figure 1.2 Example of remote access using direct connections.

- Connections to a service provider's (SP's) private network (see Figure 1.3)
- Connections to the Internet (see Figure 1.3)

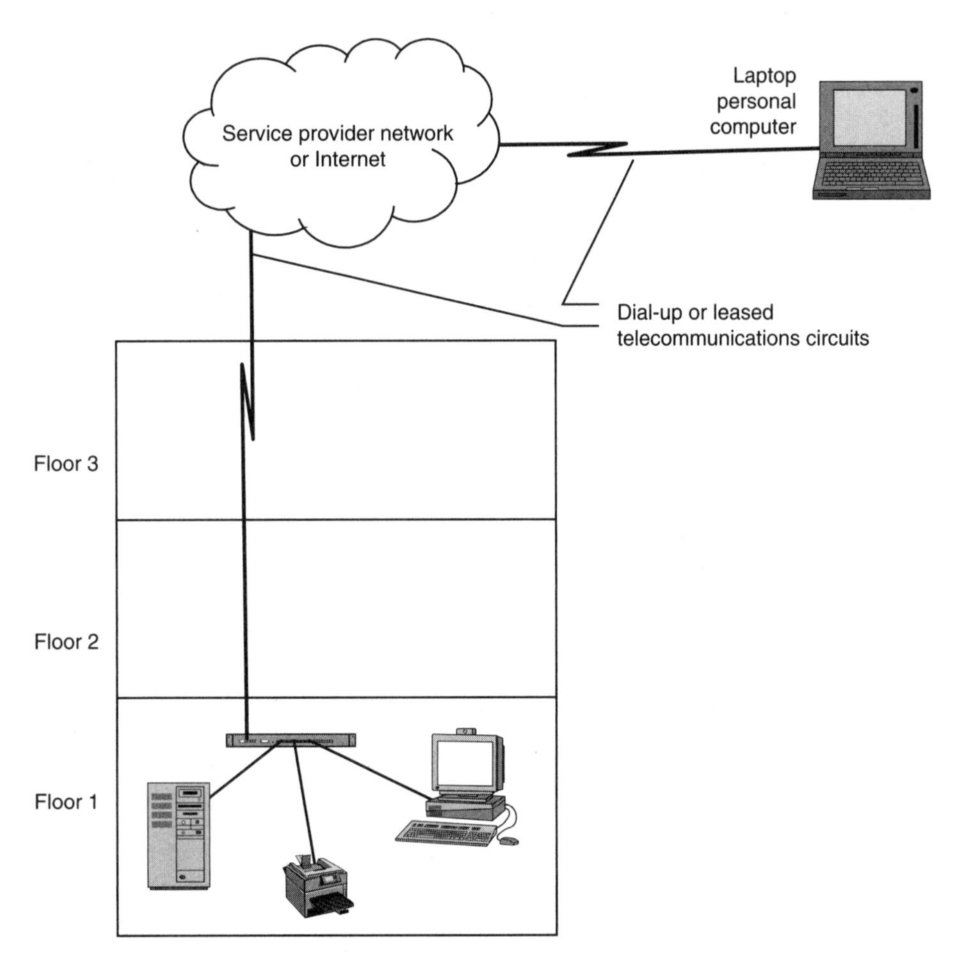

Figure 1.3 Example of remote access using service provider or Internet connections.

NOTE: In this context, extended distance refers to a separation that does not allow for the installation of cabling or wireless equipment between the PC and the LAN. Examples of extended distances are those spanning a city, a region, a country, or multiple countries.

Linked Local Area Networks (LANs)

A linked LAN consists of two interconnected LANs in the same building (see Figure 1.4) or in different buildings on the same campus (see Figure 1.5).

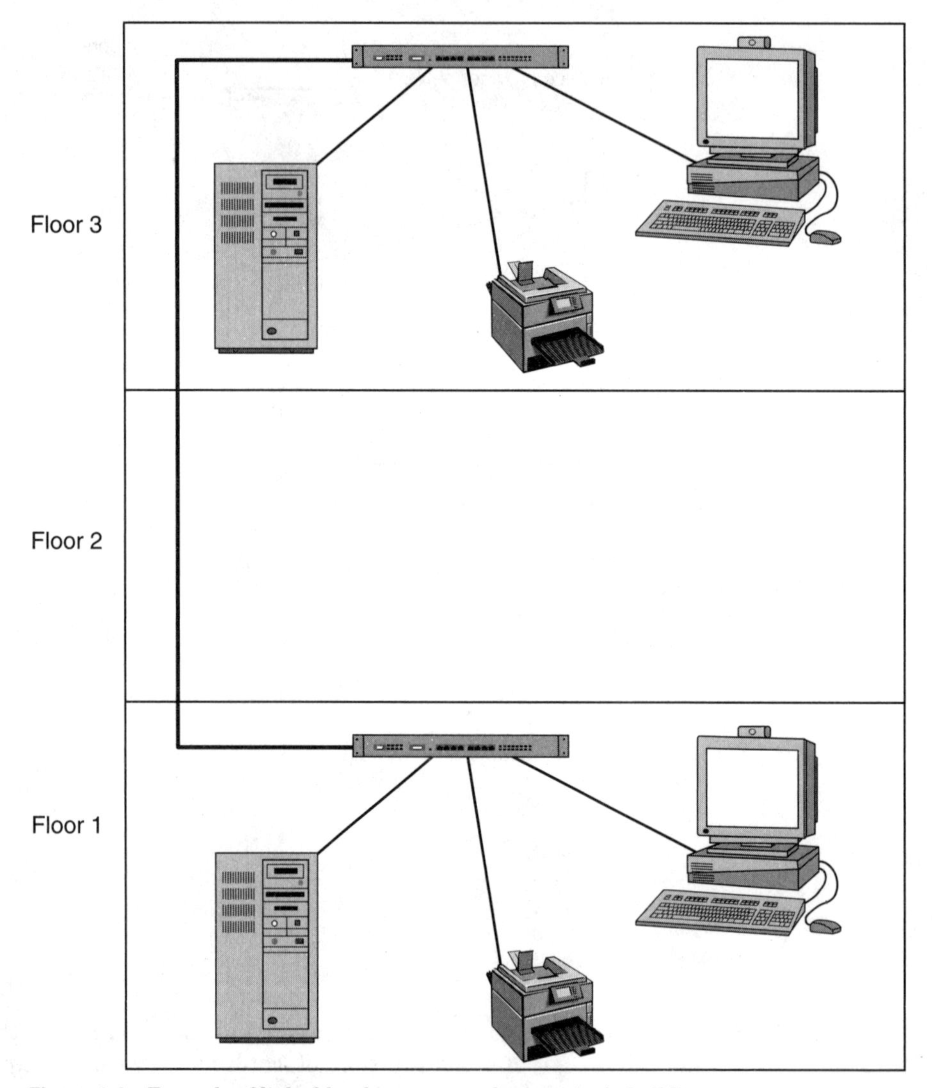

Figure 1.4 Example of linked local area networks in a single building.

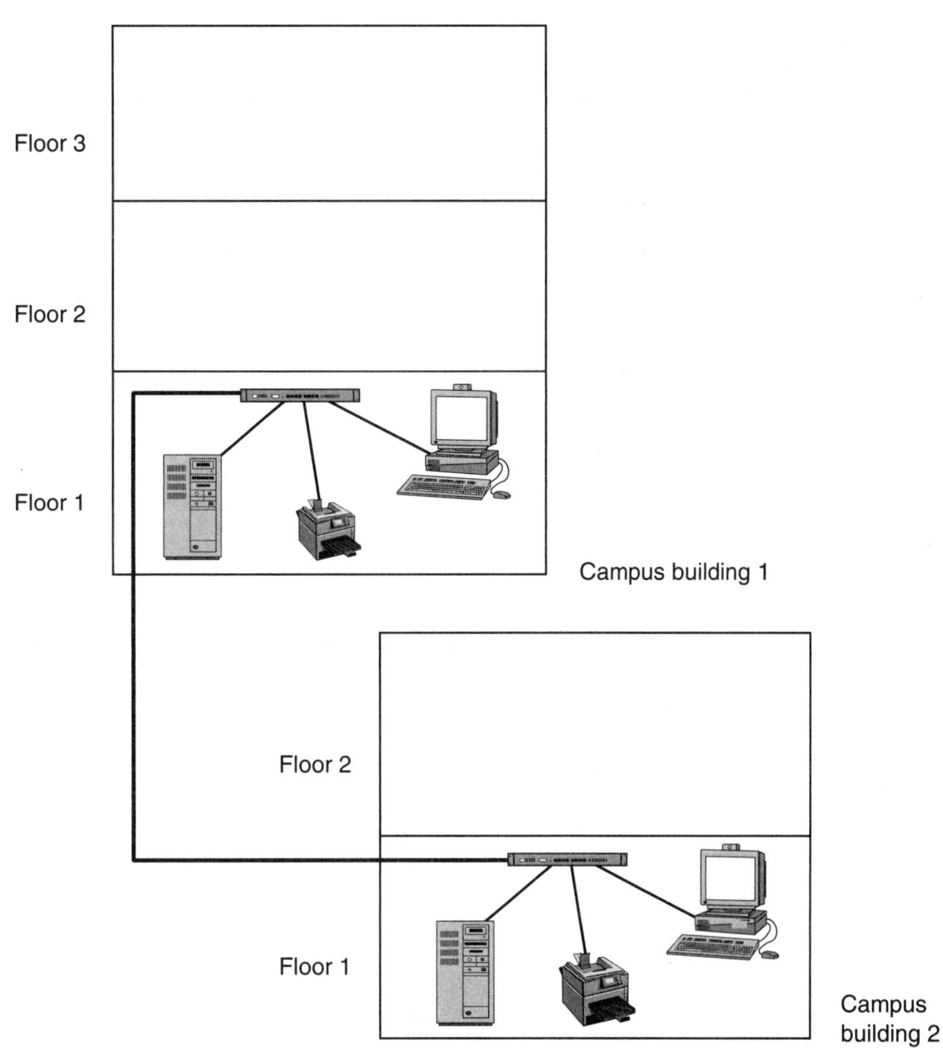

Figure 1.5 Example of linked local area networks on a campus.

Backbone network

A backbone network connects multiple LANs in the same building (see Figure 1.6)—or in a group of buildings on the same campus (see Figure 1.7)—to a single network called the backbone. The backbone network makes it possible for each group or department to have its own LAN, while allowing data to be exchanged between LANs when necessary.

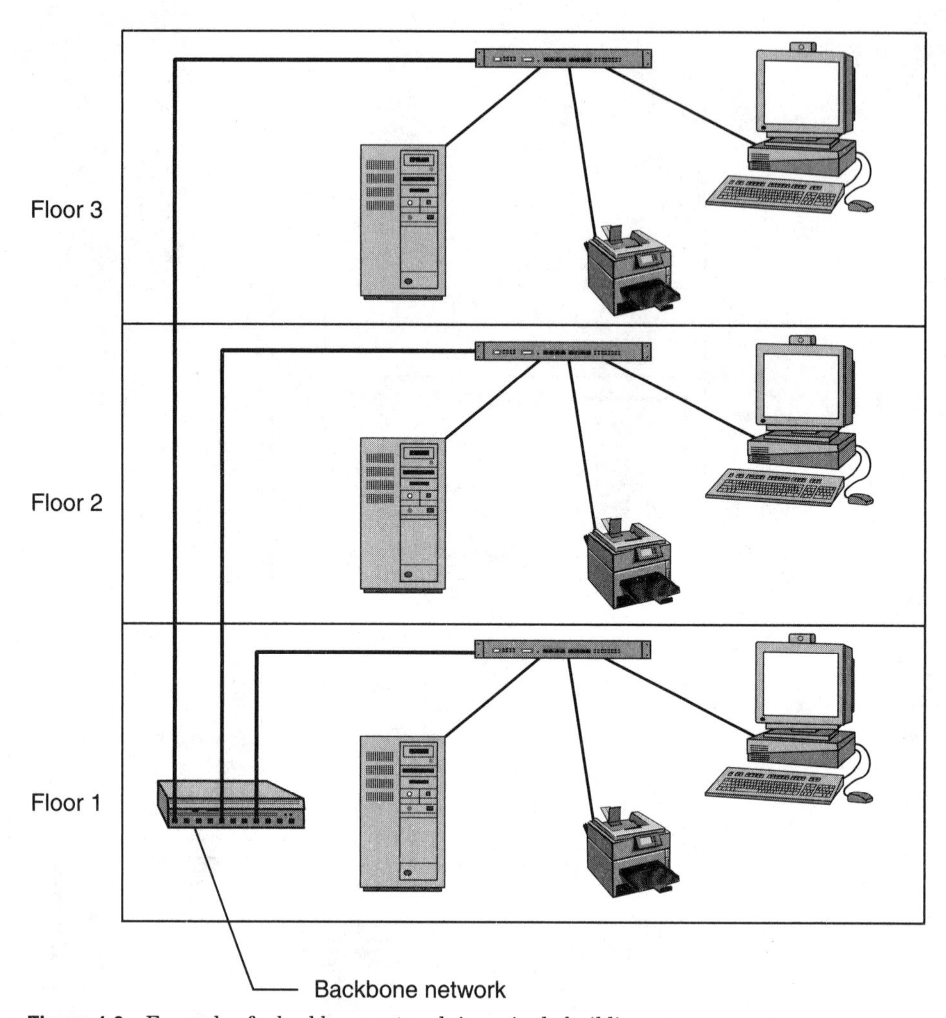

Figure 1.6 Example of a backbone network in a single building.

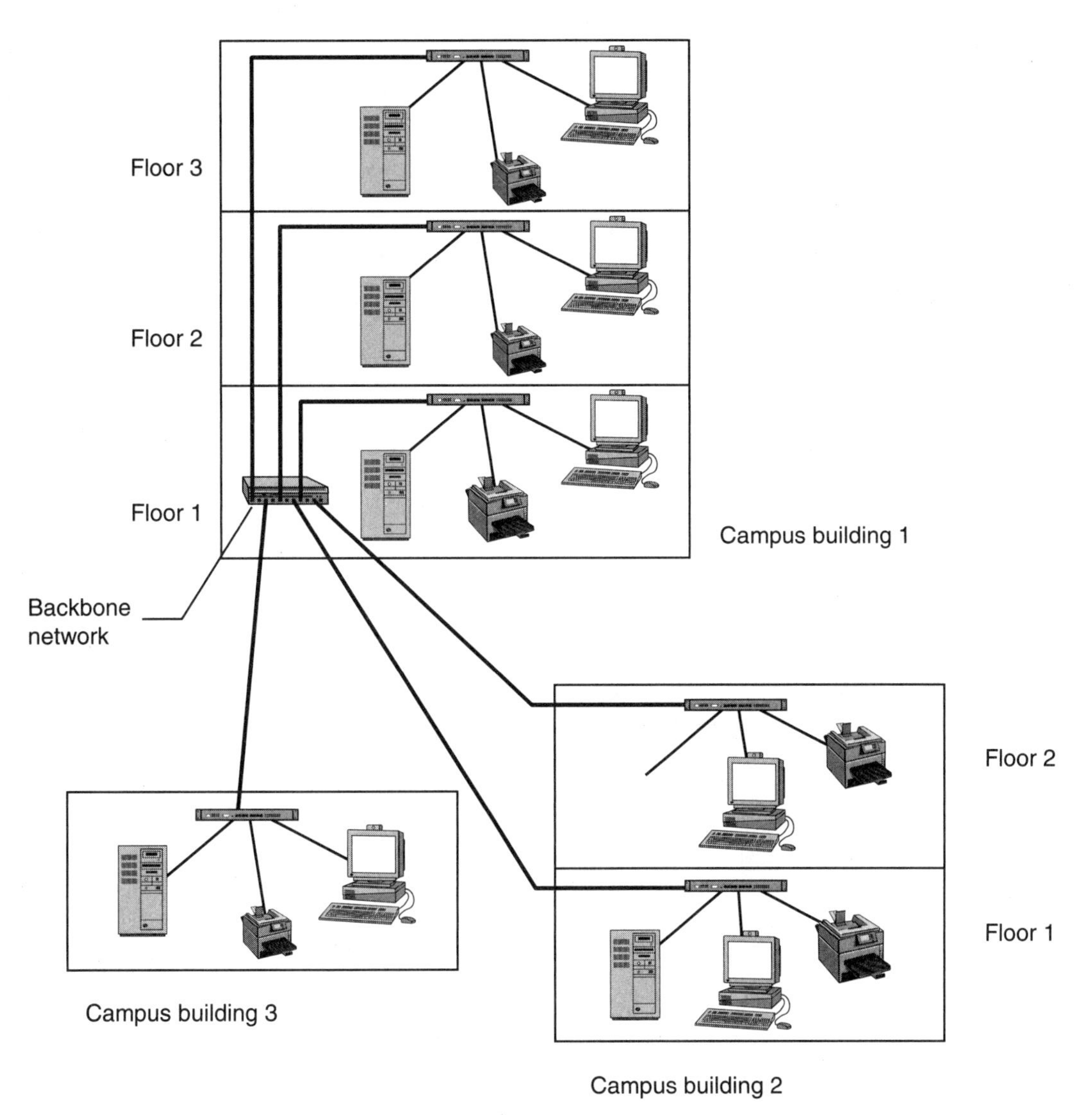

Figure 1.7 Example of a campus backbone network.

Wide Area Network (WAN)

A WAN links two or more LANs over an extended distance through any of the following:

- Direct connections between sites, using dial-up or leased telecommunications circuits (see Figure 1.8)
- Connections to an SP's private network (see Figure 1.9)
- Connections to the Internet (see Figure 1.9)

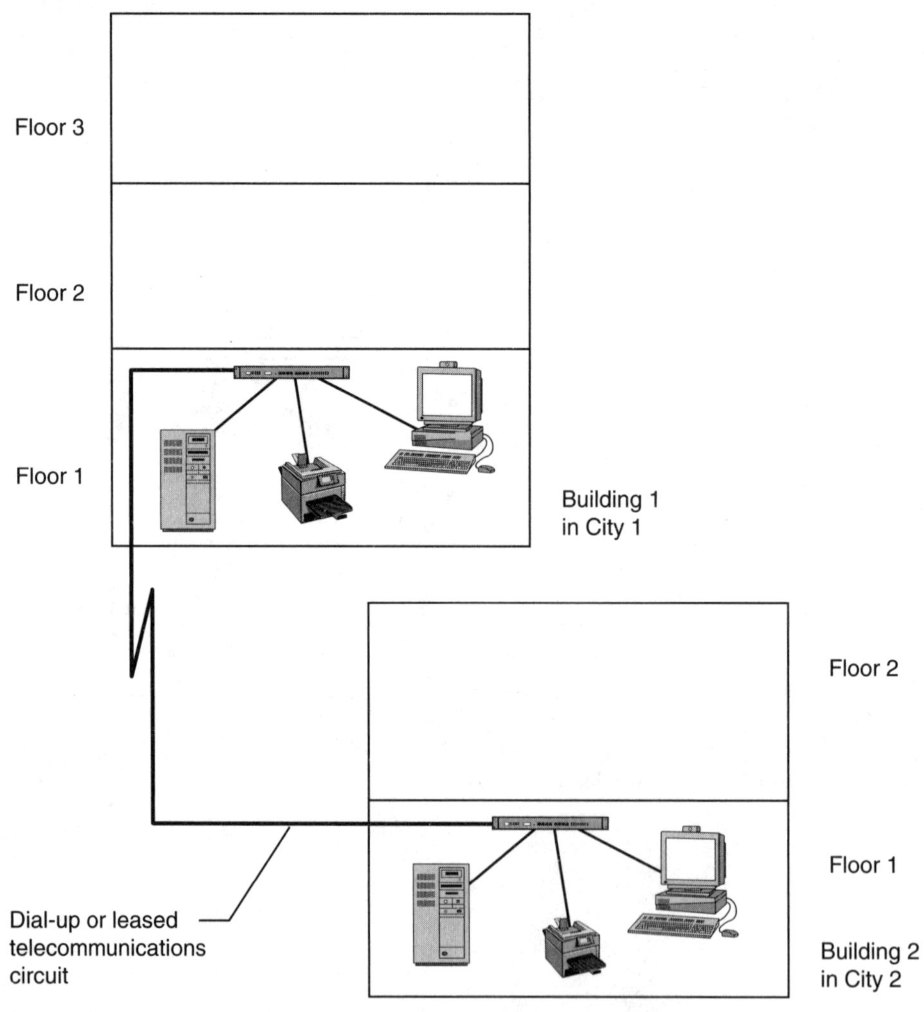

Figure 1.8 Example of a wide area network using direct connections.

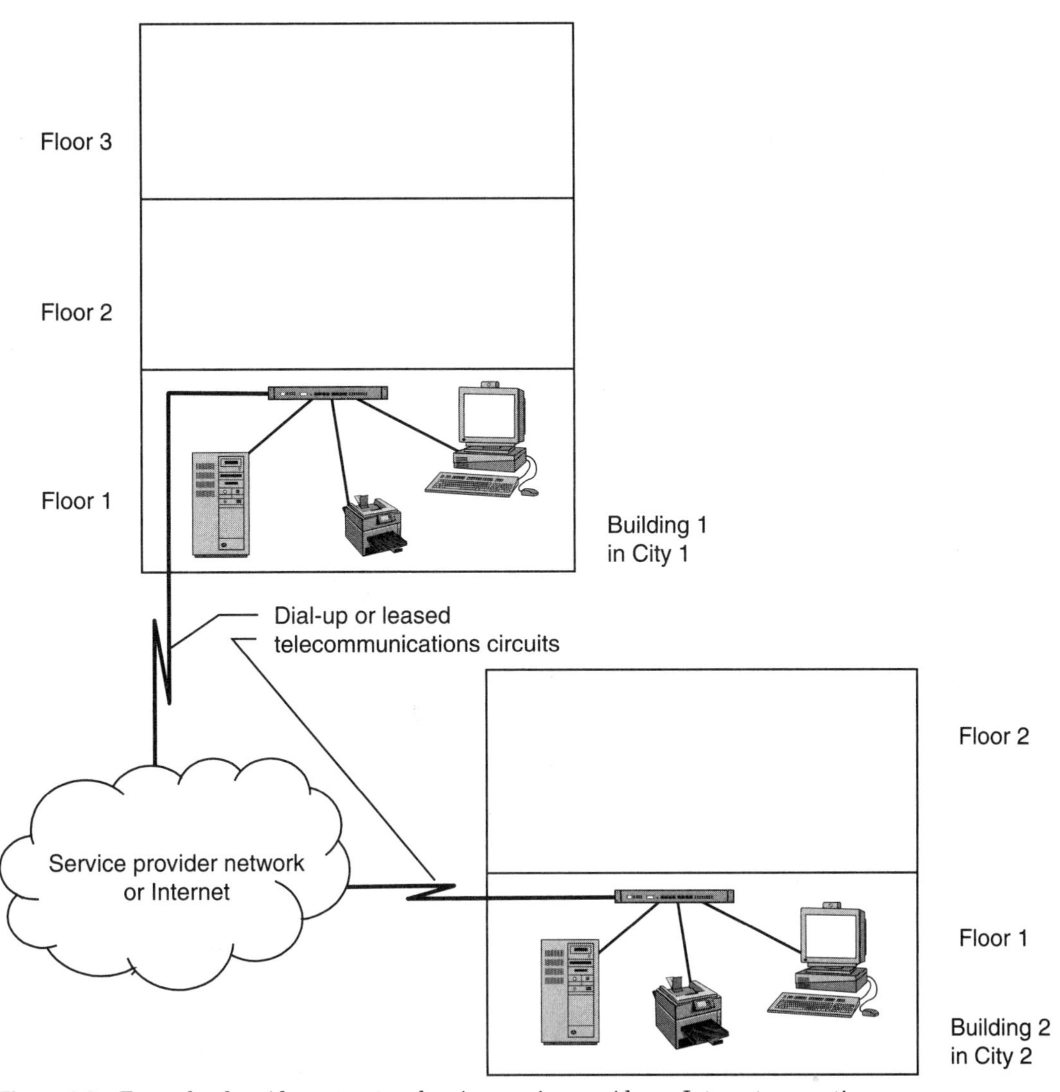

Figure 1.9 Example of a wide area network using service provider or Internet connections.

NOTE: In this context, extended distance refers to a separation that does not allow for the installation of cabling or wireless equipment between the sites to be linked. Examples of extended distances are those spanning a city, a region, a country, or multiple countries.

Design Process

Overview

A single network can have many types of designs, each focusing on a specific characteristic of the network. For example, one of the designs can detail network security, while another can detail the location or the identification of each network device. In most cases, multiple characteristics of the network must be considered, making it necessary to use both functional and physical design processes.

Functional design process

Functional design is also referred to as top-down design. In this process, the network designer begins with an assessment of the types of PCs and software applications to be supported by the proposed network.

Other factors, including the type of NOS to use and the types of files to share, are also evaluated. Based on the information gathered, the designer can generate preliminary estimates for network storage requirements, the expected level of network traffic, and the types of security and management tools needed. This information can then be used to select the appropriate technologies and products (see Figure 1.10).

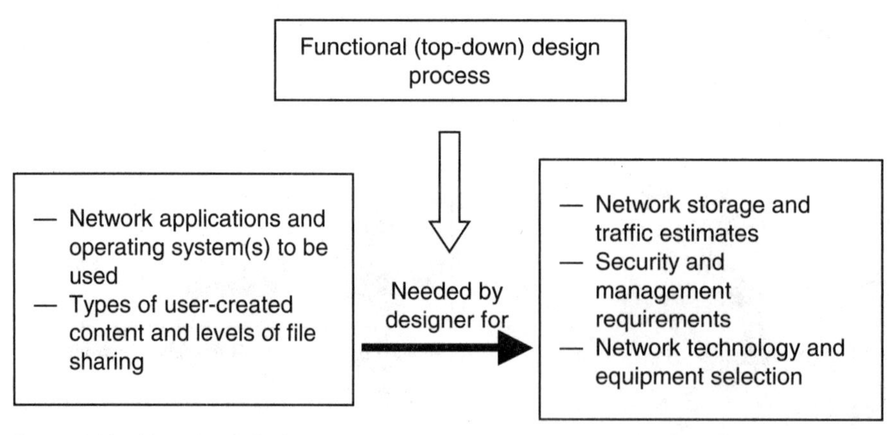

Figure 1.10 Functional design process.

Physical design process

Physical design is also referred to as bottom-up design. In this process, the network designer begins with an assessment of the site where the proposed network is to be implemented.

Included are such items as the expected location of each network device, available pathways and spaces, and distance measurements. Based on the information gathered, a network technology suitable for the site can be selected (see Figure 1.11).

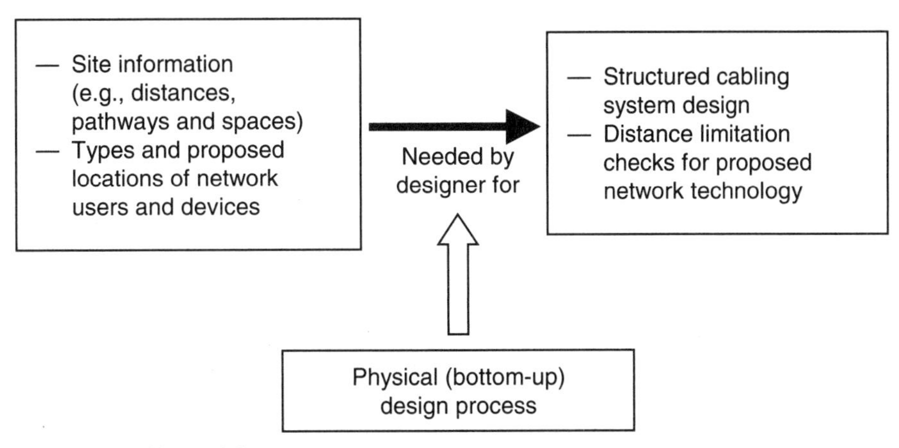

Figure 1.11 Physical design process.

For both functional and physical network design, the process requires at least an awareness—and preferably, detailed knowledge—of multiple networking and structured cabling technologies, since the final design is typically used to identify and acquire the necessary hardware, software, and cabling system. The following are examples of the type of knowledge needed when considering a specific networking technology or structured cabling system:

- Applicable standards
- Available products
- Compatibility of each product with other proposed network components

Chapter

2

Networking Fundamentals

Chapter 2 introduces the types of networks discussed in this manual. The design and operational characteristics of each type of network is presented, followed by a description of the model used to define network architectures.

Overview

General

This chapter introduces the types of networks described in this manual. The design and operational characteristics of each type are presented, followed by a description of the architecture and standards associated with most organizational networks. Since networks vary in size and scope, the following list can be used to classify networks based on design complexity:

- Local area network (LAN)—All personal computers (PCs) are connected to the same network using a structured cabling system (SCS). This is the simplest form of networking, since all users, devices, and applications are grouped together in the same location.
- Remote access—One or more PCs in other locations can connect to a LAN using telecommunications circuits. This is a more complex form of networking. Two types of connections, local and remote, must be incorporated into the design of the LAN.
- Internetwork—Two or more LANs are linked using a SCS, telecommunications circuits, or both. This is the most complex form of networking. It involves the design of multiple LANs, each with or without remote access, and the design of the internetwork that will connect the LANs.

Since many organizations provide some or all users access to the Internet, it is common to include connections to the Internet for all types of networks. This is sometimes described as outbound Internet access because the request for Internet-based data originates from within the organizational network. The term inbound Internet access can similarly be used to describe instances where the organizational network provides data in response to Internet-based requests.

Many variables influence a network's design, including the number of users and the types of applications. In turn, the design of the network can influence the design of a cabling system for a given location. The following are examples of the types of variables that influence, or are influenced by, the design of a network under consideration for a given site.

- User requirements influence the selection of network software applications.
- Software applications influence the selection of PCs to be provided to users.
- The capabilities of the PCs influence the selected design of the organizational network.
- The network design influences the selection of the cabling system (and the telecommunications circuits, if required).
- The cabling system influences the selection of pathways and spaces at the site where the network is to be implemented.

Introduction

General

Networks are implemented to enable the sharing of resources and the exchange of information between users. As the number of resources and users increases, the network design must periodically be reviewed. The outcome of this review can result in modifications or an expansion of the network, making network design an ongoing process within most organizations.

An organizational network can be defined as an interconnected system of computers, peripheral devices, and software. The network makes it possible to transfer all types of messages between users and devices. At a physical level, a network can be identified by its connected devices and the medium used to connect them—a cabling system, a wireless system, or telecommunications circuits.

In addition to the medium, a network provides other services necessary for communications between connected devices, including:

- Access control—Needed in cases where two or more devices attempt to use a shared resource (e.g., a common communications channel) at the same time.
- Synchronization—Ensures that a receiving device is listening when a sending device is transmitting.

- Flow control—Adjusts the rate at which data is transferred from sender to receiver (e.g., a receiving device can be occupied with other tasks, in which case it will alert the sending device to transmit at a slower rate).
- Error control—Verifies that a message was successfully transferred (and in cases where it was not, to request a retransmission).

History

Although this manual uses the LAN as the starting point for organizational networking, note that a LAN is only one type of data network. The earliest form of networking for the purposes of message exchange was the telephone network, which continues to be an essential organizational resource for both voice and data traffic.

Data networks designed exclusively for computing environments followed the introduction of business computing in the 1950s. Prior to that time, computers were used mostly for research and national defense purposes. Milestones in the history of computer networking in the United States include the following:

- 1950s—Geographically dispersed university computer terminals are connected for defense-related research.
- 1960s—The first large-scale commercial computer network is created for an airline reservation application. The Advanced Research Projects Agency Network (ARPANET) links computers from different manufacturers, forming what is later described as the origin of the Internet.
- 1970s—The specifications for a networking technology called Ethernet are published.
- 1980s—Large numbers of desktop and laptop PCs create a demand for PC-based LANs in many organizations.
- 1990s—The World Wide Web (WWW) service on the Internet and the PC-based WWW browser are introduced.

The advances in computing and manufacturing currently make it possible for organizations to provide each user with one or more personal computing devices. This is in contrast to earlier times, when all users had to share the processing and storage capabilities of a single system. Therefore, the evolution in computing can be described as a migration from centralized to decentralized computing, with networks used to combine the capabilities of multiple dispersed systems.

Mainframe computers. The first computers developed for organizational use were mainframes, available since the 1950s. A mainframe is characterized by its centralized approach to processing and storage—all computing operations take place in a central unit. Hundreds of users can simultaneously issue commands or enter data using devices called terminals, which are a combination of a monitor and a keyboard.

At the time of their introduction, mainframes were considered affordable only by large organizations. In most cases, one mainframe was acquired and placed in the main office to serve the needs of the entire organization (see Figure 2.1). This made it necessary to develop a way for both local users as well as staff in various offices throughout the country to access the mainframe. The solution was to use cabling to link local terminals and telecommunications circuits to link remote terminals—the original forms of local and wide area networking.

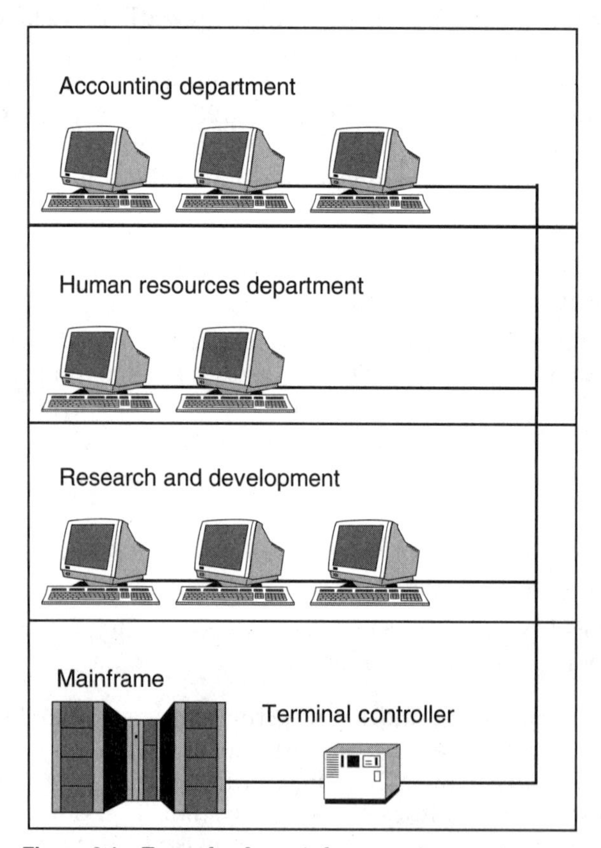

Figure 2.1 Example of a mainframe environment.

NOTE: The user devices shown in Figure 2.1 are terminals, not PCs.

Minicomputers. In the 1970s, the advances in computing made it possible to develop smaller versions of mainframes. Called minicomputers, these devices functioned in the same manner as mainframes, but on a smaller scale. The lower cost of minicomputers made it possible for small businesses, branch offices of large organizations, and individual departments to acquire a computer system (see Figure 2.2).

From a networking perspective, minicomputers operate like mainframes. Terminals are connected to a central processing and storage unit using cabling or telecommunications circuits. Compared to mainframes, fewer simultaneous terminal connections (also called sessions) are permitted, due to the lower capabilities of minicomputers.

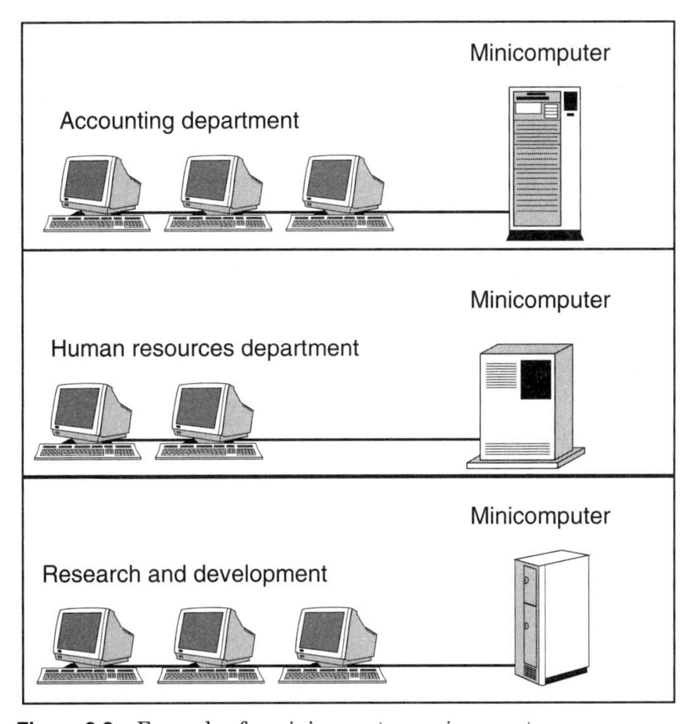

Figure 2.2 Example of a minicomputer environment.

NOTES: The user devices shown in Figure 2.2 are terminals, not PCs.

Although this manual focuses on PC-based networks, note that mainframes and minicomputers can also be connected to the LANs and internetworks described in this manual.

Personal Computers (PCs). Due to their low cost and the availability of all types of software, PCs have been acquired in large volumes by organizations since the 1980s. The advances in computing and manufacturing have made it possible to incorporate the functionality of a desktop computer into many types of devices, such as laptops, palmtops, and cellular telephones with computing capabilities (see Figure 2.3).

When first introduced, PCs provided a great deal of flexibility to users accustomed to negotiating with data processing departments for computing services from the organizational mainframe or minicomputer. The PC made it possible for any user or group to choose, purchase, install, and operate a computer without the assistance of programmers or other computing specialists. With the introduction of terminal emulation software, it became possible to replace terminals with PCs. When connected to the mainframe or minicomputer, terminal emulation software enables the PC to mimic the characteristics of any one of the many models of terminals that have been introduced over the years.

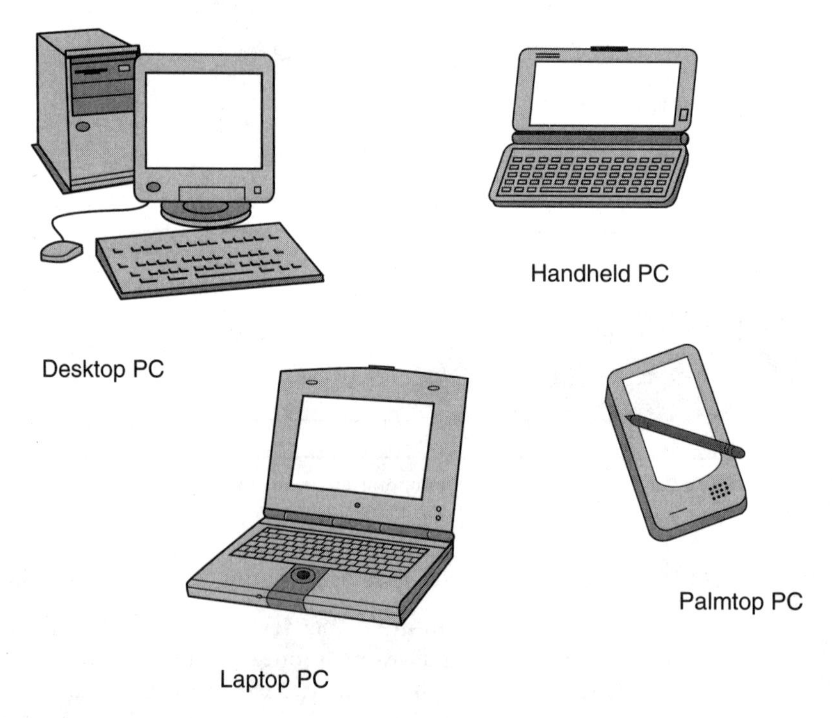

Figure 2.3 Examples of personal computers.

Local Area Network (LAN) Fundamentals

General

Local area networking for PCs was introduced in the early 1980s as a solution to the problem of managing the rapid growth in the number of PCs found within most organizations. Stand-alone PCs can cause a variety of problems in environments with many users. The following are some examples of such problems.

- Each PC stores its own data, unlike mainframes and minicomputers, both of which use centralized storage. As a result, faulty reporting and decision-making can occur if data is updated on some PCs, but not on others.
- Security is a major concern, since every PC can potentially contain valuable data, making it easier for unauthorized individuals to gain access to organizational information.
- Backups may not exist for critical data if users avoid or forget duplicating the contents of their PCs on a regular basis. The failure of a single component can result in a large loss of time, resources, and money.
- Various users or groups can install and work with different software applications on their PCs, making file sharing difficult within the organization.
- It is more difficult to justify the purchase of expensive products or services, since only one PC at a time can benefit. For example, a color laser printer or a dedicated high-speed Internet connection may not be economically feasible for every user in an organization with 50 PCs.

Implementing a LAN makes it possible to manage such concerns in the following ways:

- LANs make it possible to centralize data. All users retrieve their data from a common location, which guarantees consistency and simplifies the update process.
- Multiple levels of security can be implemented on a LAN, making it more difficult to obtain unauthorized access to data.
- LAN backup systems can be programmed to run at specific intervals, ensuring that critical data is always available from a secondary source, if needed.
- PCs can obtain their software applications from a centralized storage device connected to the LAN. A new application or an upgrade needs to be installed only once on the network. It can then be accessed from any PC. Without a LAN, the same application has to be installed separately on every PC.
- Any resource connected to the LAN (e.g., color laser printers or high-speed Internet access) can be accessed by all users.

Definition and characteristics of a Local Area Network (LAN)

The Institute of Electrical and Electronics Engineers, Inc.® (IEEE®) defines a LAN as "A datacom (data communications) system allowing a number of independent devices to communicate directly with each other, within a moderately sized geographic area."

In its basic form, a LAN is a group of PCs connected with cabling to a central network access device (see Figure 2.4). One of the PCs is used to provide network services and store shared data. This PC is called the server. The other PCs are called stations, clients, or nodes. The central network access device is commonly referred to as a hub or a switch. It is responsible for managing the communications between the stations and the server.

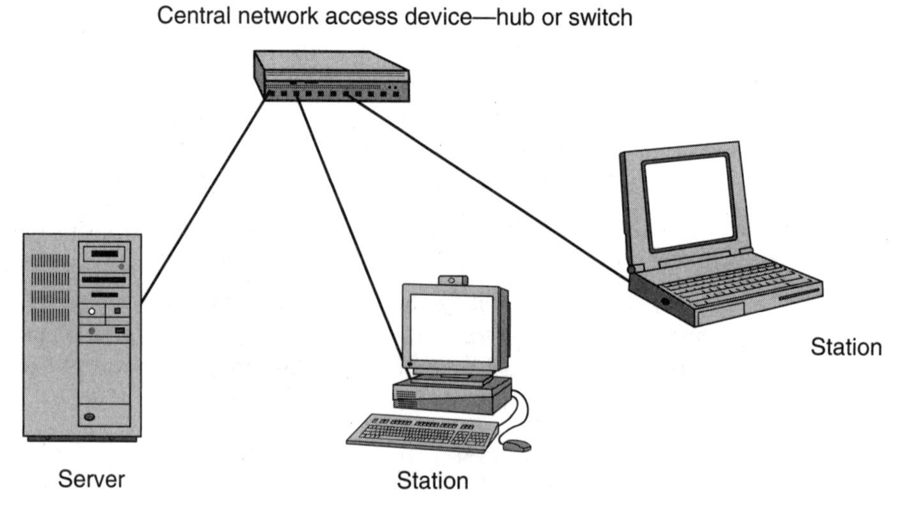

Figure 2.4 A typical local area network.

The following are some of the characteristics of a LAN:

- It is confined to a relatively small area, typically a single building or a campus (two or more buildings in close proximity to each other). Cabling systems, wireless systems, or both are used to connect stations and servers to the hubs or switches.
- There is full connectivity between devices. Any station or server can send data to (or receive data from) any other station or server.
- The operating speed is measured in millions or billions of bits transferred per second, called megabits per second (Mb/s) and gigabits per second (Gb/s).

NOTES: Mb/s and Gb/s are data communications terms used to describe the rate of data transfer. Megabyte (MB) and gigabyte (GB) are data storage terms, with one byte equal to eight bits. In some documents, the term megabytes per second (MB/s) is used to describe the transfer speed of a device, such as a disk drive. If needed, the MB/s value can be multiplied by eight to convert to the equivalent Mb/s value (e.g., 12 MB/s equals 96 Mb/s).

The term megahertz (MHz) is also used in data communications as a description of the frequency associated with a given signal or, more commonly, as a description of bandwidth—the frequency range of a communications channel. MHz and Mb/s are not interchangeable (e.g., a LAN can operate at 10 Mb/s, 100 Mb/s, or 1000 Mb/s over a cabling system with a bandwidth of 100 MHz).

Shared, dedicated, and hybrid Local Area Networks (LANs)

Communications on a LAN can take place in a shared or a dedicated environment. On a shared LAN, all stations and servers share a single communications channel. A station or server with a message to send cannot transmit its message until it obtains access to the network, which is available to only one device at a time. The central device that controls network access is typically called a hub (see Figure 2.5).

NOTE: Access control is described in detail in Chapter 5: Network Communications.

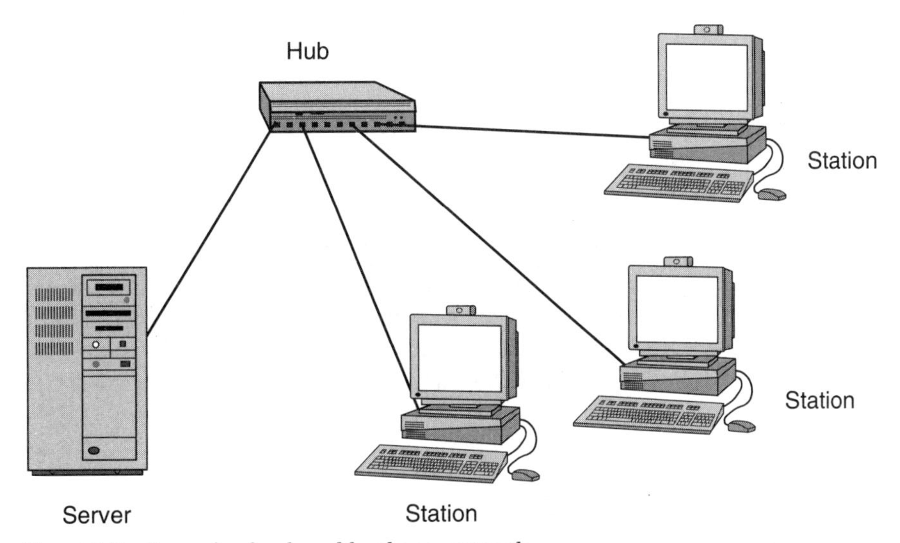

Figure 2.5 Example of a shared local area network.

NOTE: In Figure 2.5 all connections share a single communications channel.

A dedicated LAN—also called a switched LAN—provides each station or server with its own communications channel. This eliminates the need to wait for access to a shared channel, making it possible for any number of stations and servers to send their messages at the same time. The central device that controls network access is typically called a switch (see Figure 2.6).

NOTE: Switching is described in detail in Chapter 6: Switching and Virtual LANs (VLANs).

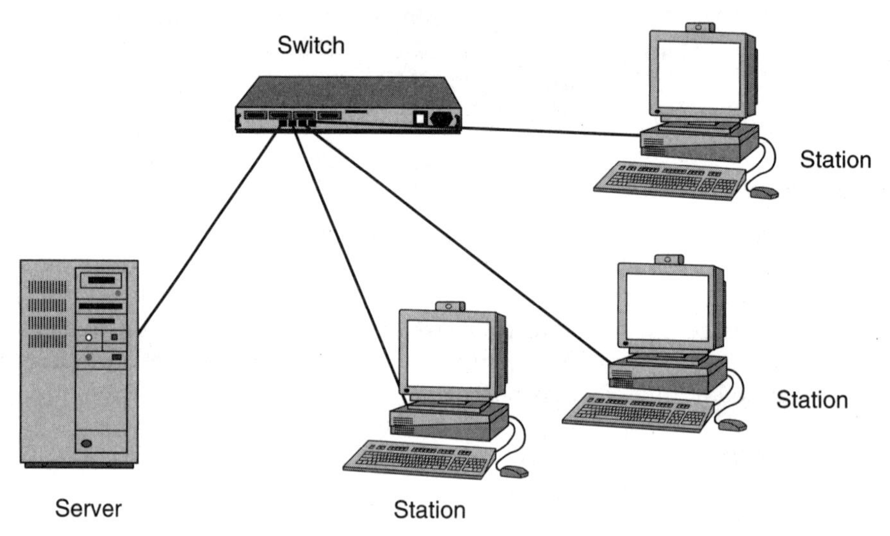

Figure 2.6 Example of a switched local area network.

NOTE: In Figure 2.6 each connection has its own dedicated communications channel.

A hybrid LAN provides a single shared and one or more dedicated communications channels, which costs less than a fully switched environment. The shared channel is typically used to connect stations. It is common to assign each dedicated channel to a server, since servers are usually the busiest devices on the network. The central device that controls network access can be described as a hybrid hub/switch (see Figure 2.7).

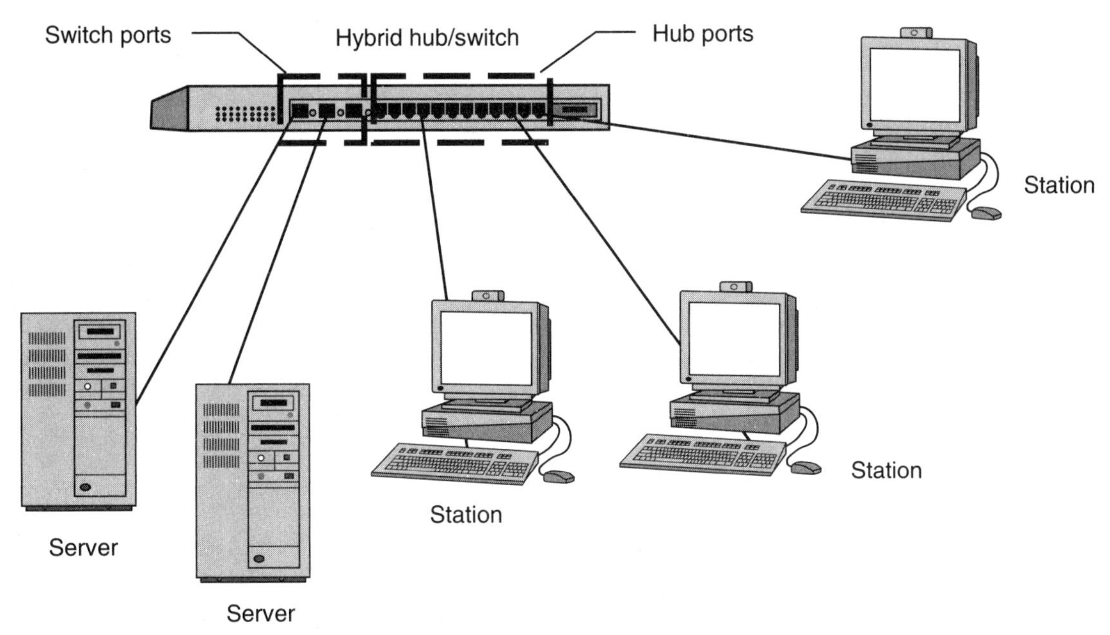

Figure 2.7 Example of a hybrid local area network.

NOTE: In Figure 2.7 all hub port connections share a single communications channel and each switch port connection has its own dedicated communications channel.

Remote Access Fundamentals

General

Remote access technologies enable a station to connect to a LAN over extended distances, using telecommunications circuits. After the connection is established, the remote station can access all LAN resources in the same manner as a local station. From a user's point of view, the only difference between remote and local LAN access is the slower response time of a typical remote connection, caused by the lower rate of data transfer over the telecommunications circuit.

A remote station can connect to a LAN using either remote control or remote node technologies. In both cases, specialized communications software must be installed on the end devices, which is used to send and receive messages over the telecommunications circuit.

Remote control

When remote control is used to access LAN resources, the remote station connects to a local station and controls its operations. The telecommunications circuit is used mostly to send commands from the remote station and receive screen displays from the local station. The local station—on behalf of the remote station—makes all requests for LAN resources (see Figure 2.8). As a result, remote control is very efficient in minimizing traffic, making it practical to connect to a LAN over relatively slow telecommunications circuits.

The main disadvantage of the remote control approach is the need to operate a local station each time a remote station requires access to the LAN. If many remote users need to access the network at the same time, a large number of stations have to be acquired, connected to the LAN, and set aside for the exclusive use of remote stations.

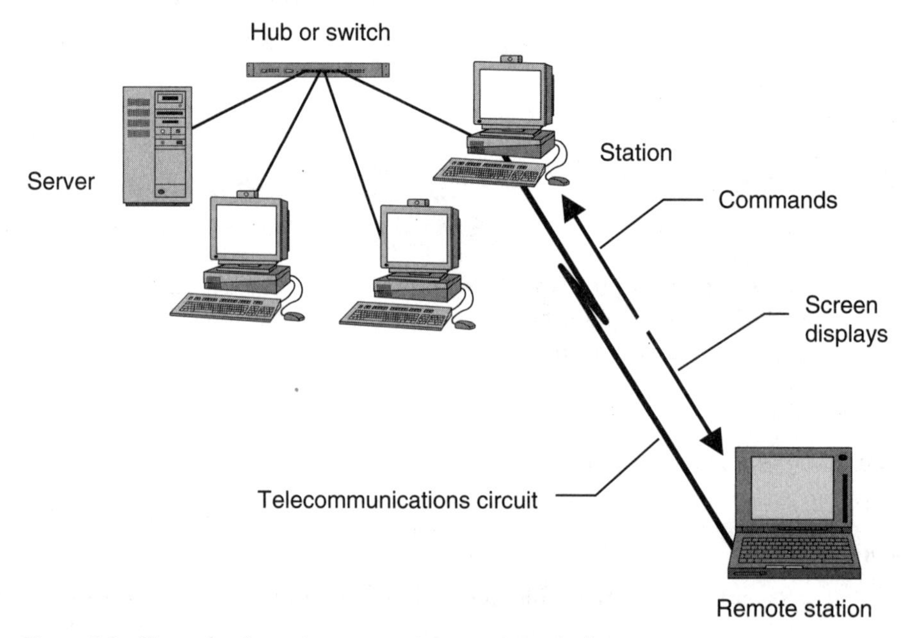

Figure 2.8 Example of remote access using remote control.

Remote node

When remote node is used to access LAN resources, the remote station communicates with a specialized LAN resource commonly referred to as a remote access server or remote access services (RAS). The RAS can be software installed on an existing server or a dedicated device used exclusively to manage many simultaneous remote access connections (see Figure 2.9). The remote node approach eliminates the need to assign local stations to remote users, as is the case with remote control. Remote node is appropriate for organizations with many remote users, since a single RAS resource can be configured to manage the requests of one hundred or more remote stations simultaneously connected to the LAN.

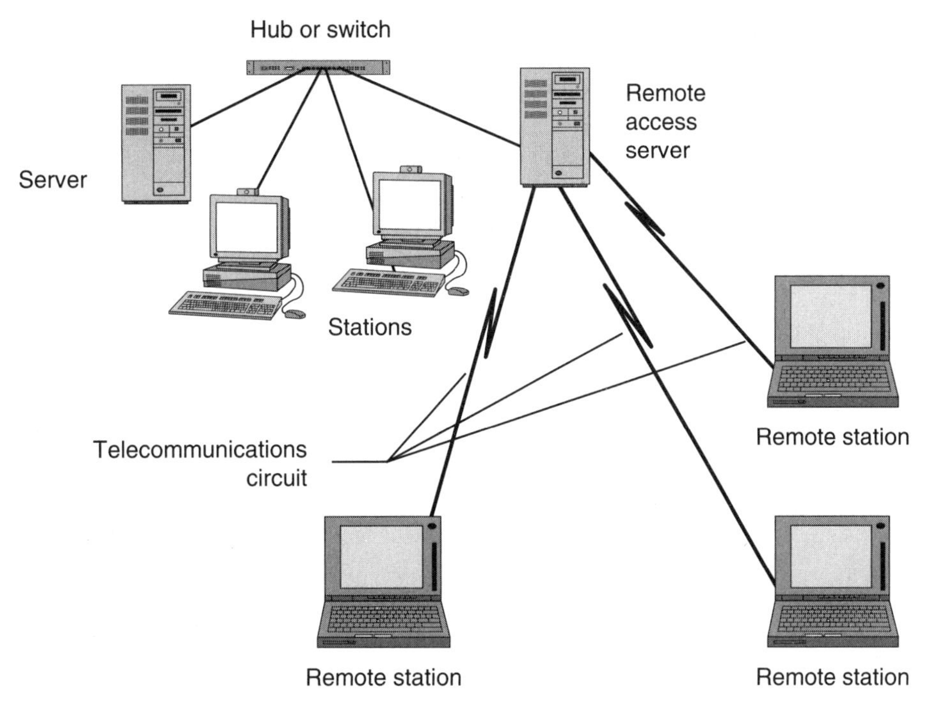

Figure 2.9 Example of remote access using remote node.

Internetwork Fundamentals

General

An internetwork connects LANs in the same way a LAN connects PCs. An internetwork can be used to link two or more existing LANs. It can also be created to divide a large LAN into two or more smaller ones to improve response time and reliability. The largest internetwork in existence is the Internet, which is global in scope and capable of serving as the universal resource for message transfer between all types of LANs (or between remote stations and LANs).

In many cases, a single device can be used to implement multiple LANs and the internetwork that links them. When this occurs, it is impossible to associate a physical device with any single network, since one unit manages all communications on and between several LANs. The term broadcast domain is often used in such cases to describe the communications boundary between a LAN and the internetwork to which it is connected.

When a device on a LAN needs to send a message to all other devices on the same LAN, it issues a broadcast. All devices that are able to receive the broadcast are said to be in the same broadcast domain as the sending device. When a LAN is connected to an internetwork, all broadcasts are confined to the LAN and prevented from entering the internetwork, since they are not intended for other LANs. Therefore, an internetwork can also be described as a collection of interconnected broadcast domains. The device typically used to link one broadcast domain to another is also the device most commonly associated with internetworking. It is called a router.

NOTE: An example of a confined broadcast application is the electronic mail (e-mail) distribution system on a network connected to the Internet. When a message is addressed to all users, it is forwarded to all members on the distribution list, not to all users on the Internet. The distribution list represents the broadcast domain for e-mail messaging.

Types of internetworks

Internetworks can be classified based on their size, scope, and complexity, as described and illustrated below.

A simple internetwork links two LANs of the same type in close proximity to each other, using cabling or wireless media. A variation of this design is used if the two LANs are geographically distant, in which case a telecommunications circuit serves as the medium between the LANs.

A second type of internetwork links two LANs that are dissimilar in architecture. Each uses forms of addressing and communications incompatible with the other. As with the simple internetwork, these two LANs can be local (see Figure 2.10) or remote (see Figure 2.11) to each other.

Internetworking becomes more complex as the number of LANs to interconnect grows. If the Internet is used as the internetworking link, additional security concerns must be addressed, since messages no longer remain exclusively within the organizational boundaries as they move from one LAN to another.

A third type of internetwork is formed when three or more LANs of similar or dissimilar architecture are connected to each other using local or remote connections. The type and number of links from one LAN to another can vary, but in all cases it is possible to send a message from any LAN to any other LAN. This type of internetwork is described as the distributed form of internetworking (see Figure 2.12).

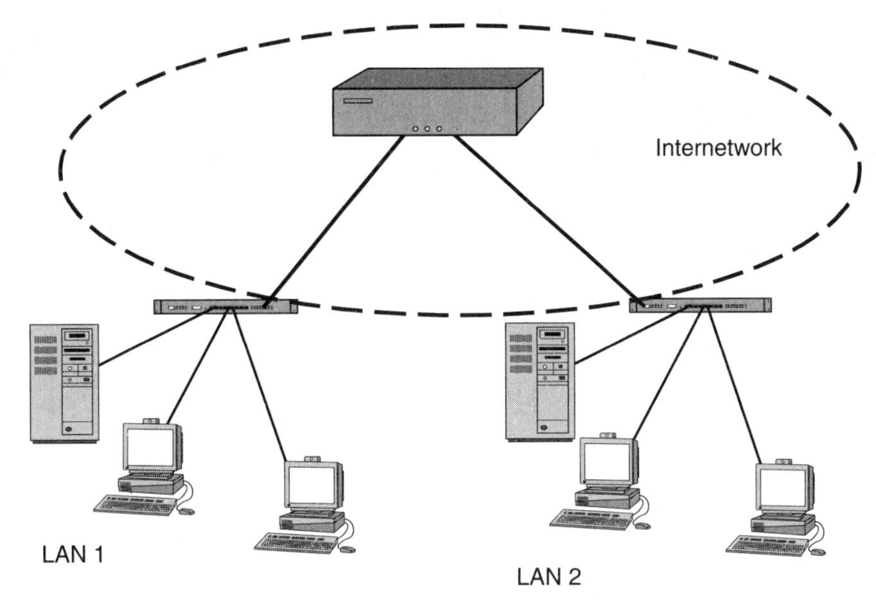

LAN = Local area network

Figure 2.10 Example of a local two-LAN internetwork.

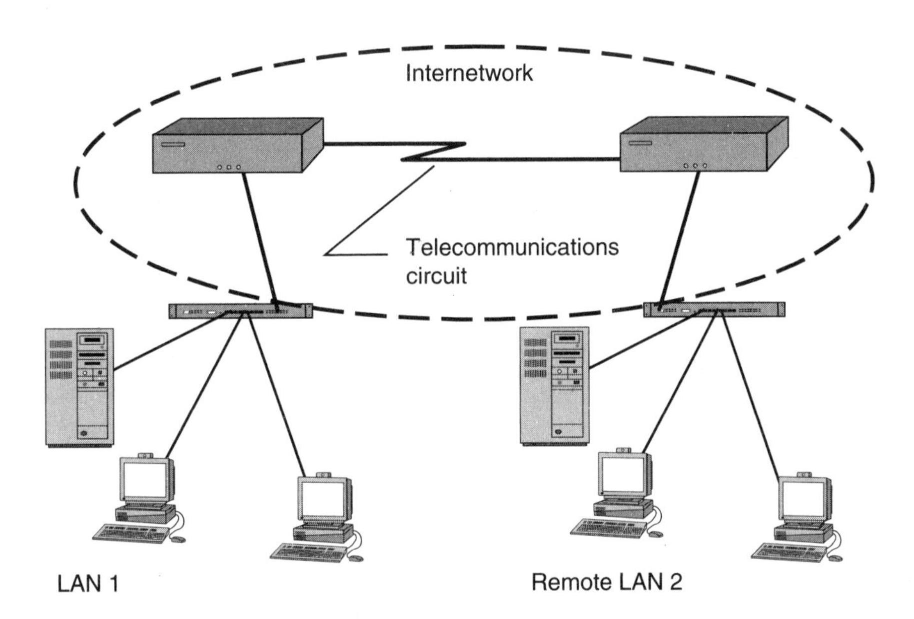

LAN = Local area network

Figure 2.11 Example of a remote two-LAN internetwork.

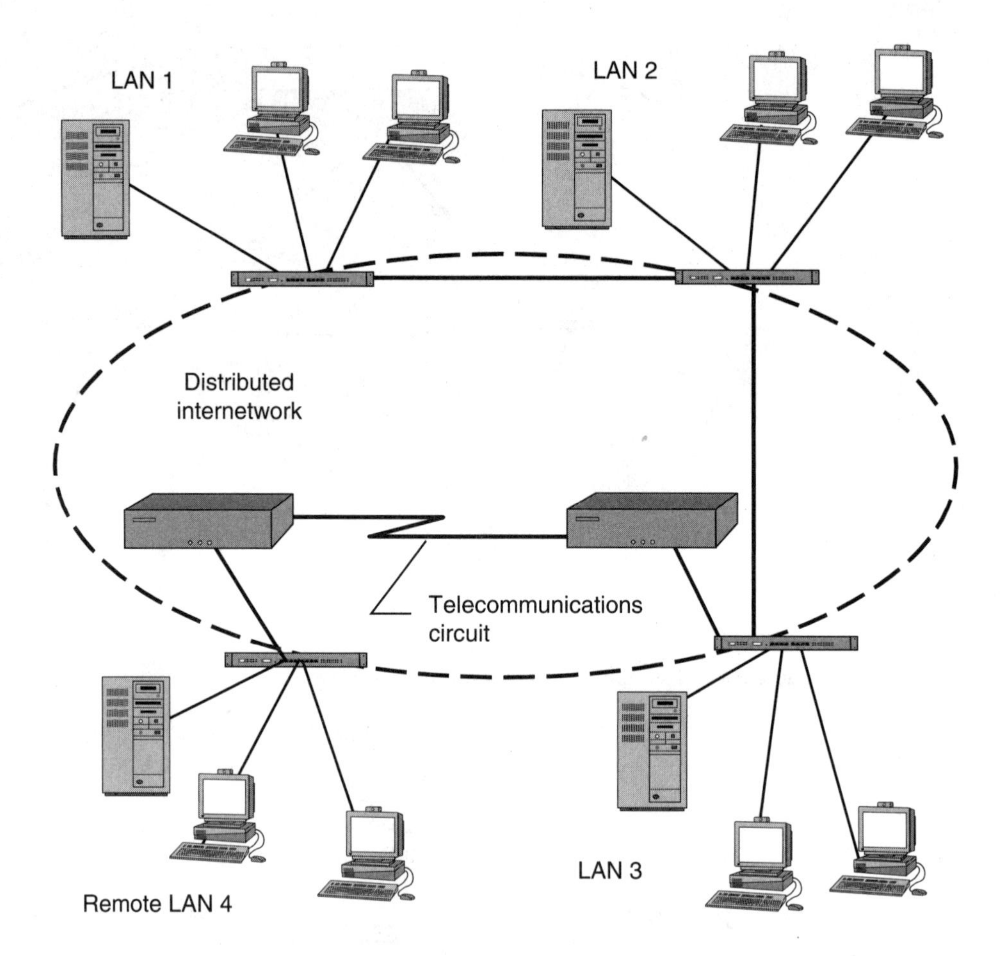

Figure 2.12 Example of a distributed internetwork.

The fourth type of internetwork uses the centralized form of internetworking. Each LAN is connected to the same backbone LAN, which forms the internetwork (see Figure 2.13). As with the distributed model, the LANs—including the backbone LAN—can be of similar or dissimilar architecture, local or remote.

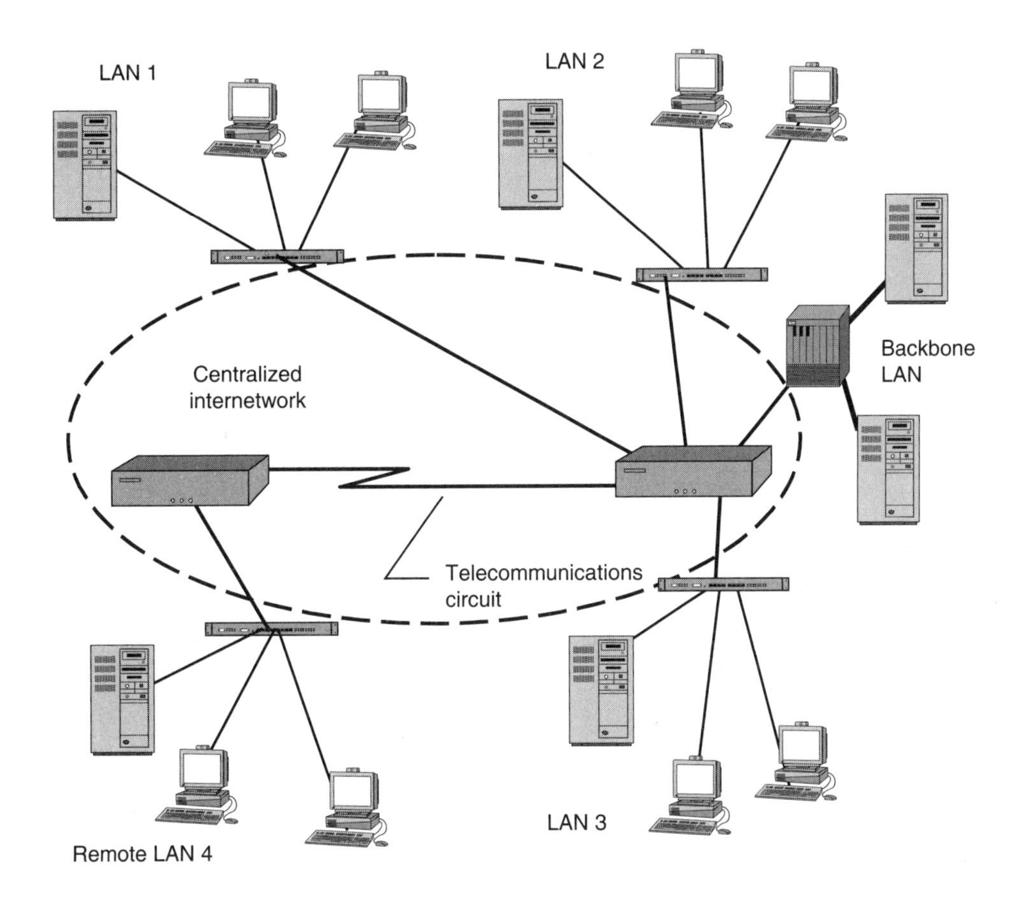

Figure 2.13 Example of a centralized internetwork.

Open Systems Interconnection (OSI) Reference Model

General

Communications on a LAN or an internetwork take place at many levels (e.g., between devices as well as between the software programs running on the same devices). In 1978, the International Organization for Standardization (ISO) introduced a framework for classifying all of the processes associated with message exchange on a network. This framework is formally called the Open Systems Interconnection Reference Model, but is commonly referred to as the OSI model.

The objective of the OSI model is to provide a structured approach for the development of all types of networks. The model specifies the sequence of processes required for message transfer between applications running on different systems.

NOTE: In this context, a system can be defined as two or more computers and the associated software, peripherals, operators, physical processes, and media that form an autonomous whole capable of processing and transferring data.

This broad definition of a system makes it possible to use the OSI model to describe any type of network, from a small LAN to the Internet.

Open Systems Interconnection (OSI) model layers

Overview. The OSI model uses an approach called layering to illustrate and explain the message exchange process. This approach divides the various functions and services provided by a network into discrete groupings called layers. In the OSI model, seven layers are used, as illustrated in Figure 2.14.

Layer 7	Application layer
Layer 6	Presentation layer
Layer 5	Session layer
Layer 4	Transport layer
Layer 3	Network layer
Layer 2	Data Link layer
Layer 1	Physical layer

Figure 2.14 Open Systems Interconnection Reference Model.

NOTE: A layer function can be described by number or by name. For example, the terms Layer 3 switching and Network layer switching refer to the same process.

In the OSI model, each layer provides services to the layer above, while hiding from that layer the processes used to implement the services. This enables changes to be made to any layer without requiring changes to any of the other layers, as long as the inputs and outputs of the changed layer remain the same. For example, a new type of cabling can be incorporated into an existing LAN technology without modifying LAN operations in any way. This makes it possible to take advantage of new technologies without sacrificing compatibility with existing networks.

The layers in the OSI model connect to each other in vertical form, also called a stack or protocol stack. The stack defines how network hardware and software interact at various levels to transfer messages between devices on a LAN and between LANs on an internetwork. Protocol stacks have the following characteristics:

- Each layer provides a set of services.
- Services of each layer and the stack are defined by protocols.
- Lower layers provide services to upper layers.
- Service access points (SAPs) are the connection points between layers.

When a message needs to be transferred between two systems, a peer-to-peer relationship is established between the corresponding layers in the protocol stack of each system—a given layer communicates with its counterpart over the network. The message, along with any control information, is passed down from the sending layer to the layer below. This process continues until the lowest layer in the stack is reached. The data is then transmitted from the lowest layer of the sending system to the lowest layer in the receiving system, where it is passed up through the layers until it reaches the counterpart—or peer—of the sending layer (see Figure 2.15).

Application layer services. The Application layer protocols make it possible for identical or different applications running on different systems to use the services of a network to exchange information. Services defined by this layer include file transfer, message handling, and remote management. For example, various types and versions of e-mail software can use the same Application layer protocols to exchange messages over the Internet.

Presentation layer services. The Presentation layer protocols are responsible for various forms of message conversion. It negotiates and establishes a common form for data representation, which includes character code translations, data compression and decompression, or data encryption and decryption.

Session layer services. The Session layer protocols provide the services needed to synchronize and manage message transfer between network devices. For example, Session layer protocols direct devices to start, stop, restart, or abandon data transfer activity.

Transport layer services. The Transport layer protocols make it possible to introduce various levels of quality to the message transfer process. After a connection has been established between network devices, the Transport layer

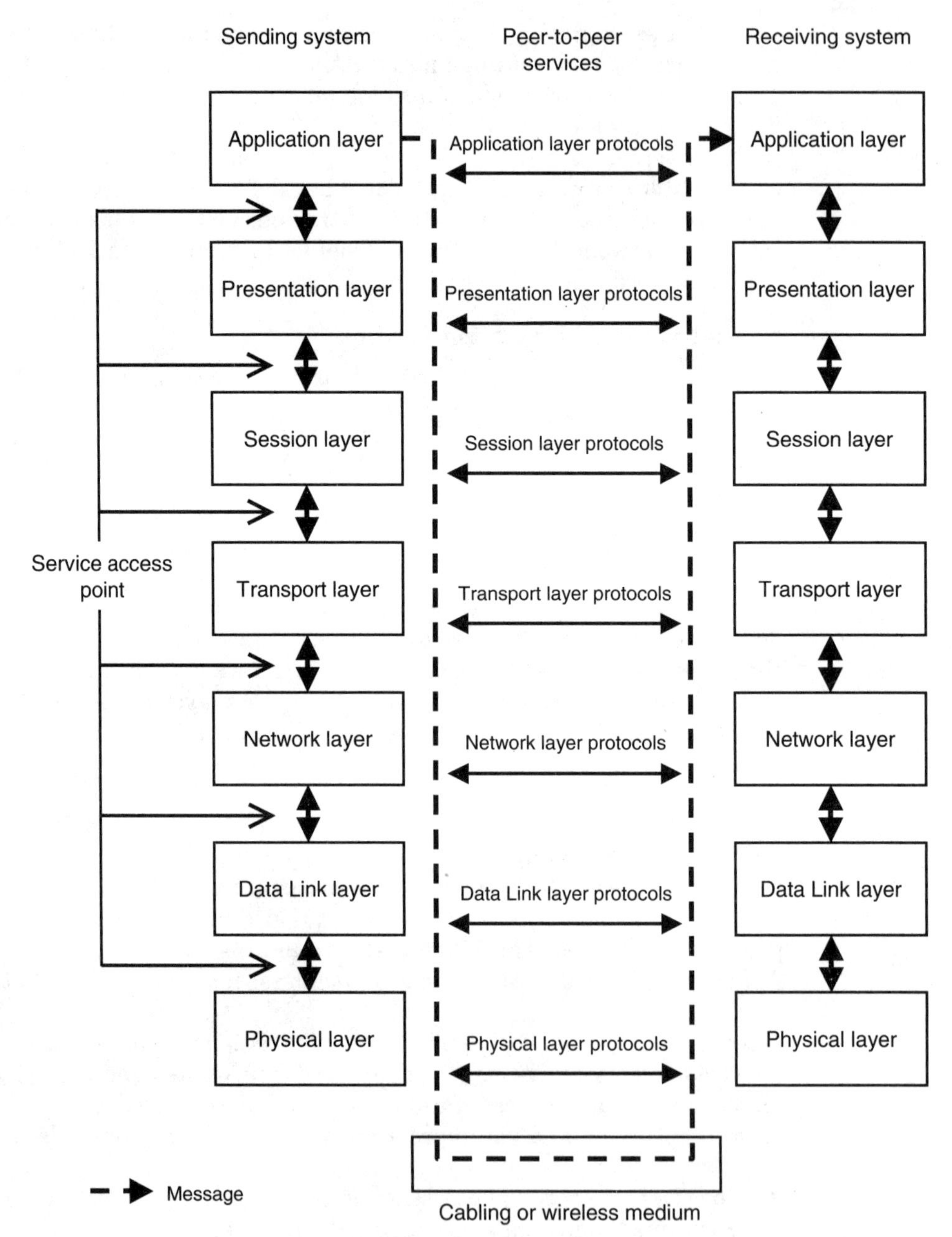

Figure 2.15 Message transfer described using the Open Systems Interconnection model.

protocols can be used to select a particular class of service, monitor the transfer for billing purposes, ensure that the appropriate service quality is maintained, and generate an alert if this quality has been compromised.

Network layer services. The Network layer protocols are responsible for internetwork message transfer (e.g., between LANs on an internetwork). When multiple routes exist, the Network layer protocols choose the most appropriate one, based on such criteria as message priority, route congestion, or route cost.

Data Link layer services. The Data Link layer protocols are responsible for intranetwork message transfer (e.g., between devices on a LAN). Some of the functions include device identification and access to shared transmission channels.

Physical layer services. The Physical layer protocols are responsible for the transfer of bits over various media.

Open Systems Interconnection (OSI) model message transfer sequence

Any message transfer that takes place between two systems follows a four-step process:

1. Preparation of the message to be transferred after a request is made by a software application running on the sending system.
2. Access to the network and transmission of the message.
3. Retrieval of the message by the receiving system.
4. Delivery of the message to the appropriate software application running on the receiving system.

Examples 2.1 and 2.2 illustrate the role of each layer in the sending and receiving systems when a message is transferred.

EXAMPLE 2.1 Message output at the sending system.

MESSAGE ↓	A software application running in the sending system directs a message to the Application layer.
Application layer ↓	The Application layer is responsible for: ▪ Organizing the message into blocks of data.
(AH)•Data ↓	An application header (AH) is added to: ▪ Identify the sending application. ▪ Identify the destination application.

(continued on next page)

EXAMPLE 2.1 Message output at the sending system. *(continued)*

↓

Presentation layer

The Presentation layer is responsible for:

- Translating the character code used by the sending system to that used by the receiving system, if necessary.
- Compressing the data to improve transfer efficiency, if necessary.
- Encrypting the data for improved security, if necessary.

↓

(PH)•AH•Data

A presentation header (PH) is added to:

- Provide details on any encoding, compression, and/or encryption used.

↓

Session layer

The Session layer is responsible for:

- Marking the first and last block of data.
- Including a special marker with the last block of data to allow the destination system to send a reply (if necessary).

↓

(SH)•PH•AH•Data

A session header (SH) is added to:

- Indicate any markers accompanying the data blocks.

Transport layer

The Transport layer is responsible for:

- Dividing each block into segments.
- Tracking the sequence of the segments.

(TH)•SH•PH•AH•Data
↓
TPDU

A transport header (TH) is added to each segment, forming a transport protocol data unit (TPDU):

- Each TPDU includes a sequence number and verification bits for error detection.

(continued on next page)

EXAMPLE 2.1 Message output at the sending system. *(continued)*

	▪ A copy of each TPDU is kept by the sending device, to be used if retransmission is required. When receipt of the message is acknowledged by the receiving device, the copy is erased.
↓	
Network layer	The Network layer is responsible for: ▪ Breaking the TPDU into fragments if necessary, to conform to the limitations of the network.
↓	
(NH)•TPDU ↓ NPDU	A network header (NH) is added to each TPDU or fragment, forming a network protocol data unit (NPDU): ▪ Each NPDU includes a sequence number and a destination address.
↓	
Data Link layer	The Data Link layer is responsible for: ▪ Adding a header and a trailer to each NPDU, forming a frame. ▪ Creating a copy of the frame in the sending device, in case retransmission is required.
↓	
F•A•C•NPDU	A data link header is added: ▪ Contains framing (F), address (A), and control (C) information.
↓	
F•A•C•NPDU•FCS•F	A data link trailer is added: ▪ Contains a frame check sequence (FCS) for error detection and optionally, additional framing information (F).
↓	
Physical layer	The Physical layer is responsible for: ▪ Signal encoding and transmission.
↓	
BITS ON THE MEDIUM	

EXAMPLE 2.2 Message input at the receiving system.

BITS ON THE MEDIUM

The signal is retrieved from the medium by the receiving system.

↓

Physical layer

The Physical layer is responsible for:

- Converting the incoming signal into a sequence of bits.

↓

F·A·C·NPDU·FCS·F

↓

Data Link layer

The Data Link layer is responsible for:

↓

NPDU

- Removing the header and trailer information.
- Using the FCS to check if the contents were modified after transmission.

↓

(NH)•TPDU

↓

Network layer

The Network layer is responsible for:

↓

TPDU

- Removing and inspecting the network header.
- Verifying that the values for destination address and sequence number are correct.
- Waiting until all of the packets that form a TPDU arrive and then assembling the TPDU.

↓

(TH)•SH•PH•AH•Data

↓

Transport layer

The Transport layer is responsible for:

- Removing and inspecting the transport header.
- Using the frame check verification bits in the TPDU to check if the contents were modified after transmission.

(continued on next page)

EXAMPLE 2.2 Message input at the receiving system. *(continued)*

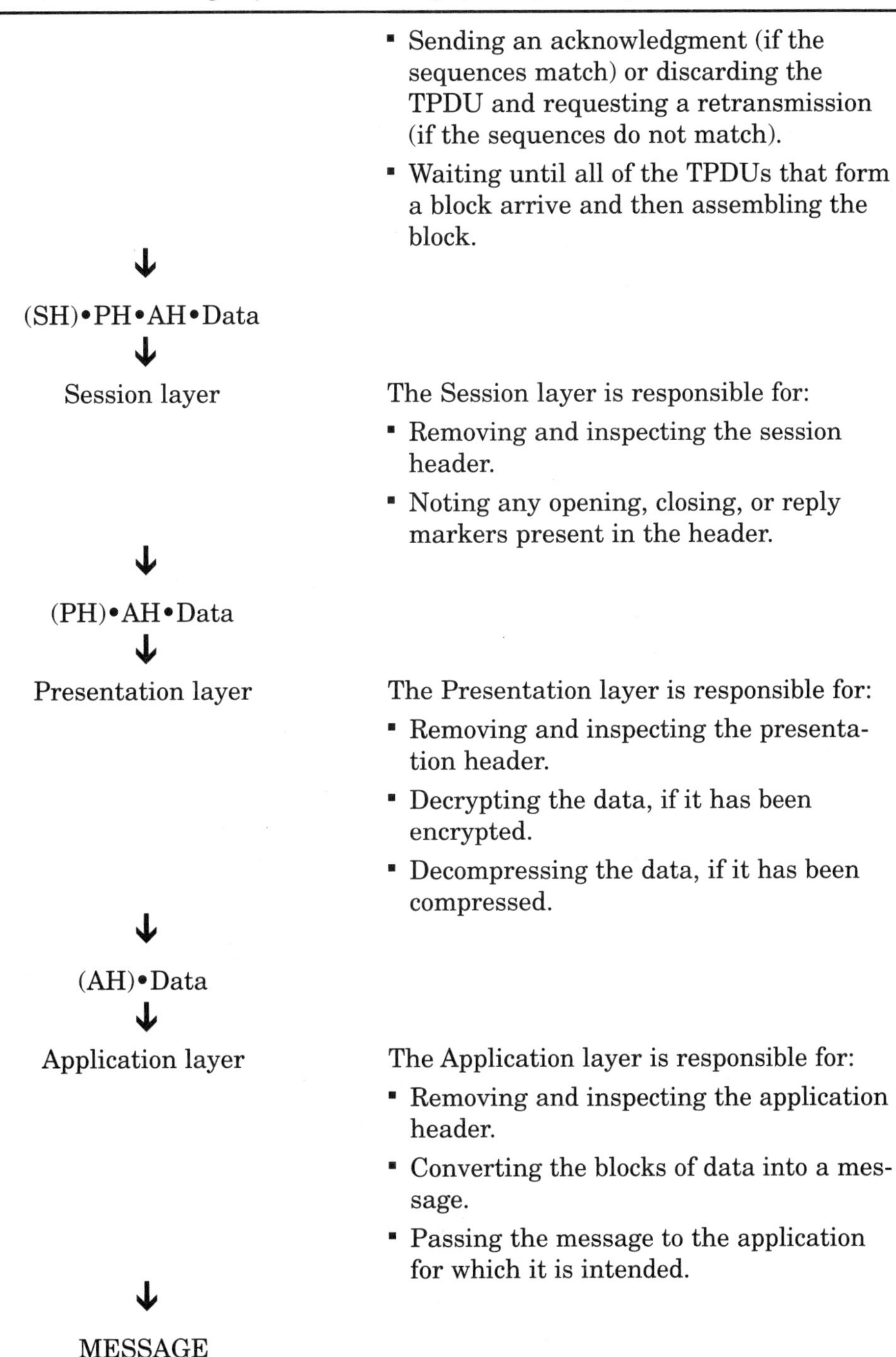

- Sending an acknowledgment (if the sequences match) or discarding the TPDU and requesting a retransmission (if the sequences do not match).
- Waiting until all of the TPDUs that form a block arrive and then assembling the block.

↓

(SH)•PH•AH•Data

↓

Session layer

The Session layer is responsible for:

- Removing and inspecting the session header.
- Noting any opening, closing, or reply markers present in the header.

↓

(PH)•AH•Data

↓

Presentation layer

The Presentation layer is responsible for:

- Removing and inspecting the presentation header.
- Decrypting the data, if it has been encrypted.
- Decompressing the data, if it has been compressed.

↓

(AH)•Data

↓

Application layer

The Application layer is responsible for:

- Removing and inspecting the application header.
- Converting the blocks of data into a message.
- Passing the message to the application for which it is intended.

↓

MESSAGE

Chapter

3

Network Components

Chapter 3 describes the components commonly found in local area networks (LANs) and internetworks. The general characteristics of each component are presented, and where appropriate, the Open Systems Interconnection (OSI) model is used to classify the component.

Overview

General

This chapter describes the components commonly found in local area network (LAN) and internetwork designs. The characteristics of each type of component are presented, and where appropriate, the Open Systems Interconnection (OSI) model is used to classify the component.

An organizational network can be described as a collection of hardware and software components, configured to provide a set of services to users. The role of the network is to provide secure and managed access to resources that users do not have in their personal computing environments (also described as nonlocal resources). Examples of nonlocal resources include a printer not attached to a user's computer, a data file stored on another computer, and a Web site on the Internet.

The components used to provide network services can be grouped into the following classes of products:

- Stations
- Servers
- Shared peripherals

- Network access devices
- Cabling systems
- Wireless systems
- Telecommunications circuits
- Security
- Management

Due to the wide variety of components in each product class, this chapter focuses on stations, servers, shared peripherals, and network access devices. The remaining product classes are described in the following chapters:

- Cabling and telecommunications circuit components are described in Chapter 4: Network Connections.
- Security components are described in Chapter 8: Network Security.
- Management components are described in Chapter 9: Network Management.

Stations

General

The term station is used to describe the device with which a user accesses the network. Stations can be local or remote.

- A local station connects to the network with cabling or wireless media.
- A remote station connects to the network with a telecommunications circuit.

At a minimum, a station provides its user with the necessary interfaces for data input and display. The most common type of local station is a desktop personal computer (PC), capable of storing, retrieving, and manipulating data whether or not it is connected to a network. Similarly, laptop PCs typically serve as remote stations for mobile users.

When an organization has a large number of PCs to be networked (e.g., several thousand students and staff on a university campus), management and security can be difficult. Unless all PCs are identical in terms of components and configuration, it is necessary to track and update multiple types of devices on an ongoing basis. The presence of storage devices such as hard disk or compact disc (CD) drives makes it possible for users to transfer unauthorized material from the network to their stations and vice versa.

Modified versions of desktop PCs have been developed to make networks more secure and easier to administer. A variety of terms are used to describe such stations, including:

- Thin client.
- Network appliance.
- Managed PC.
- Network computer.
- Net PC.
- Diskless workstation.

Although the descriptions vary, all of these devices make it easier for a network administrator to monitor, control, and change the configuration of any station from a centralized location using management software. The network replaces the need for physical access to a station for such tasks as:

- Taking inventory of the components in the station, both hardware and software.
- Installing or updating all types of software, including the station operating system.
- Scanning for viruses and removing them before they can affect other network users.
- Backing up files from the station to a network backup device.
- Powering on or powering off the station.

Management software can also be used to notify network administrators of component failure in a station. In some cases, the software can also detect and report any attempt to remove the components in a station, reducing the possibility of theft or unauthorized modifications.

In cases where data security is critical, the stations do not contain storage devices—no floppy disk, CD, or hard disk drives—making it more difficult to remove data from the network or place unauthorized material on the network. These types of stations operate like mainframe or minicomputer terminals. All software, including the station's operating system, is stored on the network and transferred to the station when it is powered on.

Station infrastructure

A variety of hardware and software components are used to transform a desktop or laptop PC into a local or remote station, including:

- Network interface cards (NICs).
- Modems.
- Media converters.
- Uninterruptible power supplies (UPSs).
- Station software components.

Each of these components is described in the following sections.

Network Interface Card (NIC)

The hardware component that connects a station to the network medium is called a NIC. Typically, the circuitry of a NIC is packaged in the form of an adapter that is installed in an expansion slot within a PC (see Figure 3.1). NIC circuitry can also be incorporated into a PC by its manufacturer, making the device network-ready—capable of being connected to a network at any time.

Some NICs can be configured to improve the performance or security of data transfer between the station and the network. To improve performance, a technique known as traffic prioritization can be used. Prioritization makes it possible to assign different levels of priority to applications running simultaneously on a station. For example, a NIC can be configured to transfer data to and from a videoconferencing application before processing any other messages. Some NICs also perform additional error checking, a task usually relegated to the device's central processor. To improve network security, a NIC can be configured to apply encryption to all messages before transmission.

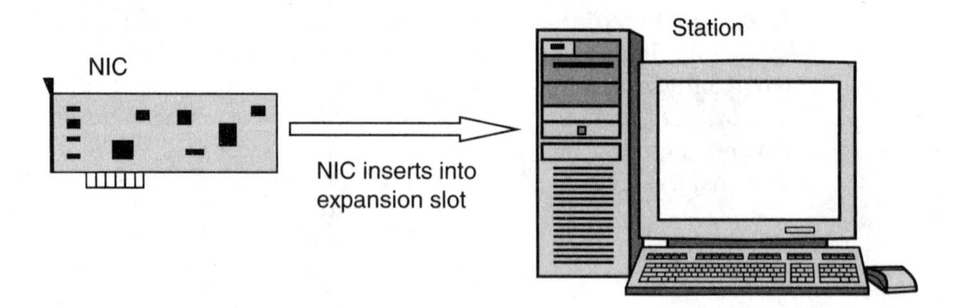

Figure 3.1 Network interface card.

Modem

A modem is a hardware device commonly used by remote stations for network access over public switched telephone network (PSTN) telecommunications circuits, also called dial-up lines. When sending data, the modem converts the digital signals produced by the remote station to an analog signal suitable for transmission over the PSTN. The reverse conversion takes place when the remote station receives data.

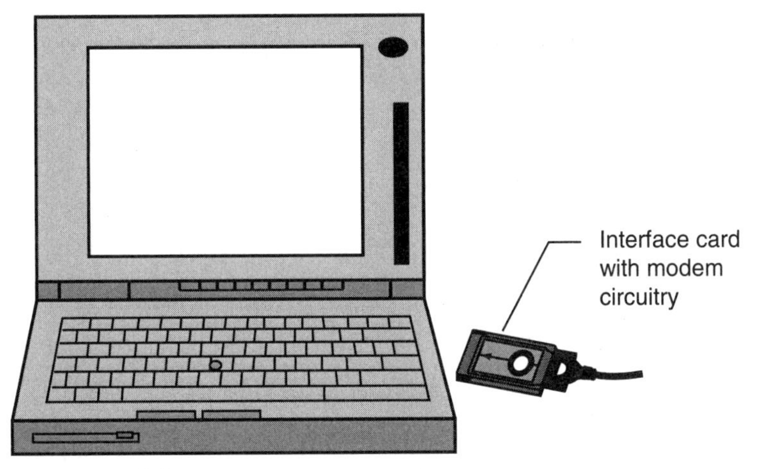

Figure 3.2 Modem.

It is common for laptop computers to be equipped with internal modems. If this is not the case, a separate interface card with modem circuitry can be inserted into most laptops when needed (see Figure 3.2). Often, such cards are equipped with both modem and NIC functionality and have two connectors. When the laptop is used as a:

- Local station (it is in the same premises as the network) the NIC connector is used to link to the network medium.
- Remote station (it is at another location) the modem connector is used to link to a remote access server (RAS) over the PSTN.

As with other networking technologies, modem performance continues to evolve, making it possible to transfer greater amounts of data over a dial-up line in a given time period. Since the PSTN infrastructure is global in scope, the organization responsible for formalizing modem standards is the International Telecommunication Union-Telecommunication (ITU-T). Some of these standards, published as *Series V: Data Communications over the Telephone Network*, are described in Table 3.1.

TABLE 3.1 Modem Standards

Standard	Specification
V.22	Specifications for data transfer at 1.2 kb/s.
V.22 bis	Specifications for data transfer at 2.4 kb/s.
V.32	Specifications for data transfer at 9.6 kb/s.
V.32 bis	Specifications for data transfer at 14.4 kb/s.
V.34	Initially issued as the specifications for data transfer at 28.8 kb/s. Subsequently updated to 33.6 kb/s.
V.90	Specifications for data transfer at 56 kb/s in one direction (downstream) and 33.6 kb/s in the other direction (upstream).
V.92	Specifications for data transfer at 56 kb/s downstream and 48 kb/s upstream.
V.42	Specifications for error detection and correction.
V.42 bis	Specifications for data compression at a maximum ratio of 4:1.
V.44	Specifications for data compression at a maximum ratio of 6:1.
V.17	Specifications for fax communications at 14.4 kb/s.

NOTES: The data transfer rates listed in Table 3.1 represent maximum attainable values. The actual transfer rates achieved are affected by a number of factors, the most important of which is the quality of the local loop—the link from the modem to the nearest PSTN switching center.

The extension bis is used by the ITU-T to denote the second in a series of related standards.

Media converter

In some cases, the available LAN medium does not match the transceiver of the NIC. One example would be a NIC equipped with a connector and circuitry suitable for copper cabling that requires connection to optical fiber cabling. Instead of replacing the NIC, a hardware device called a media converter or a media translator can be attached externally to the copper cabling connector on the NIC and linked to the optical fiber medium (see Figure 3.3).

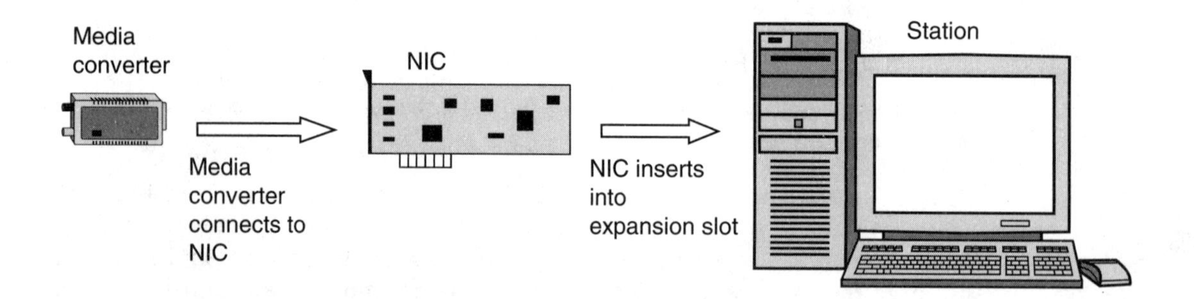

NIC = Network interface card

Figure 3.3 Media converter.

Uninterruptible Power Supply (UPS)

A station can optionally be equipped with a hardware device called a UPS to prevent data loss in the event of a power failure. All network devices can benefit from UPS protection, which provides power to a connected device through one or more batteries—making it possible to close files and exit applications in a normal manner during a power failure.

Station software components

Different types of software are used to transform a stand-alone PC into a network station. Such software components include drivers, utilities, dialers, and clients.

Drivers are needed to link a NIC to its station's operating system, making it possible for the station to use the NIC.

Various utilities are typically supplied by the NIC manufacturer to perform diagnostic tests on the NIC and the network communications channel.

Dialers, also referred to as remote client software, are used by remote stations to connect to the network over telecommunications circuits. In addition to message transfer, dialer software can provide various security tools (e.g., data encryption to conceal the contents of messages and user authentication to verify the identities of remote users).

The term client software is used to describe the parts of the station operating system responsible for network communications. Some operating systems are equipped with networking modules, requiring only activation and configuration. In other cases, the client software is an additional application that must be separately installed and configured.

Other terms used to describe client software include shell, requester, and redirector. All of these terms refer to the way such software functions. After activation, the client software intercepts and examines each user command to determine whether it is intended for a network device or a device connected to the station. Depending on the type of request, the client software either sends the command to the network device through the NIC or to the local device through the appropriate module in the station operating system.

Servers and Shared Peripherals

General

The term server is used to describe a device that manages one or more resources accessed by multiple users on a network. The types of resources that can be accessed through a server are varied. They include software applications, shared peripheral devices, and user-created files.

A server can be specialized in form and components. Mainframes or minicomputers can be servers. Alternatively, a server can be a PC of the same type used as a station.

A server can manage multiple types of resources or it can be dedicated to a single resource. For example, a dedicated print server can be configured to control the operation of multiple printers. When a station sends a request for a file to be printed, the print server processes the request and instructions are issued to the appropriate printer. When the printing is finished, the print server notifies the station that submitted the request. If the printer runs out of paper before completing the printing, the print server issues an alert to a designated network administrator.

On a small LAN, a single general-purpose file server can manage network printing in addition to storing all shared software applications and user-created files (see Figure 3.4).

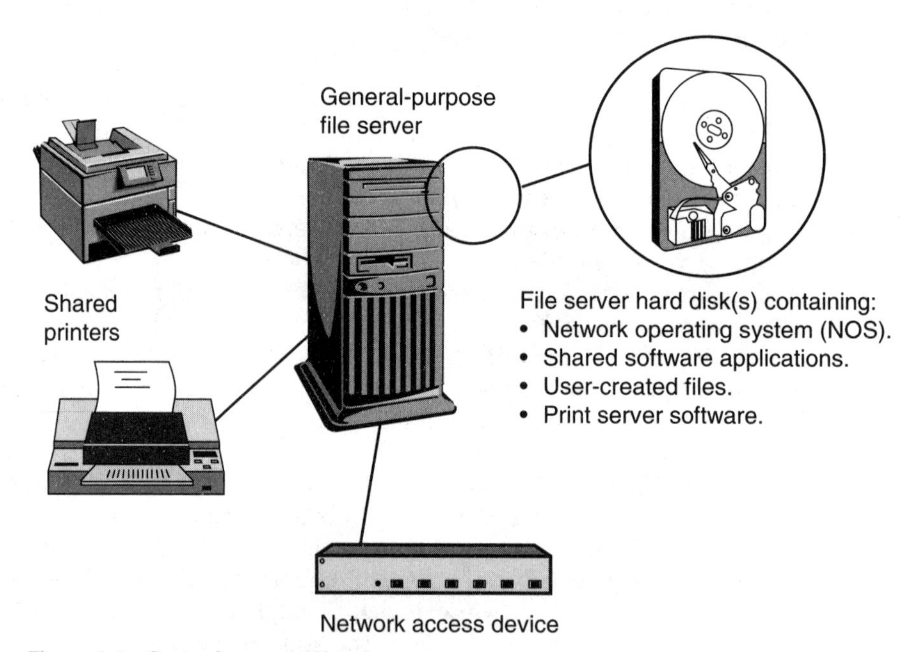

Figure 3.4 General-purpose file server.

On a large LAN, multiple dedicated servers may be required to manage a single application (e.g., remote access, e-mail distribution, an organizational Web site, or print services [see Figure 3.5]).

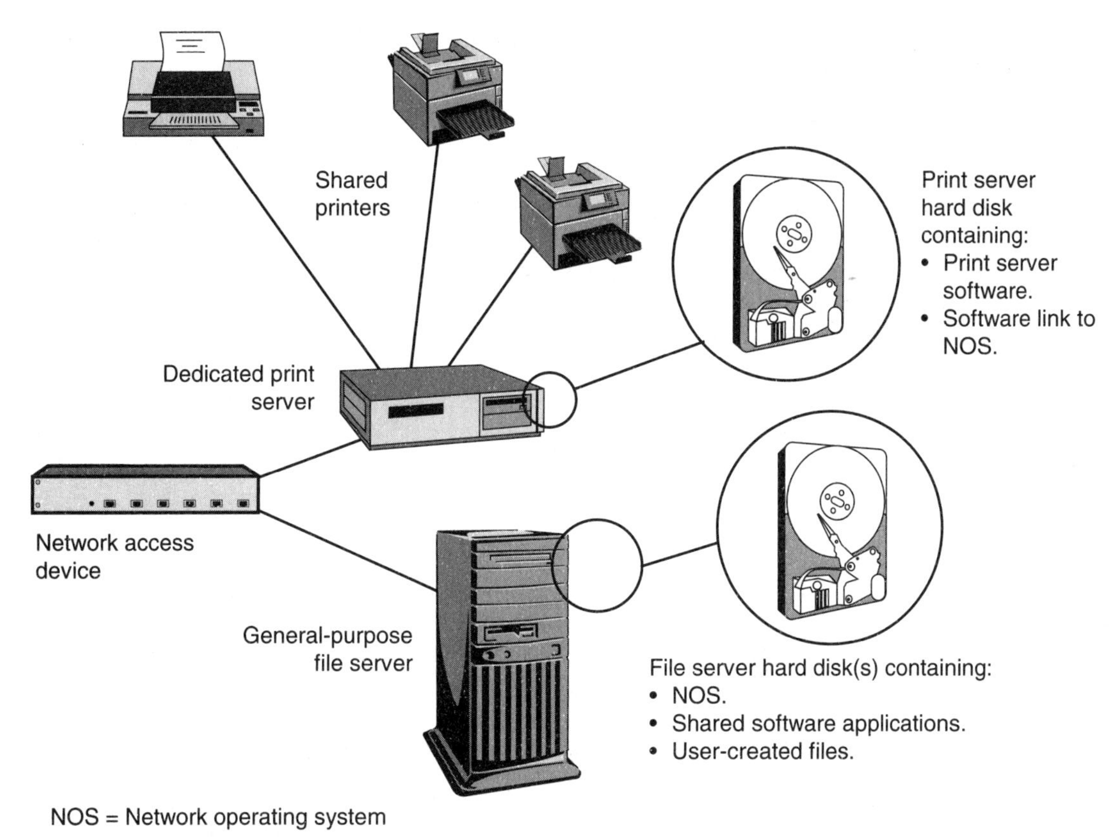

Figure 3.5 Dedicated print server.

NOTE: Figure 3.5 shows a dedicated print server and the printers as separate devices. It is also possible to integrate a NIC and print server functionality into a network printer. In such cases, the same device is both a printer and a print server, connected directly to the network medium using its NIC.

Server infrastructure

A variety of hardware and software components can be used to transform a PC into a server, including:

- NICs.
- Remote access interface cards.
- Network operating system (NOS) software.

Each of these components is described in the following sections.

Network Interface Card (NIC)

As is the case with stations, a NIC is required to connect a server to the network medium. However, NICs designed for server use can provide additional functions to reduce the response time and increase the reliability of a server.

Link aggregation. Link aggregation makes it possible to place multiple NICs in a server and configure them to operate as a single connection (see Figure 3.6). In cases where a single NIC is routinely subjected to high traffic loads, this technique enables a server to respond more quickly to station requests.

A second benefit of using link aggregation is incremental growth. For example, a server equipped with a single NIC that operates at 100 Mb/s can be equipped with a second similar NIC and configured to use link aggregation to double its network operating speed to 200 Mb/s at minimal cost. Without link aggregation, an upgrade to more expensive equipment that operates at the next available operating speed of 1000 Mb/s would be required.

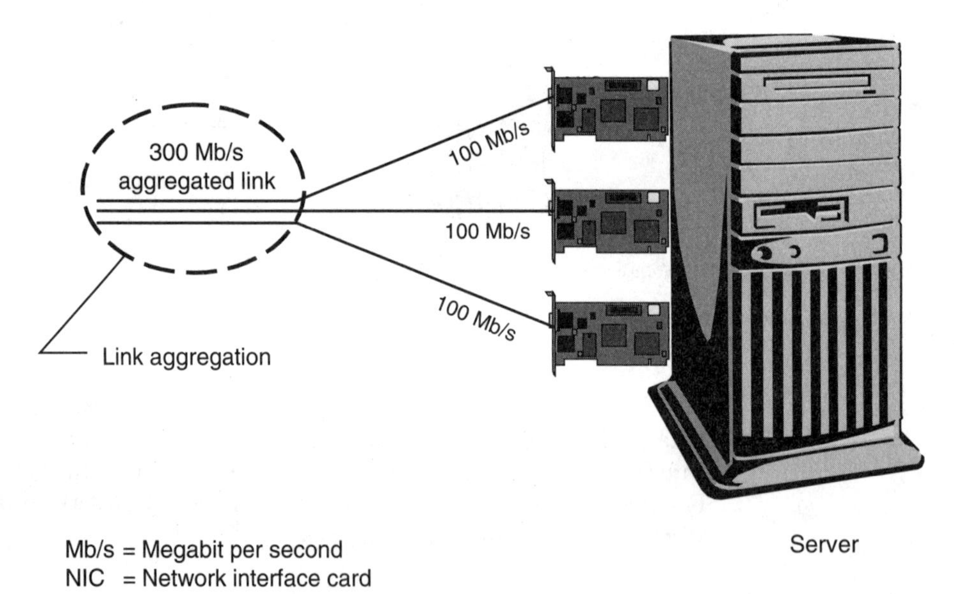

Figure 3.6 Link aggregation.

Redundant links. Two NICs can also be configured to operate in redundant (or backup) mode when placed in a server. In such cases, one NIC is designated as the primary interface, responsible for all network communications to and from the server during normal operations. The second NIC continuously monitors the status of the primary NIC, ready to take over with no disruption in communications in the event of a fault condition in the primary NIC.

Remote access interface card

Various types of remote access interface cards, also called wide area interface cards (WICs), can be used to connect an RAS to one or more telecommunications circuits. The most common type is the multiport modem card, capable of managing the communications sessions of multiple modems in simultaneous operation. In some cases, all of the modem circuitry is found on the interface card. In other cases, external modems are connected to one or more interface cards.

As with print servers, an RAS can be part of a general-purpose file server or it can be a dedicated unit, responsible only for managing the network connections of remote stations (see Figure 3.7).

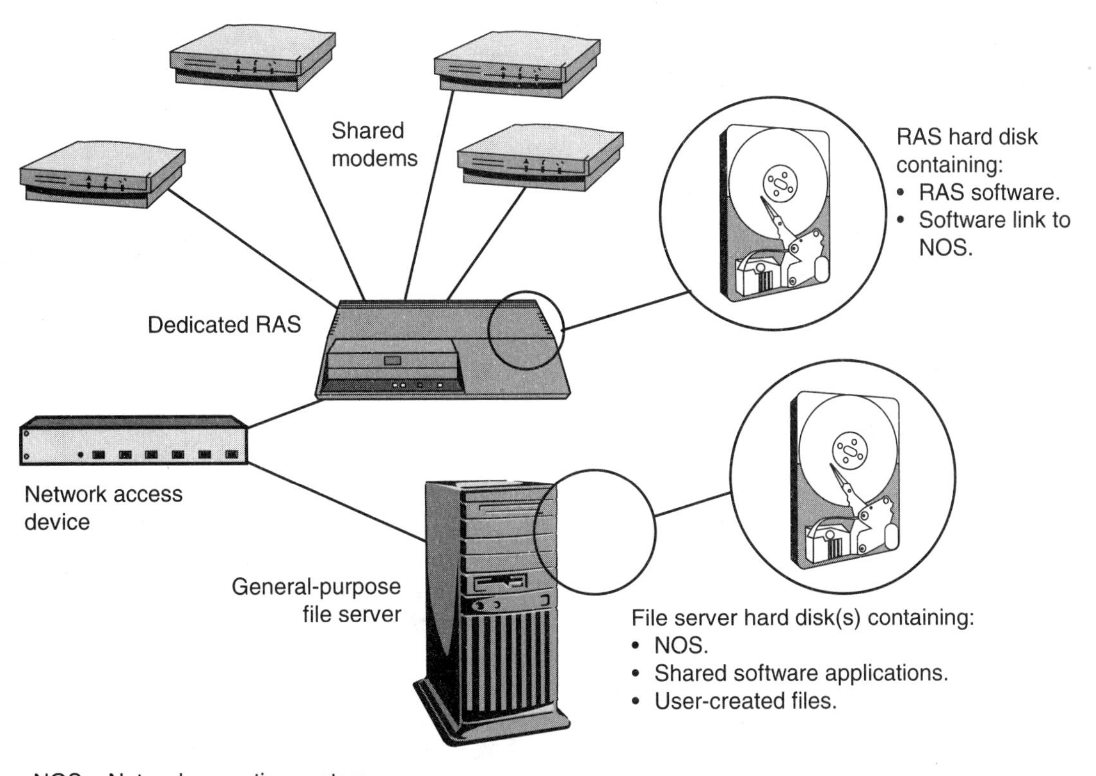

Figure 3.7 Dedicated remote access server.

Network Operating System (NOS) software

The NOS software is an integrated set of programs designed to control and coordinate all access to network resources. It enables sharing of software applications, peripheral devices, and user-created files by authorized network users. Typically, when each server on a network is powered on, it loads a separate copy of the NOS. If multiple servers are present, the NOS software in each server will communicate with the others to synchronize network activities.

In addition to controlling the operations of its server, the NOS software creates and maintains an updated directory listing of all resources on the entire network. This directory service provides users and administrators with a global view of the organizational network, making it possible—with appropriate authorization—to find and access the resources controlled by any server.

In most cases, an organization will standardize on a particular brand of NOS software, installing the same version on all of its server(s). In large organizations with a diverse set of equipment and software applications, it may be necessary to use multiple types or versions of NOS software.

In addition to enabling secure access to network resources, the NOS software provides network administrators with tools to monitor and manage operations. The NOS software can:

- Limit the amount of network hard disk space assigned to each user.
- Maintain log files of specific network activities such as remote access or printing.
- Initiate backups at specific intervals.
- Issue an alert if a network device or component experiences a fault condition.

Shared peripherals

The term peripheral device or peripheral describes any equipment that can be connected to and controlled by a PC. Examples of peripherals include printers, scanners, modems, and mass storage devices (e.g., compact disc-read only memory [CD-ROM] or tape drives).

Multiple network users can access a peripheral if the device is connected to the network, either directly or indirectly. A direct connection is possible if the peripheral can be equipped with a NIC and driver software compatible with the type of network and NOS software in use. With an indirect connection, the peripheral is connected to a station or a server on the network and its resources are made available through the NOS software.

Two common network peripherals are printers and backup units, since both types of devices are essential for users and network administrators, respectively. Dedicated print and backup servers can be used if network printing or backup activities significantly increase the response time of a general-purpose file server.

Print server operation. A print server operates as a centralized printing resource, controlling the printing activities of multiple printers. It is not always necessary to connect all network printers to a print server. In some cases, a print server can control printers connected to both itself and to one or more stations—as long as those stations are powered on and connected to the network.

Print servers commonly use a technique called spooling, a term derived from the phrase simultaneous peripheral operation online. A spool is a combination of hardware and software that intercepts any file sent to a network printer and stores it temporarily on a network storage device. These print files are organized and sent to the appropriate printer on a first-in, first-out basis. If necessary, the printing sequence can be changed by an administrator or an authorized user (e.g., a document that is several hundred pages long can be scheduled to begin printing after office hours, to allow all users fair access to a printer).

Spooling print files eliminates two common problems associated with printers:

- A spooled file is stored on a hard disk, and can be forwarded to a printer in small increments. This is useful when the file to be printed is larger than the memory available in the printer, which can occur with large image files. Without spooling, a printer may be unable to process and print such files without an increase in its memory capacity.
- A spool can accept data at a faster rate than a printer, since data can be written to a hard disk faster than it can be printed. This makes user applications more responsive, especially when large files (or a large number of files) need to be printed, because the time needed to send a file to the printer is reduced due to the presence of the spool.

NOTE: Some network printers can be equipped with an internal hard disk or additional memory to store the spooled files before printing.

Backup server operation. A backup server automates the task of duplicating network data for recovery purposes in the event of data loss from a primary storage device. On most networks, both servers and stations are equipped with hard disk drives, which makes it necessary to perform backups on both types of devices.

The backup server is a combination of hardware and software, connected to the network with a NIC. On large networks, the hardware can be a tape library device that uses multiple drives and magnetic tape media with sufficient total capacity to copy all necessary data from stations and servers. The backup software must be compatible with the NOS software used on the network, since the backup process requires authorization to access and copy files on servers controlled by the NOS software (see Figure 3.8).

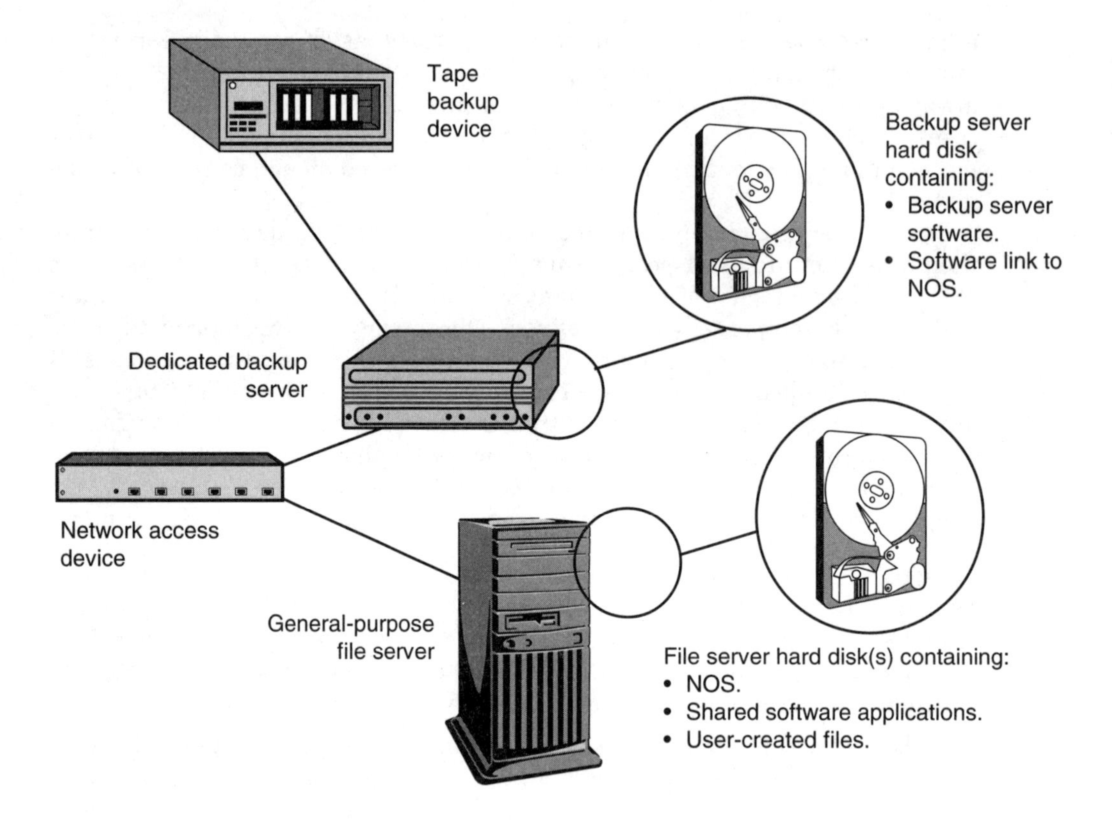

Figure 3.8 Dedicated backup server.

Backups typically take place at specific intervals (e.g., after office hours every night). In cases where the data is considered critical, a backup server can be configured to operate continuously, monitoring the network for new or changed files and backing them up immediately. In all cases, three general types of backups can be performed.

- A full backup, which copies all files.
- An incremental backup, which copies all new and changed files since the last backup (full or incremental).
- A differential backup, which copies all new and changed files since the last full backup.

In addition to these general forms of backup, it is also possible to configure a customized backup, which includes or excludes any number of individual files or groups of files on the network.

Network Access Devices

General

The term network access device is used to describe equipment that interconnects stations, servers, and shared peripheral devices on a LAN. It also describes devices used to link one LAN to another on an internetwork. In some cases, a single network access device can do both, if it is equipped with multiple LAN and internetwork interfaces.

The OSI model can be used to classify network devices in terms of functionality. The bottom three layers of the model—Physical, Data Link, and Network—are associated with most network access device operations.

- The Physical layer deals with signaling over a shared medium. It is associated with LAN access devices such as hubs (see Figure 3.9).
- The Data Link layer deals with communications between devices in the same broadcast domain. It is associated with LAN access devices such as switches (see Figure 3.10).
- The Network layer deals with communications between LANs, where each LAN is defined as a single broadcast domain. It is associated with internetwork access devices such as routers (see Figure 3.11).

The upper four layers of the OSI model—Transport, Session, Presentation, and Application—deal with network services for messages. When systems using different message formats need to exchange messages, devices called gateways are used. For example, a PC-based LAN and a mainframe connected to the same internetwork require a gateway to exchange messages. The gateway is responsible for message translation before the lower layer network access devices can process the message (see Figure 3.12).

NOTE: The term gateway is also used for identifying equipment connected to wide area network (WAN) telecommunications circuits.

The remainder of this section describes hubs, switches, and routers in detail.

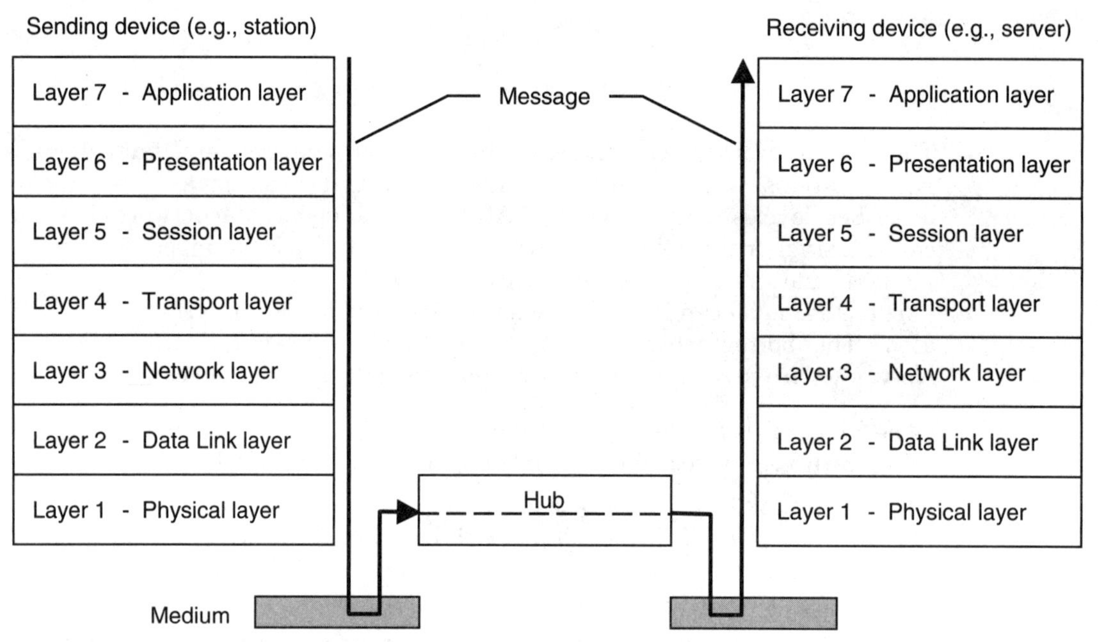

Figure 3.9 Hubs and the Open Systems Interconnection model.

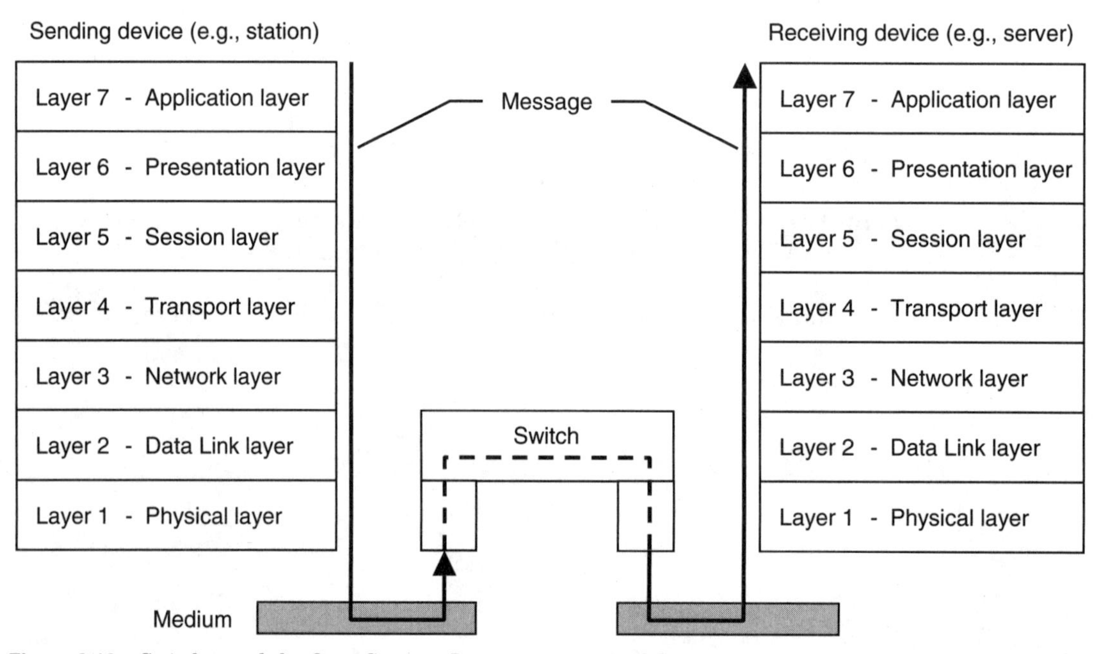

Figure 3.10 Switches and the Open Systems Interconnection model.

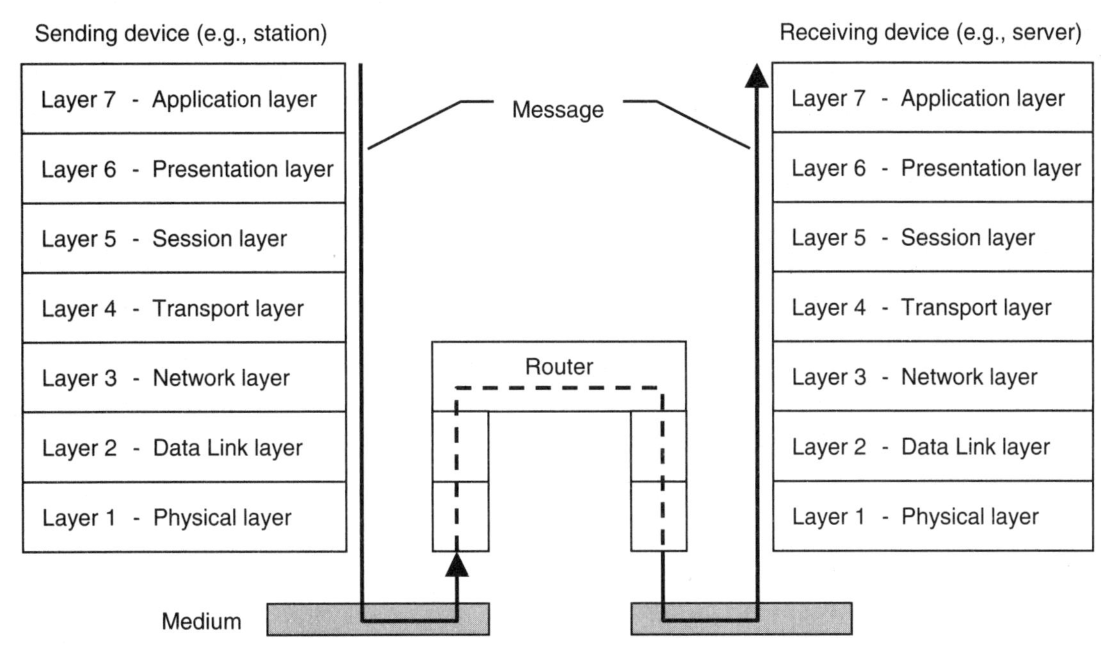

Figure 3.11 Routers and the Open Systems Interconnection model.

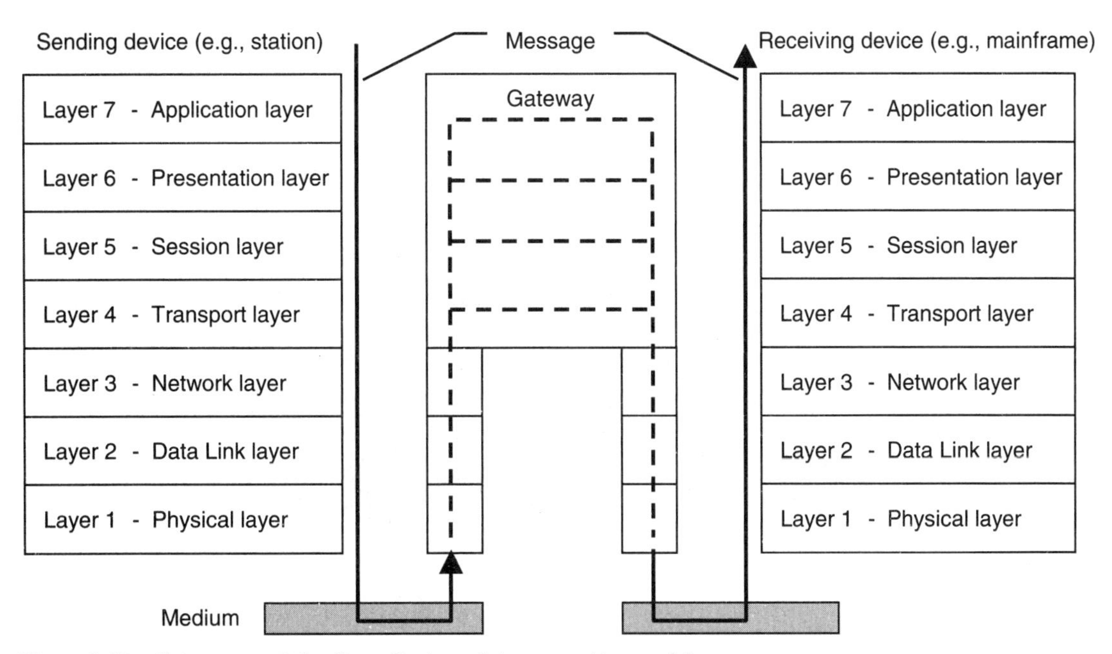

Figure 3.12 Gateways and the Open Systems Interconnection model.

Hub

A hub is a network access device that provides a centralized point for LAN communications, media connections, and management activities. By monitoring all attached stations, servers, and shared peripheral devices, the hub can inform administrators of network traffic levels, error conditions, or device failures. In most cases, a hub is also capable of disconnecting a device experiencing a fault condition from the LAN, improving network reliability.

Before the introduction of hubs, a LAN was formed by connecting network devices to a common length of cable—the cable became the shared communications medium (see Figure 3.13). However, this configuration is not reliable, since any single media failure can disable, or crash, the entire network. Functionally, a hub is the equivalent of a short length of cable in a box, to which all stations, servers, and shared peripheral devices are attached. Each connection point on a hub is called a port (e.g., a hub capable of linking eight devices is described as an 8-port hub).

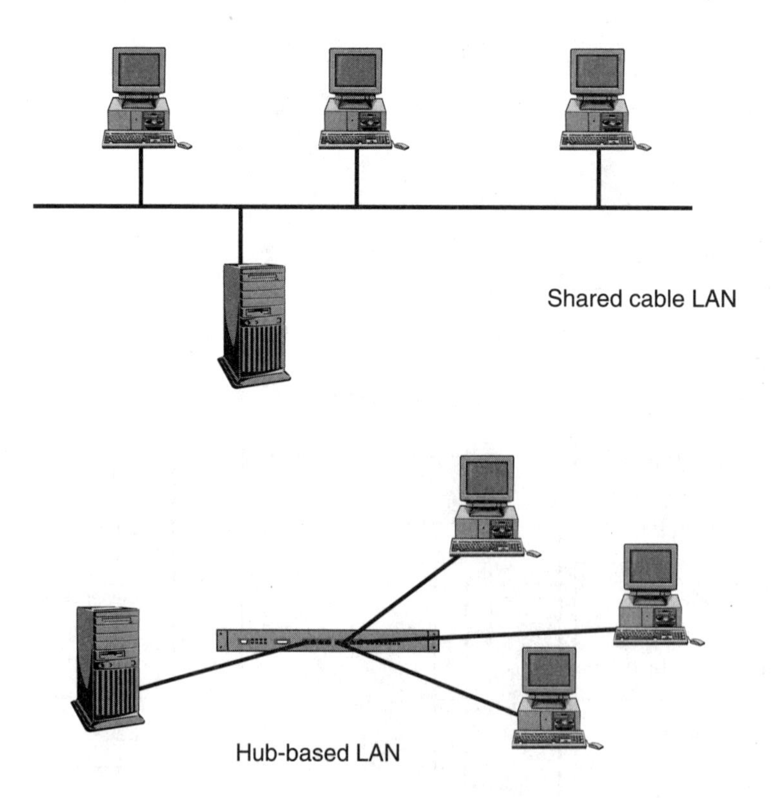

Figure 3.13 Shared cable versus hub-based local area networks.

NOTE: Because a hub concentrates all cabling into a common device, other terms used to describe hubs include multiport repeaters, concentrators, wiring concentrators, and multistation access units (MAUs).

Since a hub is centralized, a message does not travel directly from the sending device to the receiving device. The message must first travel from the sending device to the hub, which transmits the message to all connected devices. To maximize the physical span of the network (also called the maximum network diameter), most hubs amplify or regenerate incoming signals before retransmitting them. The circuitry used to amplify or regenerate signals is called a repeater. Therefore, a hub can also be described as a multiport repeater—an incoming message on one port is transmitted out of all other ports.

All devices connected to a hub share a single communications channel. If two or more devices attempt to transmit at the same time, their messages collide and become unreadable, or corrupted. Therefore, only one device at a time can be allowed to transmit its message, which is then received by all devices connected to the hub. Various techniques are used to allow devices to contend equitably for access to the shared communications channel—the most common of which is called collision detection. In collision detection, all devices that communicate through a common hub are sometimes described as sharing a single collision domain, although the term contention domain can also be used. Similarly, the term network diameter can also be described as the collision diameter.

In addition to linking stations, servers, and shared peripheral devices, hubs can be connected to each other to expand the physical span of the network, the number of connected devices, or both (see Figure 3.14). The connections between hubs can use the same medium as the other devices or the hubs can be equipped with media converters to increase the network diameter. For example, a hub can be equipped with 24 ports for connecting to stations, servers, and shared peripheral devices using copper cabling and a single port for connecting to another 24-port hub using optical fiber cabling.

The term stackable hub describes hubs that are in close proximity, capable of being connected to each other using a short length of specialized cable, and functioning together as a single unit. Such units allow for incremental LAN expansion—additional stations, servers, and shared peripheral devices can be purchased and added to the LAN as required. If more ports are needed, a second or third hub can be stacked on top of the existing unit(s).

Both the maximum number of hubs that can be connected to each other in a single shared domain and the maximum network diameter depend on the type of LAN technology used. The two most common LAN technologies are called Ethernet and token ring.

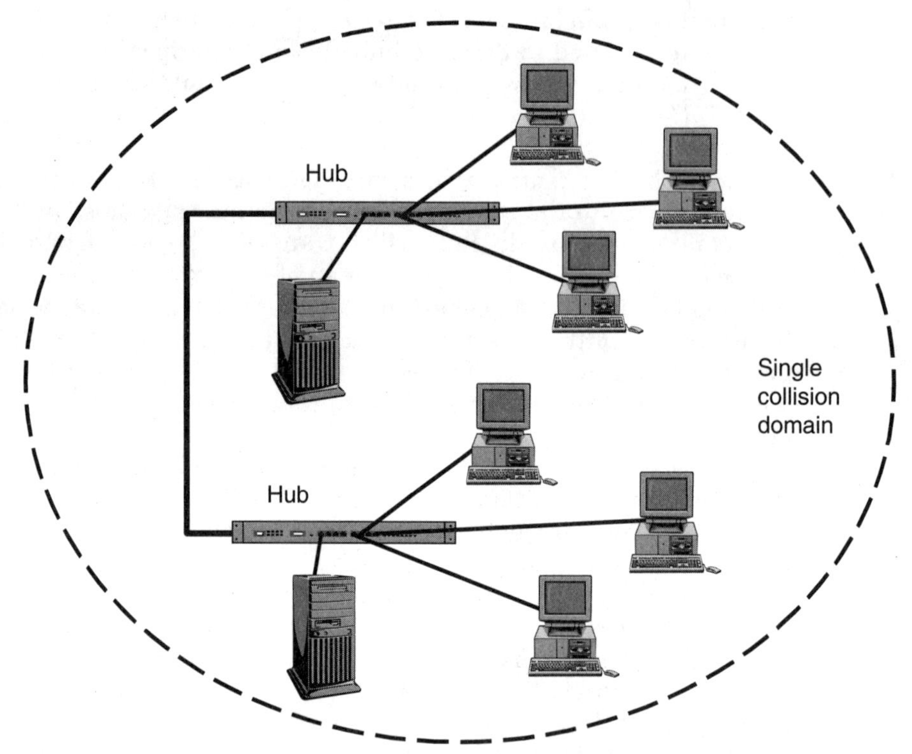

Figure 3.14 Collision domain created with multiple linked hubs.

Managed hub. A managed hub is capable of being monitored and controlled through network management software. Network administrators can observe traffic conditions and collect operational status information on the hub as well as all devices connected to the hub. If needed, diagnostic routines can be initiated to discover the cause of a fault condition. All of these actions can be performed remotely through a centralized management console, which saves time by eliminating the need to physically inspect the unit(s).

NOTE: Additional management issues are discussed in Chapter 9: Network Management.

Passive hub. Most hubs are active units—devices that require electrical power to operate. Active hubs can amplify, retime, or regenerate incoming signals to minimize errors and message corruption. Some LAN technologies such as token ring permit the use of passive hubs, which do not require electrical power to operate. Such devices function as simple signal distribution units, where an incoming signal on one port is directed to another port with no amplification, retiming, or regeneration. For this reason, the use of passive hubs reduces the maximum network diameter on a LAN.

Switch

A switch is a network access device that provides a centralized point for LAN communications, media connections, and management activities—like a hub. Unlike a hub, where all ports connect to a single communications channel, each port on a switch is a separate communications channel representing a different collision domain. This makes it possible for multiple devices on a switch-based network to transmit simultaneously, as long as they are connected to different ports on the switch.

NOTE: A port on a hub is referred to as a shared port, whereas a port on a switch can be described as a dedicated port.

Before the introduction of switches, less complex devices called two-port bridges were used to divide a single congested LAN into two separate collision domains. The purpose of a bridge is to improve network response time by reducing the number of devices sharing a single communications channel. A typical two-port bridge functions as follows:

1. Before the installation of the bridge, two or more hubs are connected to each other, forming a single collision domain (see Figure 3.15).
2. When the bridge is installed, each of its two ports is connected to a different hub, forming two collision domains (see Figure 3.16).
3. During operation, the bridge inspects the destination device address of all messages transmitted on each collision domain. If a message needs to pass from the hub connected to one bridge port to the hub connected to the other bridge port, the bridge performs the transfer and forwards the message. If the message does not need to cross the bridge, the message is described as discarded or filtered.
4. The hubs, stations, servers, and shared peripheral devices connected to the network are not aware of the existence of the bridge. For this reason, these types of bridges are sometimes described as transparent bridges, since they can be incorporated into a network without making any changes to other components.

A simple switch functions in a similar manner to the two-port bridge described previously, but it is typically equipped with many more ports. Therefore, the term multiport bridge can also be used to describe a switch. During operation, a switch inspects the destination device addresses of messages arriving on all ports and forwards each message out of the port connected to the destination device.

It is possible to connect hubs as well as stations, servers, or shared peripheral devices to a port on a switch. Both switches and hubs can be deployed in various configurations to form a hybrid LAN, made up of both shared and dedicated communications channels.

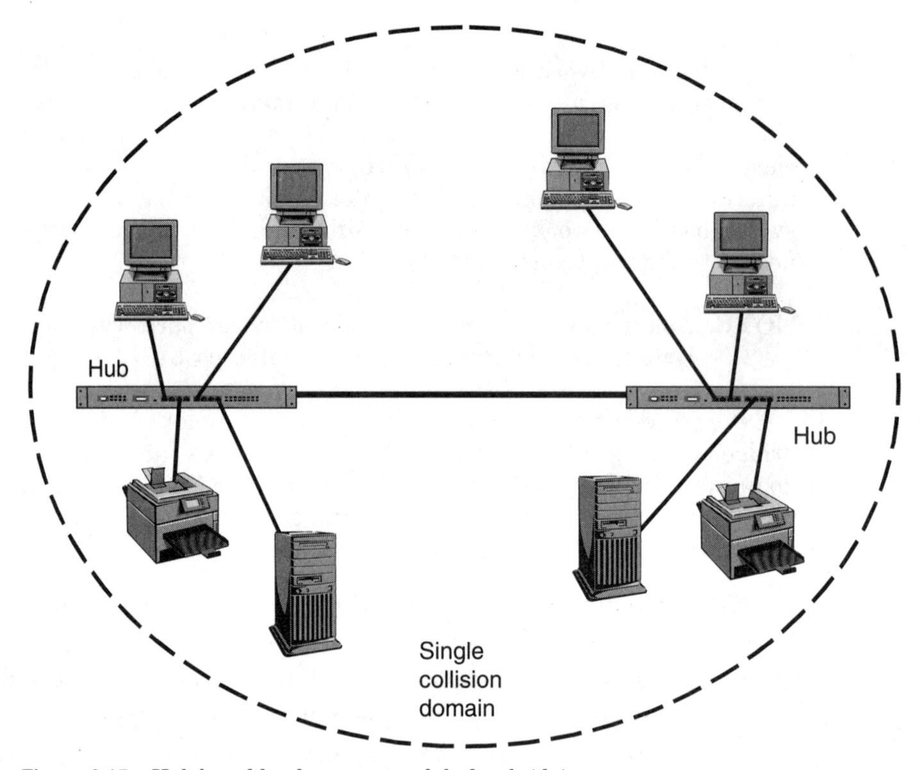

Figure 3.15 Hub-based local area network before bridging.

NOTE: It must be emphasized that a simple switch creates multiple collision (or contention) domains, not multiple broadcast domains. On a simple switch, any incoming message addressed to all devices will be forwarded out of every port—other than the port the message arrived at.

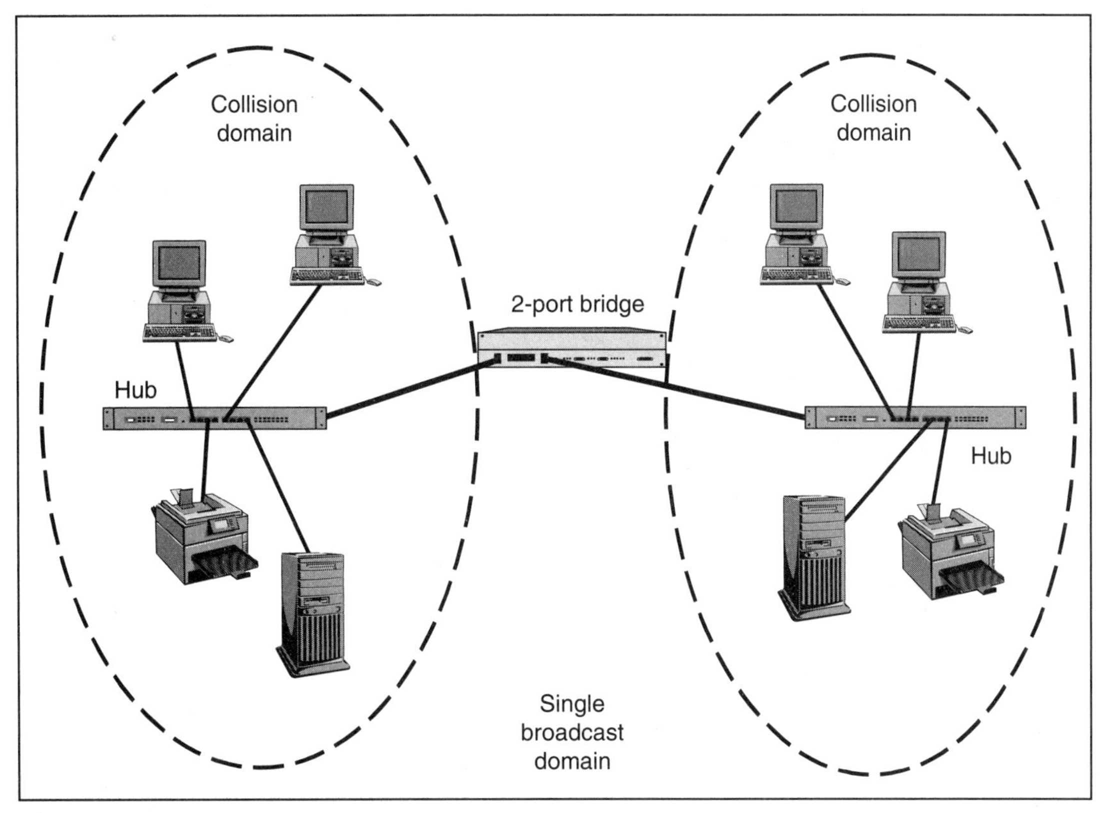

Figure 3.16 Hub-based local area network after bridging.

Although simple switches make it possible to add more devices to a LAN by creating multiple collision domains (see Figure 3.17), they cannot contain the subsequent increase in broadcasts. At some point, the level of broadcast traffic will increase network response time significantly, making it impractical to continue adding devices to the LAN. When this happens, a router can be used to create multiple broadcast domains, in the same way a simple switch is used to create multiple collision domains.

NOTE: Switches that are more complex can be used in place of routers to control broadcast traffic, using a technology called virtual LAN (VLAN). Switch operations are described in detail in Chapter 6: Switching and Virtual LANs (VLANs).

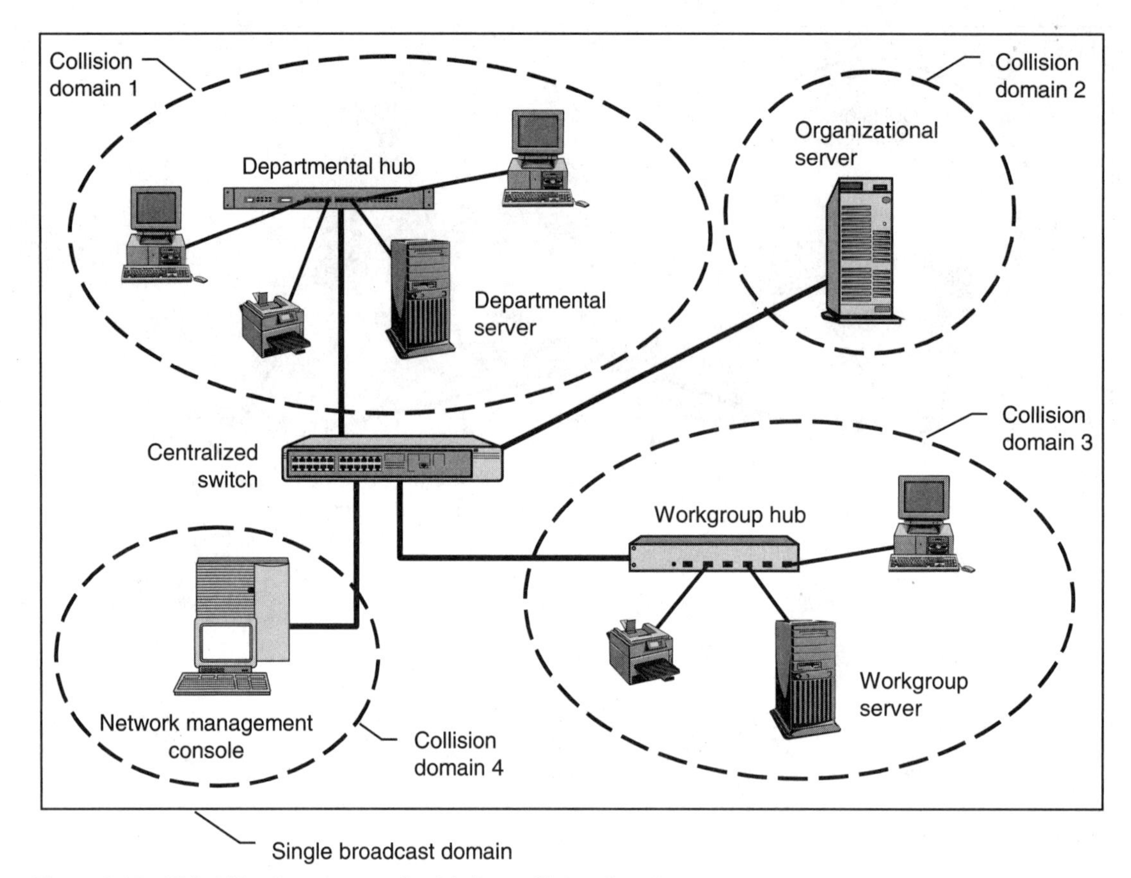

Figure 3.17 Hybrid local area network with four collision domains.

Router

A router is a network access device that provides a centralized point for communications between networks—where a network is defined as a single broadcast domain containing any number of interconnected hubs, switches, or both (see Figure 3.18). Like switches, routers forward messages based on the addressing information contained in the message. However, the address inspected by a router is different from the address inspected by a switch. Expressed in terms of the OSI model, the addressing information used by a router is the Network layer address, whereas a switch will process messages based on Data Link layer addressing information. Addressing is described in detail in Chapter 5: Network Communications.

It is possible to connect multiple networks to a single router. In such cases, the router makes it possible to maintain each network as a separate broadcast domain. A message addressed to all devices within one network will not be transferred to any of the other networks. Each of the networks connected to

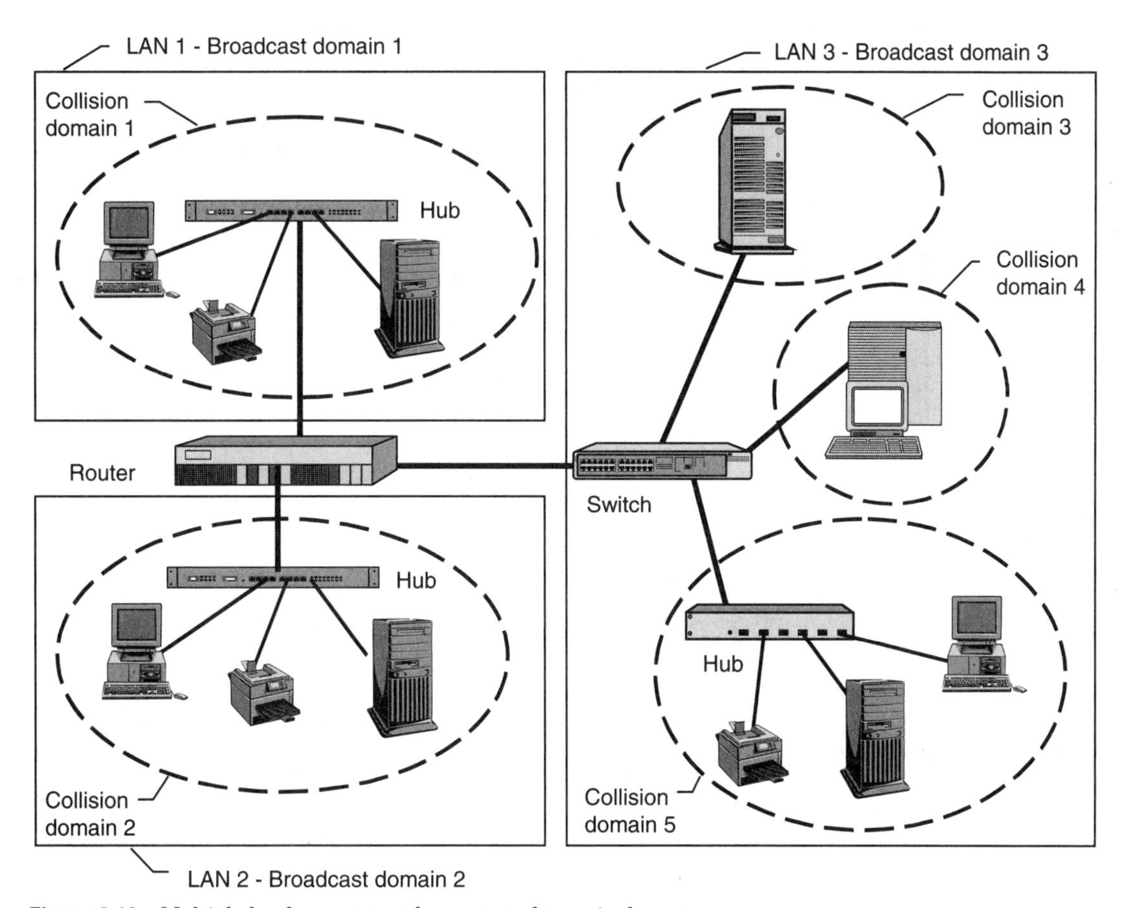

Figure 3.18 Multiple local area networks connected to a single router.

the router can use a different technology (e.g., Ethernet and token ring networks can exchange messages through the router).

Organizations with multiple offices and many LANs to interconnect must use multiple routers linked together to form an internetwork (see Figure 3.19). On a complex internetwork, one or more of the routers can be linked to several other routers, creating a mesh configuration—multiple possible paths between any two networks connected to different routers. In such environments, every router must be aware of the existence and status of the available paths to a destination network. A message to be routed to another network can be sent over a specific path based on its contents, the level of traffic on a given path, the costs assigned to various paths, or other criteria.

The software used to control and coordinate router operations is often described as the internetwork operating system (IOS) software, to distinguish it from the NOS software used on LANs.

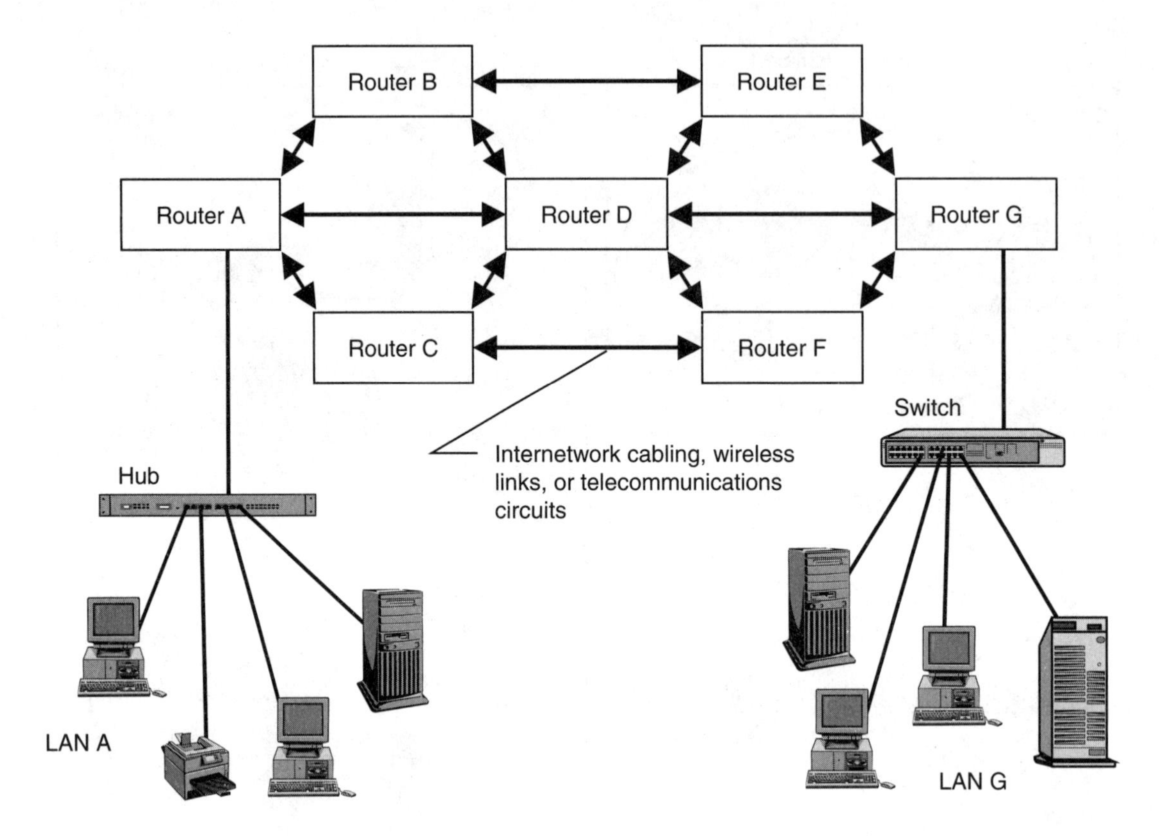

Figure 3.19 Multiple routers in an internetwork.

NOTE: In Figure 3.19, LANs connected to routers B–F have been omitted to simplify the illustration.

Routing and switching functionality can be incorporated into a single device. Such hybrid units are typically modular in form, making it possible to link various types of LANs as well as other routers to a centralized unit. To obtain Internet connectivity, an organizational router can also be linked to a router owned by an Internet service provider (ISP [see Figure 3.20]).

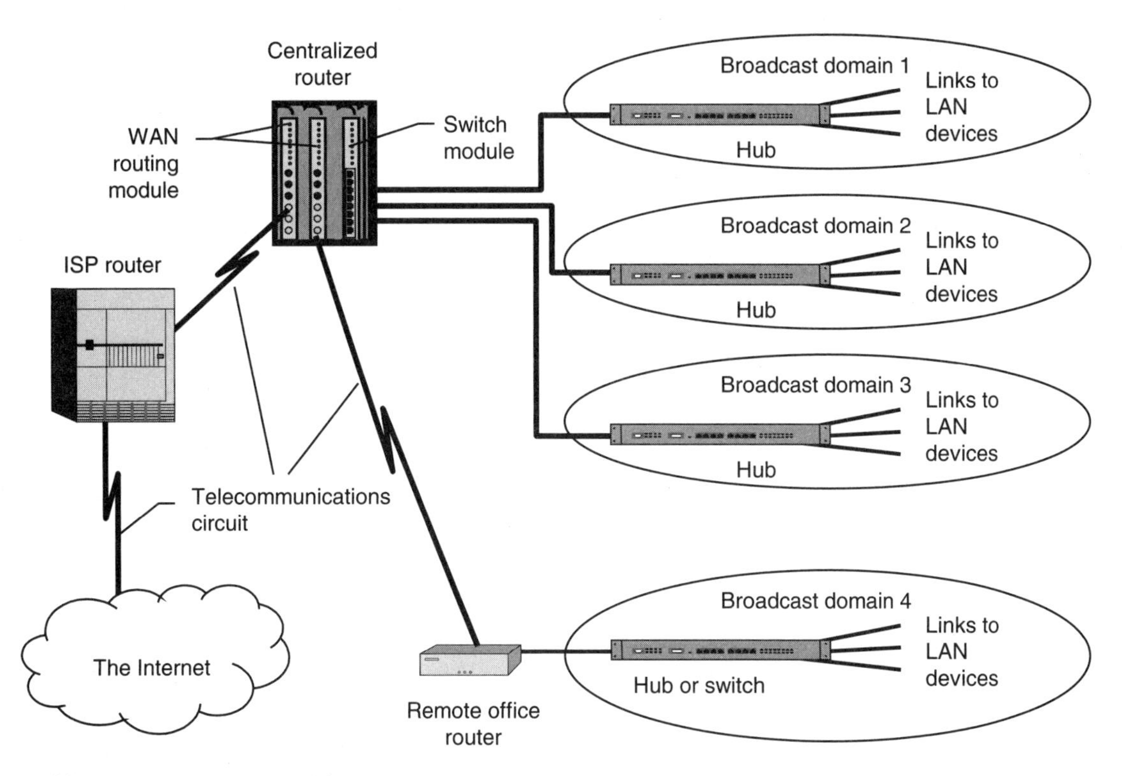

Figure 3.20 Centralized router connected to the Internet.

Chapter

4

Network Connections

Chapter 4 describes the cabling and telecommunications technologies used to link devices on local area networks (LANs) and internetworks. When the devices to be linked are in close proximity to each other—in the same building or in different buildings on the same campus—one or more types of cabling media can be used to provide network connections. In cases where the distance between the devices to be linked extends beyond the campus, telecommunications circuits provide the necessary connections.

Overview

General

This chapter describes the technologies used to link devices on local area networks (LANs) and internetworks. When the devices to be linked are in close proximity to each other—in the same building or in different buildings on the same campus—one or more types of cabling media can be used to provide network connections. In cases where the distance between the devices to be linked extends beyond the campus, telecommunications circuits provide the necessary connections.

NOTE: Wireless technologies can also be used to link network components over a variety of distances.

For both single building and campus connections, the physical span of the network is used to determine the scope of the cabling infrastructure, as follows:

- A small network typically requires only a single-floor cabling infrastructure, also referred to as horizontal cabling (e.g., a departmental LAN serving a group of users located in a common area on one floor of a building).
- A mid-sized network spanning various floors of a building requires both horizontal and vertical (or backbone) cabling.
- A large network spanning two or more multi-floor buildings on a campus requires horizontal, in-building (or intrabuilding) backbone, and campus (or interbuilding) backbone cabling.

Various types of cabling media can be used for horizontal and backbone connectivity, including the following:

- Copper cabling
  - Coaxial
  - Shielded twisted-pair (STP-A)
  - Screened twisted-pair (ScTP)
  - Unshielded twisted-pair (UTP)
- Optical fiber cabling
  - Singlemode optical fiber
  - Multimode optical fiber

For temporary connections over extended distances, dial-up telecommunications circuits can be used (e.g., remote access by mobile users equipped with laptop computers). If a connection needs to be available at all times, dedicated (or leased) telecommunications circuits are required (e.g., the connection between an organizational network and the Internet). Multiple technologies and data transfer rates are available for both dial-up and leased connections, including:

- Dial-up telecommunications circuits
  - Public switched telephone network (PSTN)
  - Integrated services digital network (ISDN)
- Leased telecommunications circuits
  - T-carrier
  - X.25
  - Frame Relay (FR)
  - Digital subscriber line (DSL)

– Synchronous optical network (SONET)
– Asynchronous transfer mode (ATM)

NOTE: ATM is a complex LAN and internetworking technology that can be used to connect devices in a single building, on a campus, or over extended distances.

Cabling Infrastructure

General

A network cabling infrastructure is a combination of products and design practices that allows for the transport of signals between connected devices in a building or campus. In many cases, a cabling infrastructure is implemented without knowledge of the specific network technologies or devices that are to be connected (e.g., during the initial construction of a multi-tenant commercial building). Alternatively, an existing network implemented at the same time as the cabling infrastructure may require replacement due to inadequate performance. When network devices are replaced, it may be necessary to verify if the new equipment can function using the cabling infrastructure in place.

To provide compatibility with existing and future network technologies, standards have been developed to guide manufacturers, telecommunications distribution designers, telecommunications cabling installers, and buyers of cabling infrastructures. The standards make it possible to implement a cabling infrastructure using a structured approach, where a common set of design and performance guidelines ensures compatibility between many types of products and networks. The term structured cabling system also describes this type of cabling infrastructure.

Structured cabling standards—described in detail later in this chapter—make it possible to implement a cabling infrastructure capable of accommodating a wide variety of LAN and internetwork technologies. From the perspective of such standards, the cabling infrastructure in any building or campus is described as a collection of the following elements:

- Horizontal cabling
- Backbone cabling
- Work areas
- Telecommunications rooms (TRs)
- Equipment rooms (ERs)
- Entrance facilities (EFs)

NOTE: The term cabling describes both cable and the components used to provide connections to the cable (e.g., cross-connect panels, telecommunications outlet/connectors, and various types of cordage [see Figure 4.1]).

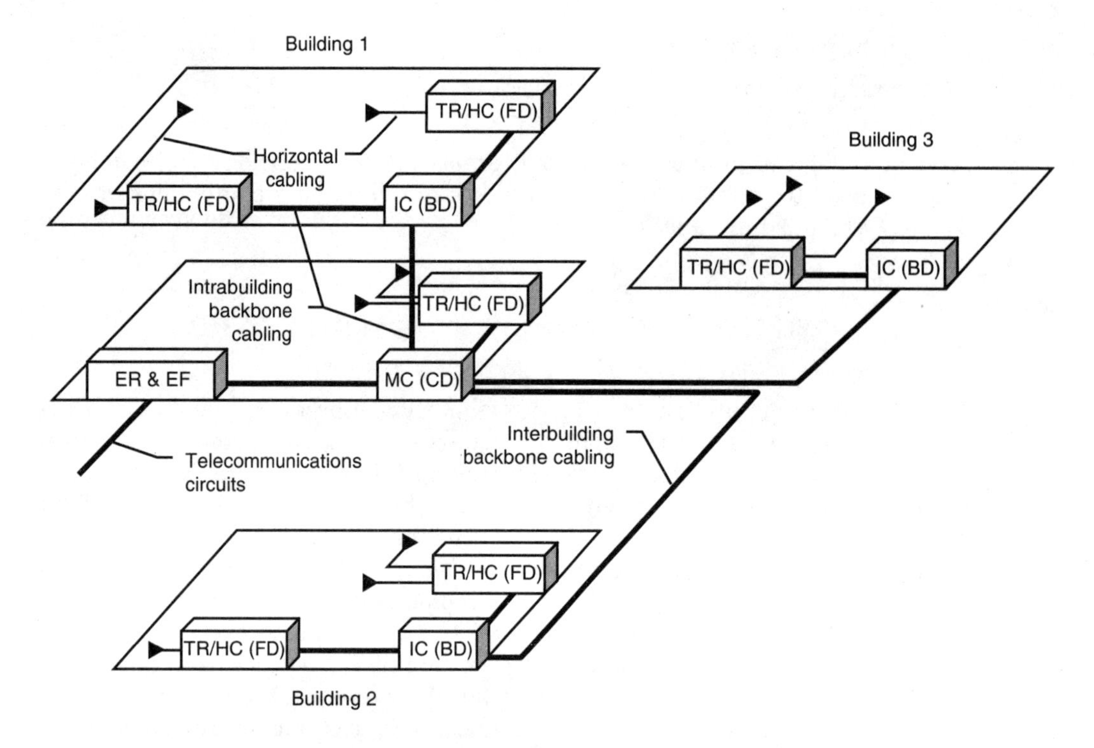

Figure 4.1 Campus cabling infrastructure.

NOTES: The terms ER, EF, and TR are used to describe telecommunications spaces. These spaces contain telecommunications equipment, the associated telecommunications cabling, and accessories.

The terms main cross-connect (MC), intermediate cross-connect (IC), and horizontal cross-connect (HC) are used to describe the functional cross-connection of telecommunications cabling. Any of these cross-connection types can be located in any of the telecommunications spaces.

The international standards' counterparts to MC, IC, and HC are campus distributor (CD), building distributor (BD), and floor distributor (FD), respectively.

Media types

The availability of many types of networking technologies and the variety of building and campus designs has made it difficult to select a single type of cabling for all network implementations. Criteria such as performance, cost, and expected lifespan are commonly used to evaluate various media. It is also common to use more than one type of medium to link various devices within a building or campus.

Three types of media can be considered for use as network cabling:

- Coaxial copper cable
- Twisted-pair copper cable
- Optical fiber cable

Coaxial cable. Coaxial cable is commonly referred to as coax. While there are many different types of coax, they are all similar in construction. Such cables consist of a central core—either a solid conductor or stranded wire—surrounded by a layer of insulating material. A metallic wire braid, a metalized plastic tape, or a metallic sleeve envelops this insulation. An outer jacket covers the entire assembly. This design inherently minimizes inbound interference from external sources as well as outbound radiation of electromagnetic energy from the cable (see Figure 4.2).

NOTE: Transmission over coaxial cable relies on signal transmission on the center conductor with respect to a zero potential reference provided on the outer shield. This type of transmission is sometimes referred to as unbalanced.

An advantage associated with coaxial cabling is that due to its construction, coax cabling is less susceptible to interference than twisted-pair cable.

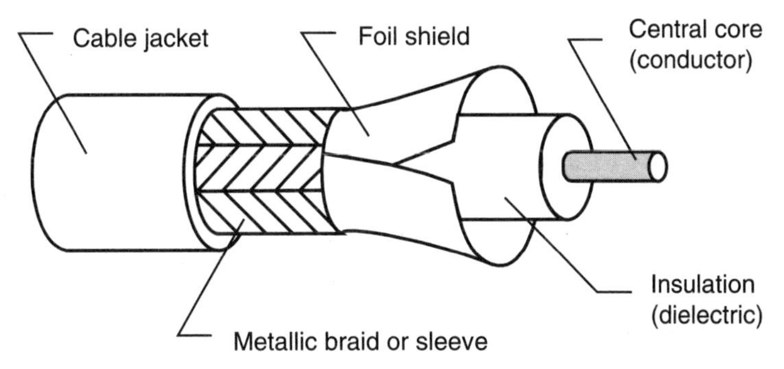

Figure 4.2 Coaxial cable.

A disadvantage associated with coaxial cabling is that coax cables are application dependent. A variety of coax cables are manufactured, each with a specific purpose and transmission characteristics. This restricts flexibility and prevents using a single type of coax to support multiple types of network technologies.

Twisted-pair cable. To form a twisted-pair, two individually insulated copper wires are physically twisted together. This twisting prevents undesirable electromagnetic interference (EMI) between separate communications channels and systems.

Typically, a twisted-pair consists of two copper conductors. Each conductor is covered with a thermoplastic insulation such as polyvinyl chloride (PVC), fluorinated ethylene propylene (FEP), or polyethylene (PE) insulation (see Figure 4.3).

Twisting the two wires together exposes each wire equally to potential interference from various sources and causes much or all of this interference to cancel. Immunity to interference is an important feature for LANs using differential signaling, where data is transmitted as the difference in voltage levels between the two conductors. Interference above a given threshold will cause errors, which must be resolved by various network error-correction mechanisms.

A twisted-pair cable consists of multiple twisted-pairs covered by an overall sheath or jacket. Varying the number of twists for each pair relative to the other pairs in the cable—known as the lay—can greatly reduce interference between signals on the different pairs. Such interference is referred to as crosstalk.

In addition to its sheath, the cable may have an overall shield, each pair may be individually shielded, or both. Shielding can increase the cable's immunity to external noise, also referred to as EMI. In instances where an excessive amount of electromagnetic energy radiates from the cable, shielding may be required to comply with appropriate regulations. Such radiation is referred to as radio frequency interference (RFI).

NOTE: The term electromagnetic compatibility (EMC) can also be used to describe the combined EMI and RFI performance of a component or system.

Figure 4.3 Twisted-pair.

There are three common types of twisted-pair copper cable:

- UTP
- ScTP
- STP-A

NOTE: Twisted-pair and quad cables are sometimes referred to as balanced cables.

Unshielded Twisted-Pair (UTP) cable. Historically, UTP cable was referred to as telephone wire. This is no longer an appropriate designation. Technical advances have transformed UTP into a high-quality medium capable of accommodating very high data transfer rates (see Figure 4.4).

UTP performance is frequently expressed using a scale based on categories. The American National Standards Institute/Telecommunications Industry Association/Electronic Industries Alliance (ANSI/TIA/EIA) has developed a numerical scale to express UTP performance. Category 3 is considered the minimum acceptable level of performance for networking. Category 5e is the ANSI/TIA/EIA-568-B.1, *Commercial Building Telecommunications Cabling Standard, Part 1: General Requirements*, recommended minimum twisted-pair cabling category. A future standard is expected to define the performance characteristics of Category 6.

NOTE: UTP should not be confused with the four-wire cable referred to as quad. Quad or inside wiring (IW) consists of four individually insulated wires covered by an overall jacket. Some quad cables exhibit significant crosstalk interference and are not considered to be an acceptable network medium. Such cables are primarily used for telephone service, typically in residential units.

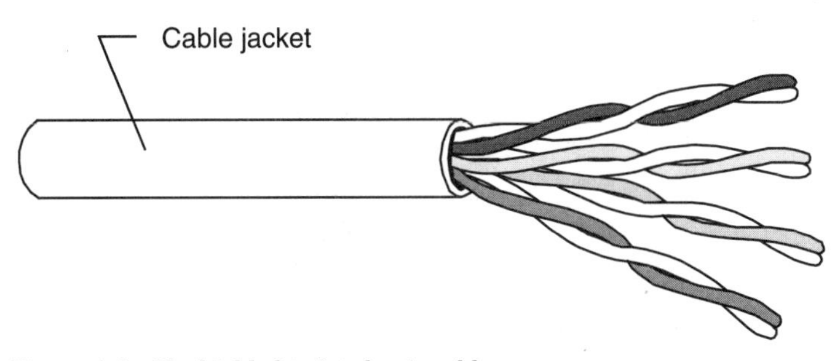

Figure 4.4 Unshielded twisted-pair cable.

Advantages associated with UTP cabling include the following:

- It has a large installed base and is a familiar technology.
- Most network technologies readily operate over UTP cabling systems.

Disadvantages associated with UTP cabling include the following:

- High-quality UTP cabling systems require strict adherence to recommended installation practices and procedures.
- UTP is potentially more sensitive to EMI than other media.
- A UTP communications channel between any two devices is limited to 100 m (328 ft) for most network technologies.

Screened Twisted-Pair (ScTP) cable. ScTP cable—sometimes referred to as foil twisted-pair (FTP) cable—is very similar in construction to UTP cable. However, ScTP cables have an additional overall foil shield and drain wire. The foil shield blocks both inbound and outbound electromagnetic energy through the cable jacket. The drain wire is placed in contact with the foil and is used as the bonding element for the shield (see Figure 4.5).

An advantage associated with ScTP cabling is that it provides better EMC performance than UTP in environments where a high level of immunity or low levels of emission are important.

Disadvantages associated with ScTP cabling include the following:

- Any break or improper grounding of the foil shield reduces the cable's overall effectiveness.
- It is a heavier and bulkier cable than UTP, with a more labor-intensive installation.

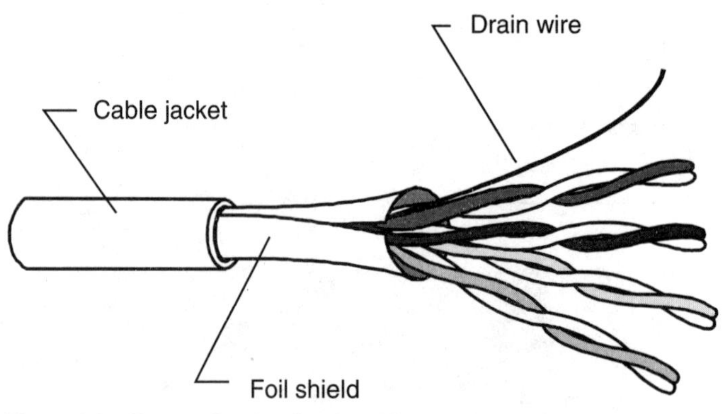

Figure 4.5 Screened twisted-pair cable.

Shielded Twisted-Pair (STP-A) cable. STP-A cables have each wire pair as well as the combined grouping of all pairs covered with a layer of shielding to minimize interference-related problems (see Figure 4.6).

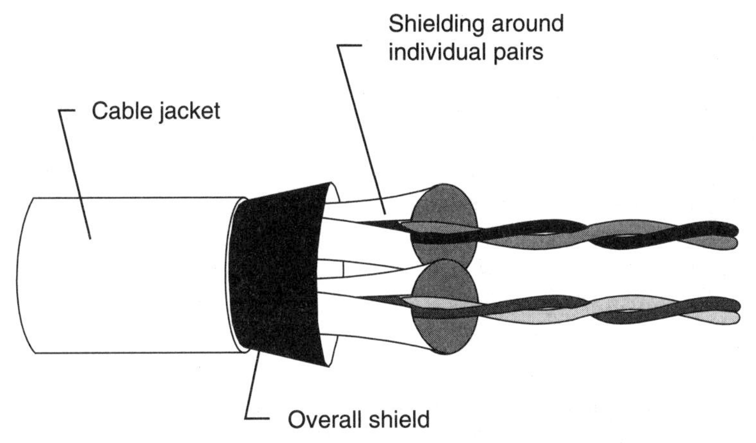

Figure 4.6 Shielded twisted-pair cable.

Advantages associated with STP-A cabling include the following:

- It provides better performance than UTP in environments where a high level of immunity, a low level of emissions, or both are critical.
- Provides better crosstalk performance than UTP and ScTP.

Disadvantages associated with STP-A cabling include the following:

- Any break or improper grounding of the shield reduces the cable's overall effectiveness.
- It is a heavier and bulkier cable than UTP or ScTP, with a more labor-intensive installation.

NOTE: STP-A should not be confused with ScTP. The two cable types have different construction and performance characteristics. One cannot be substituted for the other.

Optical fiber cable. Optical fiber cable uses glass fibers to transport signals between network devices. The signals are in the form of light pulses rather than the electrical pulses used with all types of copper cable.

NOTE: When plastic is used in place of glass, the cable is referred to as plastic optical fiber (POF).

Optical fiber strands are thin filaments of glass consisting of an inner core and an outer cladding. Both the core and the cladding are made of glass, each with a different refractive index. Signals are in the form of light pulses that travel through the core of the optical fiber. When a light pulse strikes the cladding, it is reflected back into the core, because the glass used in the cladding has a lower refractive index than the core. This mechanism transports all light pulses emitted by the device at one end of the optical fiber link to the device at the other end.

The diameter of the core varies with the type of optical fiber. Singlemode optical fiber has a core diameter less than or equal to 10 μm while multimode optical fiber typically has a core diameter of 62.5 μm or 50 μm. The cladding diameter for both singlemode and multimode optical fiber is 125 μm. These dimensions are summarized and explained in Table 4.1 (see Figure 4.7).

TABLE 4.1 Optical fiber core and cladding diameters.

	Singlemode	Multimode
Core	Less than or equal to 10 μm	62.5 μm or 50 μm
Cladding	125 μm	125 μm

NOTES: The symbol μm represents a unit of length known as a micron or micrometer. It is equal to one-millionth of a meter (0.000001 meter).

Optical fiber refers to the cabling medium while the term fiber optics is used to describe various transmission technologies that use optical fiber media.

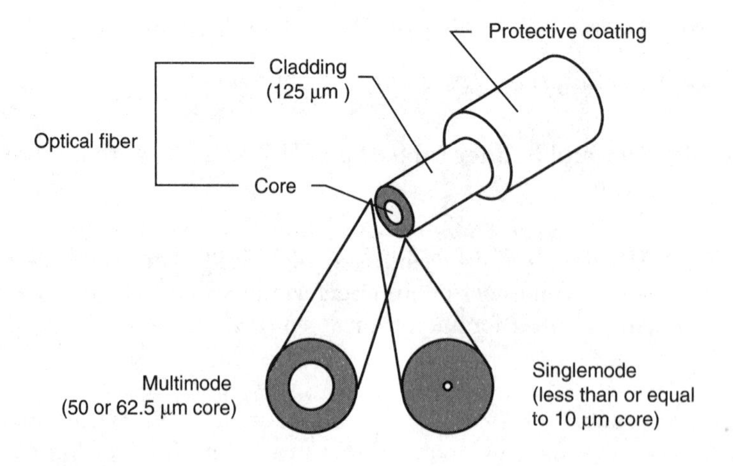

μm = Micron

Figure 4.7 Optical fiber strand.

Advantages associated with optical fiber cabling include the following:

- Light pulses traveling through glass lose less energy than electrical pulses traveling the same distance through copper. This allows network devices to be linked over longer distances with optical fiber cabling.

NOTE: The term attenuation is used to describe the decrease in the energy level of a signal.

- Optical fiber cabling is immune to EMI and RFI since signals are sent as light pulses and not as electromagnetic energy.

A disadvantage associated with optical fiber cabling is that due to the necessary conversions between light-based signaling and electricity-based computing, additional network components are required when using optical fiber cabling.

NOTE: The term optoelectronics is used to describe the components or circuitry used for such conversions.

Structured cabling standards

Multiple organizations are involved in the development of standards associated with network cabling. Some of these organizations focus on specifications for the manufacturing and testing of components (e.g., various types of cables and connectors), while others develop guidelines for the design, installation, or administration of cabling infrastructures.

To minimize duplication of effort or conflicting recommendations, organizations responsible for developing cabling-related standards are structured in a hierarchy on a worldwide basis. At the top of the hierarchy, the International Organization for Standardization/International Electrotechnical Commission (ISO/IEC) publishes the International Standard 11801:2000 – *Generic Cabling for Customer Premises*.

The ISO/IEC standard is developed using recommendations submitted by standards organizations representing individual countries or groups of countries. For example, the American National Standards Institute (ANSI) is responsible for the coordination, formalization, and adoption of national standards in the United States. Similarly, the Comité Europeén de Normalisation Electrotechnique (CENELEC) issues standards representing contributions from 19 European countries. CENELEC is sometimes referred to as the European Committee for Electrotechnical Standardization.

NOTE: This chapter uses cabling standards developed in the United States as the basis for its guidelines and examples. Applicable national standards should be observed when designing cabling systems.

Two organizations responsible for developing cabling-related standards in the United States are the Telecommunications Industry Association (TIA) and the Electronic Industries Alliance (EIA). TIA focuses on products used in the communications and information technology sectors, including cabling systems. EIA covers a broader range of products, including electronic components, consumer electronics, telecommunications, and information technology. Subsequently, a number of the standards developed by TIA and EIA have been approved by ANSI and submitted to ISO/IEC as United States national standards.

The cabling system standards developed by the TIA are commonly referenced when designing a network cabling infrastructure for a building or a campus. The topics covered in TIA standards related to cabling include design recommendations for the following:

- Spaces and pathway systems for network components and cabling
- Grounding and bonding systems for network components and cabling
- Building and campus cabling systems capable of transporting LAN and internetwork traffic
- Administration systems to manage cabling infrastructure

A series of committees exist within the TIA, each responsible for formulating standards in a particular subject matter. Examples include committee TR-34—Satellite Equipment and Systems and committee TR-8—Mobile and Personal Private Radio Standards. The committee responsible for cabling infrastructure standards is TR-42—User Premises Telecommunications Infrastructure. The following subcommittees have been formed within TR-42, each addressing a different subject area associated with cabling infrastructures:

- TR-42.1 *Commercial Building Telecommunications Cabling* (ANSI/TIA/EIA-568-B.1)
- TR-42.2 *Residential Telecommunications Infrastructure* (ANSI/TIA/EIA-570-A)
- TR-42.3 *Commercial Building Telecommunications Pathways and Spaces* (ANSI/TIA/EIA-569-A)
- TR-42.4 *Outside Plant Telecommunications Infrastructure* (ANSI/TIA/EIA-758)
- TR-42.5 *Telecommunications Infrastructure Terms and Symbols*
- TR-42.6 *Telecommunications Infrastructure and Equipment Administration* (ANSI/TIA/EIA-606-A)
- TR-42.7 *Telecommunications Copper Cabling Systems* (ANSI/TIA/EIA-568-B.2)
- TR-42.8 *Telecommunications Optical Fiber Cabling Systems* (ANSI/TIA/EIA-568-B.3)
- TR-42.9 *Industrial Telecommunications Infrastructure*

The following are some of the standards developed by various TR-42 subcommittees and jointly published by the TIA, EIA, and ANSI (in some cases, by the TIA and EIA).

Spaces and pathways systems. The following represent the standards associated with spaces and pathway systems for network components and cabling:

- ANSI/TIA/EIA-569-A
  Commercial Building Standard for Telecommunications Pathways and Spaces
- ANSI/TIA/EIA-569-A-1
  Addendum 1 – Surface Raceways
- ANSI/TIA/EIA-569-A-2
  Addendum 2 – Furniture Pathways and Spaces
- ANSI/TIA/EIA-569-A-3
  Addendum 3 – Access Floors
- ANSI/TIA/EIA-569-A-4
  Addendum 4 – Poke-Thru Fittings
- ANSI/TIA/EIA-569-A-5
  Addendum 5 – Commercial Building Standard for Telecommunications Pathways and Spaces—SP 4722-A-1 (draft). This document is an addendum to replace sub-clause 4.2, Underfloor Pathways, of ANSI/TIA/EIA-569-A.
- ANSI/TIA/EIA-569-A-6
  Addendum 6 – Commercial Building Standard for Telecommunications Pathways and Spaces—Multi-Tenant Pathways and Spaces—SP 3-2950-AD6 (draft).

Grounding and bonding systems. The following represents the standard associated with grounding and bonding systems for network components and cabling:

- ANSI/TIA/EIA-607
  Commercial Building Grounding and Bonding Requirements for Telecommunications

Cabling systems. The following represent the standards associated with building and campus cabling systems capable of transporting LAN and internetwork traffic:

- ANSI/TIA/EIA-568-B.1
  Commercial Building Telecommunications Cabling Standard, Part 1: General Requirements

- ANSI/TIA/EIA-568-B.2
  Commercial Building Telecommunications Cabling Standard, Part 2: Balanced Twisted-pair Cabling Components
- ANSI/TIA/EIA-568-B.3
  Optical Fiber Cabling Components Standard
- ANSI/TIA/EIA-758
  Customer-Owned Outside Plant Telecommunications Cabling Standard
- TIA/EIA-758-1
  Addendum No. 1 to ANSI/TIA/EIA-758
- ANSI/TIA/EIA-570-A
  Residential Telecommunications Cabling Standard

Administration systems. The following represents the standard associated with administration systems to manage cabling infrastructure:

- ANSI/TIA/EIA-606
  Administration Standard for the Telecommunications Infrastructure of Commercial Buildings. This document is to be updated by SP 4156 and issued as ANSI/TIA/EIA-606-A.

Horizontal cabling

The term horizontal cabling is used to describe cabling that links network devices in user work areas to network equipment located in a TR. Since one or more TRs are typically located on the same floor as the works areas, this type of cabling usually extends horizontally along the floors, walls, or ceilings.

ANSI/TIA/EIA-568-B.1 describes horizontal cabling as follows:

> "The horizontal cabling is the portion of the telecommunications cabling system that extends from the work area telecommunications outlet/connector to the horizontal cross-connect in the telecommunications room. The horizontal cabling includes the horizontal cables, telecommunications outlet/connectors in the work area, mechanical terminations, and patch cords or jumpers located in the telecommunications room, and may include multi-user telecommunications outlet assemblies and consolidation points."

The following are some of the characteristics of horizontal cabling. It:

- Contains the greatest quantity of individual cables in a building.
- Is not readily accessible—the time, effort, and skills required for changes can be significant.
- Should accommodate a variety of network technologies to minimize cabling changes if new components are installed.

Topology. ANSI/TIA/EIA-568-B.1 specifications for horizontal cabling (see Figure 4.8) include the following:

- Horizontal cabling is deployed in a physical star topology—each work area telecommunications outlet/connector connects to a HC in a TR.
- Horizontal cabling serving a given work area should terminate in a TR located on the same floor as the work area being served.
- Application-specific electrical components (such as impedance-matching devices) are not installed as part of the horizontal cabling. When needed, such components are placed external to the telecommunications outlet/connector.
- Horizontal cabling shall not have more than one transition point (TP) between flat undercarpet cable and a recognized horizontal cable.
- Bridged taps and splices are not permitted in copper horizontal cabling.

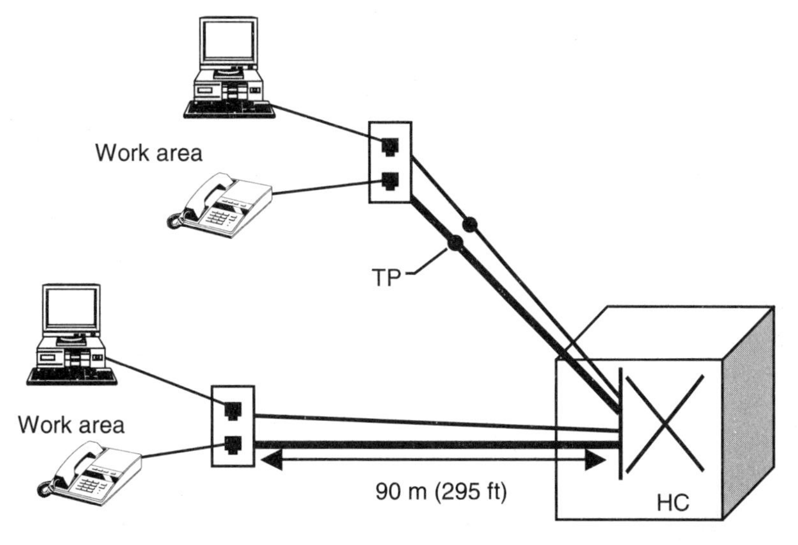

HC = Horizontal cross-connect

= Cross-connect

= Telecommunications outlet/connector

= Transition point (TP)

= 4-Pair twisted-pair

= 4-Pair twisted-pair or 2 or more strands of 50/125 μm or 62.5/125 μm optical fiber

Figure 4.8 Typical horizontal and work area cabling.

Cabling distances. Regardless of media type, the maximum horizontal distance is not to exceed 90 m (295 ft). Distance is measured from the mechanical termination of the horizontal cabling medium at the horizontal cross-connect to the termination at the telecommunications outlet/connector in the work area, and is based on the physical length of the cable.

In addition, the following allowances are made:

- 10 m (33 ft) for the total length of all cordage used in a horizontal channel, such as a patch cable, an equipment cable, and a work area cable.
- 5 m (16 ft) for the work area cable connecting a user's device to the telecommunications outlet/connector.

Recognized media. Two types of cables are recognized in the horizontal cabling system:

- Four-pair 100 Ω twisted-pair cable
- Two or more strands of 50/125 μm or 62.5/125 μm multimode optical fiber cable

Currently, 150 Ω STP-A cable is recognized. However, it is not recommended for new cabling installations and is expected to be removed from the next revision of the standard.

Hybrid cables consisting of more than one recognized cable under a common sheath may be used, providing they meet additional performance and color code specifications as defined in ANSI/TIA/EIA-568-B.2 and B.3.

Choosing media. A minimum of two telecommunications outlet/connectors shall be provided for each individual work area. One may be associated with a voice service and the other with a data service.

The two telecommunications outlet/connectors are configured as follows:

- One telecommunications outlet/connector must be supported by a four-pair 100 Ω twisted-pair cable—Category 3 or higher (Category 5e recommended).
- The second/other telecommunications outlet/connector is to be supported by a minimum of one of the following media:
  - Four-pair 100 Ω Category 5e twisted-pair cable
  - Two strands of 50/125 μm or 62.5/125 μm multimode optical fiber cable

Multi-User Telecommunications Outlet Assembly (MUTOA). ANSI/TIA/EIA-568-B.1 describes a MUTOA as a grouping in one location of several telecommunications outlet/connectors.

Using this type of assembly allows an open office layout to be changed without disturbing the horizontal cable. When MUTOAs are used, multiple horizontal cables are terminated in a common location (e.g., within a furniture cluster or similar open area). Work area cables are connected to work area devices with no additional intermediate connections. These cables originate at the MUTOA and are routed through work area pathways such as those found in modular furniture (see Figure 4.9).

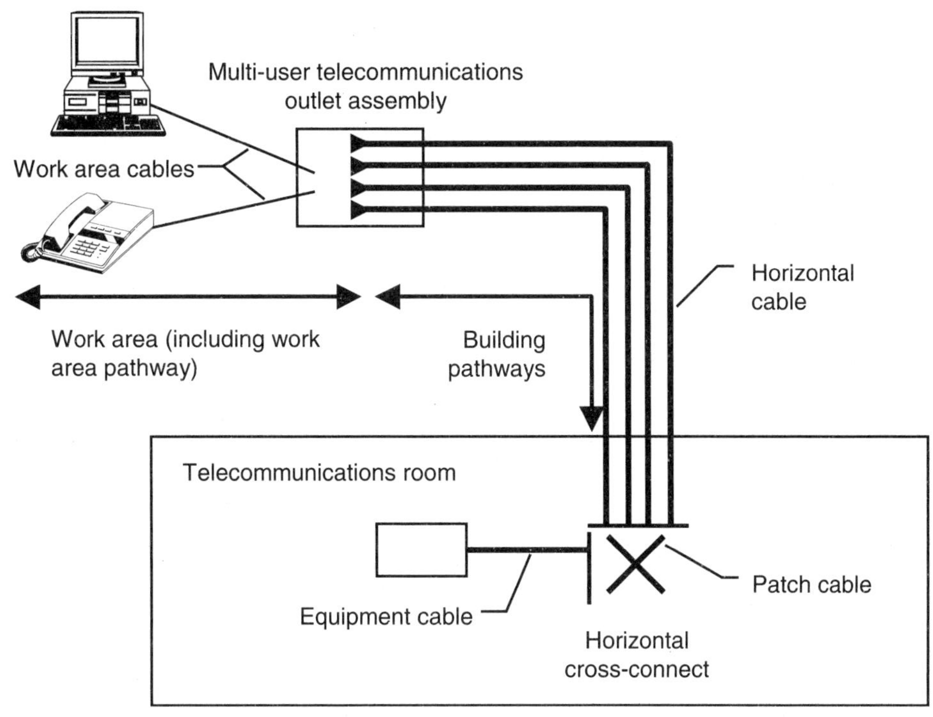

Figure 4.9 Horizontal cabling with multi-user telecommunications outlet assembly.

Planning and design. MUTOAs should be located so that each furniture cluster—a contiguous group of work areas—is served by at least one MUTOA. A single MUTOA should serve no more than 12 work areas.

Considerations when planning the use of MUTOAs should include:

- Maximum work area cable length requirements.
- Spare capacity.

Installation practices. The MUTOA shall be located in a fully accessible, permanent location such as building columns, permanent walls, and furniture permanently secured to the building structure. They shall not be placed in ceiling spaces.

Administration. Administration should follow recommendations made in ANSI/TIA/EIA-606. Label components as follows:

- Work area cables connecting the MUTOA to the work area shall be labeled at both ends with a unique cable identifier.
- The end of the work area cable at the MUTOA shall be labeled identifying the work area it serves.
- The end of the work area cable at the work area shall be labeled identifying the MUTOA and the port occupied.
- The MUTOA shall be marked with the maximum allowable work area cable length.

Horizontal distances—copper cabling. Copper cables to be used as work area cables shall meet or exceed the requirements specified in ANSI/TIA/EIA-568-B.1 and B.2. The maximum length of the work area cable is determined using the following metric formula:

$$C = (102 - H)/(1 + D)$$

$W = C - T \leq 22$ m (72 ft) for 24 AWG [0.51 mm (0.020 in)] twisted-pair or
≤ 17 m (56 ft) for 26 AWG [0.41 mm (0.016 in)] twisted-pair.

where:

C = The maximum combined length of work area cable, equipment cable, and patch cord (in meters).

H = The length of horizontal cable (in meters). ($H + C \leq 100$ m)

D = A de-rating factor for the patch cord type (0.2 for 24 AWG [0.51 mm (0.020 in)] twisted-pair and 0.5 for 26 AWG [0.41 mm (0.016 in)] twisted-pair)

W = The maximum length of work area cable (in meters).

T = The total length of patch and equipment cord in the TR.

NOTE: The above formulas must be used with values expressed in metric units.

Table 4.2 applies the above formulas using a total of 5 m (16 ft) of 24 AWG [0.51 mm (0.020 in)] twisted-pair patch cords and equipment cables in the TR.

Table 4.3 applies the above formulas using a total of 4 m (13 ft) of 26 AWG [0.41 mm (0.016 in)] twisted-pair patch cords and equipment cables in the TR.

TABLE 4.2 Maximum length of horizontal and work area cables using 24 AWG [0.51 mm (0.020 in)] patch cords and equipment cables.

Length of Horizontal Cable	Maximum Combined Length of Work Area Cables, Patch Cords, and Equipment Cables	Maximum Length of Work Area Cable
H	C	W
90 m (295 ft)	10 m (33 ft)	5 m (16 ft)
85 m (279 ft)	14 m (46 ft)	9 m (30 ft)
80 m (262 ft)	18 m (59 ft)	13 m (44 ft)
75 m (246 ft)	22 m (72 ft)	17 m (56 ft)
70 m (230 ft)	27 m (89 ft)	22 m (72 ft)

TABLE 4.3 Maximum length of horizontal and work area cables using 26 AWG [0.41 mm (0.016 in)] patch cords and equipment cables.

Length of Horizontal Cable	Maximum Combined Length of Work Area Cables, Patch Cords, and Equipment Cables	Maximum Length of Work Area Cable
H	C	W
90 m (295 ft)	8 m (26 ft)	4 m (13 ft)
85 m (279 ft)	11 m (36 ft)	7 m (23 ft)
80 m (262 ft)	15 m (50 ft)	11 m (36 ft)
75 m (246 ft)	18 m (59 ft)	14 m (46 ft)
70 m (230 ft)	21 m (70 ft)	17 m (56 ft)

Horizontal distances—optical fiber cabling. When using optical fiber, any length of horizontal, work area, patch, and equipment cable is acceptable as long as the total of the combined lengths does not exceed 100 m (328 ft) for a given horizontal cabling channel.

Consolidation Point (CP). ANSI/TIA/EIA-B.1 describes a CP as a location for interconnection between horizontal cables that extend from building pathways and horizontal cables that extend into work area pathways (see Figure 4.10).

The CP acts as an interconnection point in the horizontal cabling system. ANSI/TIA/EIA-568-B.2 and ANSI/TIA/EIA-568-B.3 compliant hardware must be used. CPs differ from MUTOAs in that they require an additional connection for each horizontal cable.

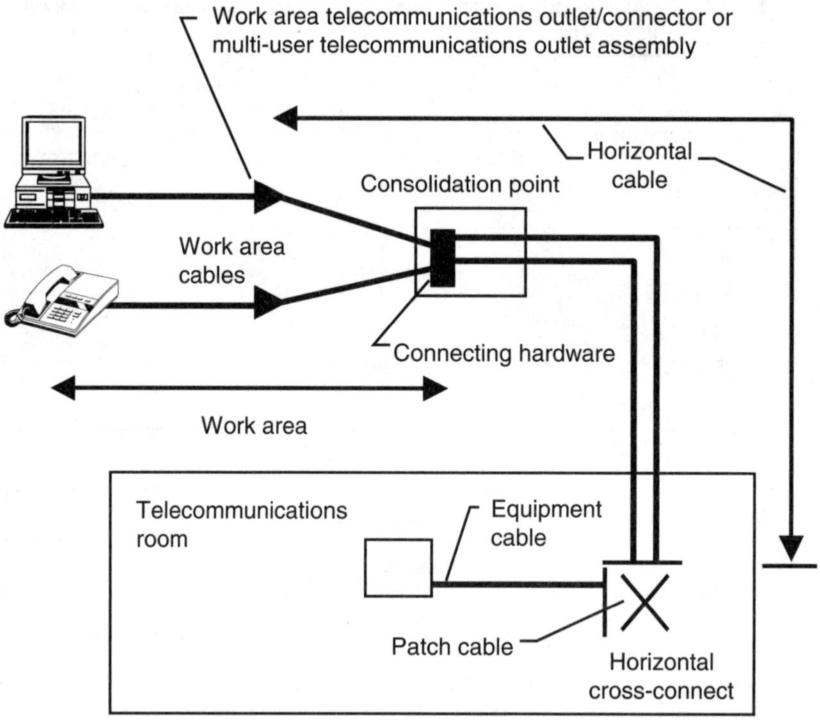

Figure 4.10 Horizontal cabling with consolidation point.

The requirements and recommendations for CPs are:

- Cross-connections shall not be permitted at a CP.
- Only a single CP shall be permitted within a horizontal cabling link.
- A TP and CP shall not be used in the same horizontal cabling link.
- Each horizontal cable extending to the work area from the CP shall be terminated at a telecommunications outlet/connector or a MUTOA.
- For twisted-pair cabling, solid conductor cable is required from the CP to the telecommunications outlet/connector.
- For twisted-pair cabling, the CP should be located at least 15 m (50 ft) from the TR to minimize performance degradation.

Planning and design. CPs should be located so that each furniture cluster is served by at least one CP. A single CP should serve no more than 12 work areas.

Spare capacity should be considered when planning the use of CPs.

Installation practices. The CP is to be located in a fully accessible, permanent location such as building columns, permanent walls, and furniture permanently secured to the building structure. Locations not to be considered for installation of the CP include obstructed areas and furniture not permanently secured to the building structure.

Administration. Administration should follow recommendations made in ANSI/TIA/EIA-606.

Any moves, adds, or changes of service not associated with rearranging the open office should be made at the horizontal cross-connect in the TR, not at the CP.

Backbone cabling

The term backbone cabling is used to describe cabling that provides connections between EFs, ERs, and TRs within a building (intrabuilding) and between buildings on a campus (interbuilding).

ANSI/TIA/EIA-568-B.1 describes backbone cabling as follows:

> "The function of the backbone cabling is to provide interconnections between telecommunications rooms, equipment rooms, main terminal space, and entrance facilities in the telecommunications cabling system structure. Backbone cabling consists of the backbone cables, intermediate and main cross-connects, mechanical terminations, and patch cords or jumpers used for backbone-to-backbone cross-connection. Backbone cabling also includes cabling between buildings."

Certain considerations should be made when selecting backbone cabling.

- The useful life of the backbone cabling system is expected to consist of several planning periods (typically, three to ten years). This is shorter than the overall life of a premises cabling system (typically, several decades).
- Prior to the start of a planning period, the maximum amount of backbone cabling required for the period should be projected. Growth and changes during this period should be accommodated without installing additional backbone cabling.
- When planning the routing and support for copper backbone cabling, areas where potential sources of EMI may exist should be avoided.

Topology. ANSI/TIA/EIA-568-B.1 provides the following specifications regarding the backbone cabling topology:

- The backbone cabling shall follow the hierarchical star topology (see Figure 4.11).
- Each HC in a TR is cabled to a MC or to an IC and then to a MC. The only exception is if requirements for a bus or ring network topology are anticipated. Then, supplemental cabling between TRs is allowed.

- There shall be no more than two hierarchical levels of cross-connect in the backbone cabling to limit signal degradation for passive systems and to simplify moves, adds, and changes. Facilities that have a large number of buildings or cover a large geographical area may elect to subdivide the entire facility into smaller areas within the scope of ANSI/TIA/EIA-568-B.1. In this case, the two-level backbone hierarchy serves multiple distinct areas.
- Backbone cabling between any two TRs shall pass through three or fewer cross-connects.
- No more than one cross-connect shall be passed through to reach the MC from any HC.
- In certain installations, a single backbone cabling cross-connect (the MC) may meet all cross-connect needs.
- Backbone cabling cross-connects may be located in TRs, ERs, or EFs.
- Bridged taps are not permitted in the backbone cabling.

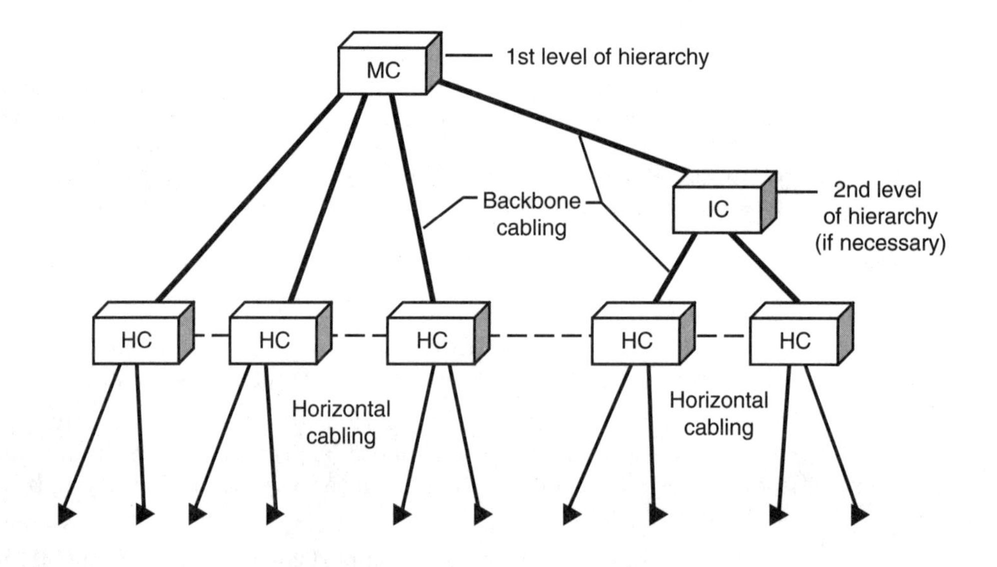

Figure 4.11 Backbone hierarchical star topology.

Recognized media. ANSI/TIA/EIA-568-B.1 recognizes three backbone transmission media, which can be used individually or in combination:

- 100 Ω twisted-pair cable
- 50/125 μm or 62.5/125 μm multimode optical fiber cable
- Singlemode optical fiber cable

Choosing media. Factors to consider when making a choice include:

- Flexibility with respect to present and future services.
- Required useful life of backbone cabling.
- Site size and user population.

When possible, service requirements should be grouped into categories such as voice, mainframe/minicomputer, or LAN technologies. Within each group, the appropriate backbone cabling should be identified and required quantities projected. Where uncertainty exists, worst-case scenarios should be used.

Cabling distances. Maximum allowable backbone distances are application dependent. Those provided in the standards are based on typical applications for a given type of cabling (see Figures 4.12, 4.13, and 4.14).

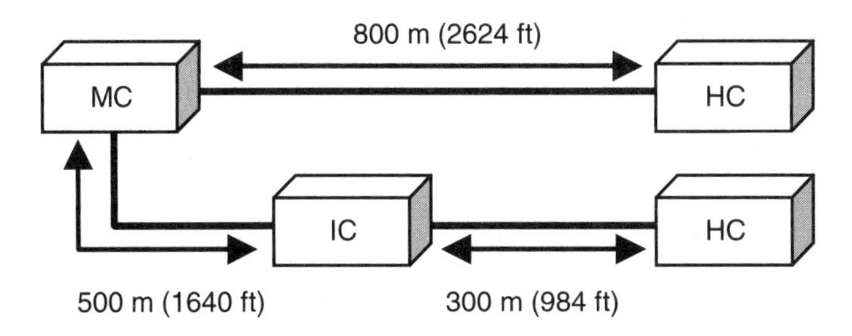

Figure 4.12 Maximum backbone distances for twisted-pair cabling.

NOTE: The distances provided in Figure 4.12 are based on voice or low-speed data applications only. Twisted-pair backbone cables used for high-speed data applications should be limited to 90 m (295 ft) between cross-connects that are connected to active equipment.

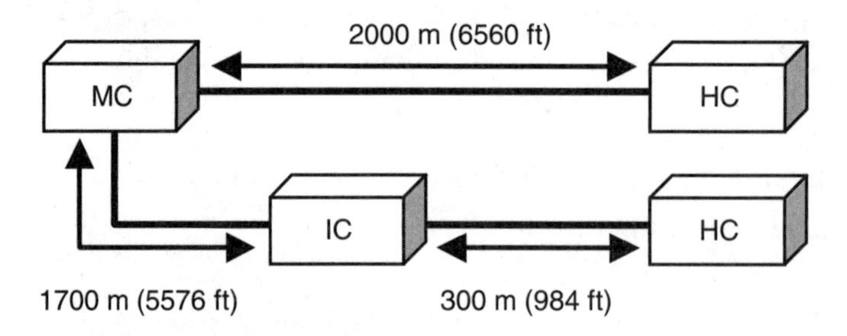

Figure 4.13 Maximum backbone distances for multimode optical fiber cabling.

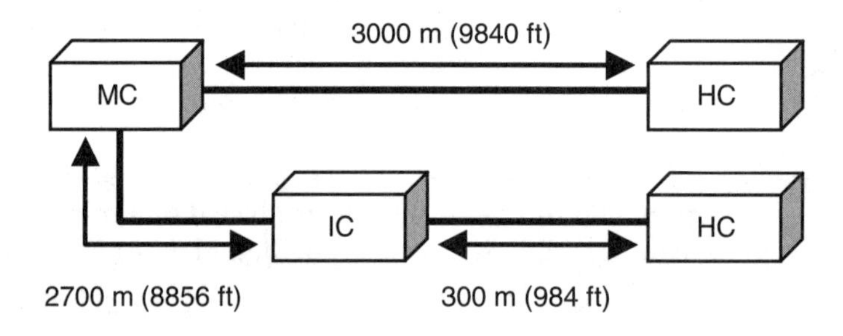

Figure 4.14 Maximum backbone distances for singlemode optical fiber cabling.

NOTE: While it is recognized that the capabilities of singlemode optical fiber allow for end-to-end distances in excess of 3000 m (9840 ft), as shown in Figure 4.14, such distances are generally considered to be outside the scope of building and campus cabling standards.

To minimize cabling lengths, the MC should be located near the physical center of a building or campus. Other backbone cabling issues include the following:

- MC to EF connections on a campus—Any cabling used to link the MC and the EF shall be included in the total interbuilding backbone distance.
- Cross-connections—In a MC, jumper wire or patch cord lengths should not exceed 20 m (66 ft). In an IC, jumper wire and patch cord lengths should not exceed 20 m (66 ft).
- Equipment cable—Cable used to link equipment to connecting hardware in MCs or ICs should not exceed 30 m (98 ft).

Centralized optical fiber cabling

It is possible to centralize all network access devices such as hubs, switches, and routers rather than distributing them throughout a building. The standards allow the use of centralized optical fiber cabling as an alternative to separate horizontal and backbone cabling systems in a single tenant application.

With proper planning and installation, a centralized optical fiber cabling system can provide a high level of security, flexibility, and manageability.

Design options. The standards provide three options for centralized optical fiber cabling systems. These are as follows:

- Pull-through cables can be used to link a centralized cross-connect in the ER directly to telecommunications outlet/connectors in each work area without any intermediate connections in the TRs.
- Splices can be used to link horizontal and intrabuilding backbone cables in the TRs.
- Connectors can be used to link horizontal and intrabuilding backbone cable in the TRs. This option (called an interconnection) provides greater flexibility, manageability, and ease of migration to cross-connection than either the pull-through or splice options.

NOTE: The term cross-connect should not be confused with the term cross-connection. A cross-connect is a facility, whereas a cross-connection (see Figure 4.15) is one possible method of connecting cables (an interconnection [see Figure 4.16] and a splice are two other methods). A single cross-connect facility (e.g., a HC) may contain any number of cross-connections, interconnections, and splices.

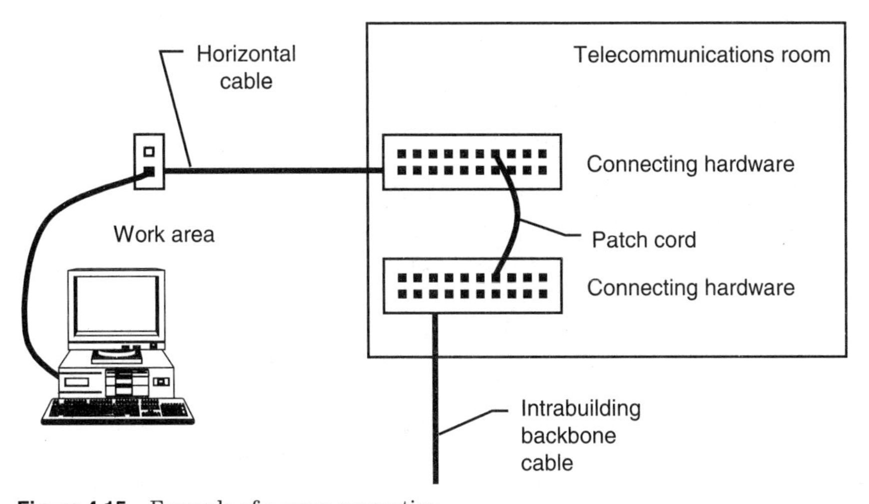

Figure 4.15 Example of a cross-connection.

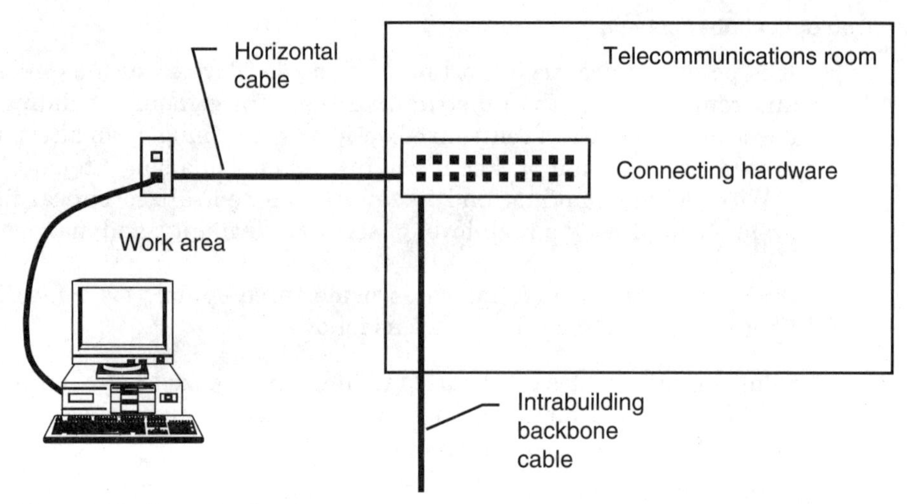

Figure 4.16 Example of an interconnection.

For all options, the design must allow for migration to a cross-connection—in total or in part. Adequate space should be available in the TR for the addition of connecting hardware required for migration to a cross-connection. The layout of termination hardware—rack- or wall-mounted—must be able to accommodate modular growth.

There must be enough cable slack available in the TR to permit movement of the cables in the case of migration to a cross-connection. The slack can be stored as either cable or unjacketed fiber (buffered or coated) with bend radius control to ensure that cable (or fiber) bend radius limitations are not exceeded. If the slack is stored as unjacketed fiber, it must be stored in protective enclosures. If it is stored as cable, it may be stored in protective enclosures or on the walls of the TR.

Cabling distances. The horizontal cable segment distance is limited to 90 m (295 ft).

When using the interconnection or splice options (see Figure 4.17), the total combined length of horizontal cable, intrabuilding backbone cable, and all cordage shall not exceed 300 m (984 ft). This distance represents the total distance from the device in the work area to the device in the ER.

When using the pull-through option (see Figure 4.17), the total cable length is not to exceed 90 m (295 ft).

The intrabuilding backbone should have sufficient spare capacity to service additional telecommunications outlet/connectors from the ER without the need to install additional backbone cables. Backbone cable planning should consider both present and future requirements. Most network technologies require two fibers for each connection to the network.

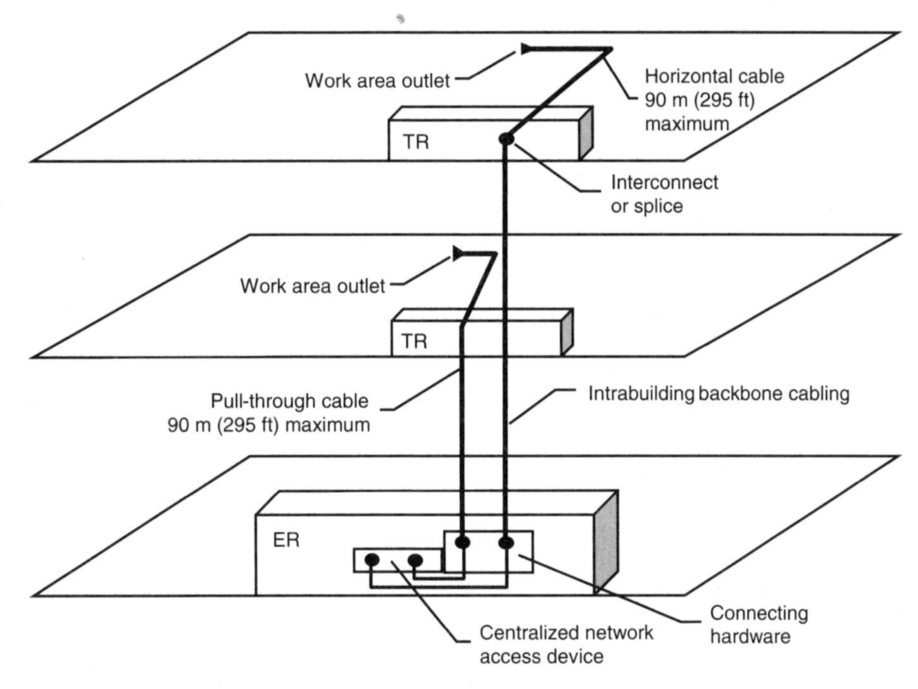

Figure 4.17 Centralized optical fiber cabling.

Telecommunications Circuits

General

Telecommunications circuits make it possible for an organizational network to grow beyond the physical boundaries of a building or a campus. Terms such as metropolitan area network (MAN), regional area network (RAN), and wide area network (WAN) are used to describe networks that span an extended geographic area.

NOTE: This document uses the term WAN to describe any network that uses telecommunications circuits to extend beyond a campus environment. This includes networks with metropolitan, regional, national, and international coverage.

The components of a typical WAN include:

- Two or more separate, independent LANs.
- Routers connected to each LAN.
- WAN access devices connected to each router.
- Telecommunications circuits connected to each WAN access device.

The combination of routers, WAN access devices, and telecommunications circuits forms the wide area internetwork linking the LANs. The combination of the internetwork and the LANs forms the WAN (see Figure 4.18).

NOTE: The telecommunications circuits and WAN access devices listed in Figure 4.18 are described in detail later in this section.

The network illustrated in Figure 4.18 is an example of a private WAN, where direct or point-to-point telecommunications circuits are used to connect organizational sites. A private network is one in which all traffic is generated by devices under the control of an organization. In cases where there are many sites to be interconnected and/or the distances between sites is substantial, it may be too expensive to set up a private WAN with point-to-point telecommunications circuits.

For example, a global organization with 50 facilities around the world can require a WAN to interconnect all sites. In such cases, two alternatives can be used in place of a private WAN:

- Connections to a WAN operated by a network service provider (NSP), sometimes referred to as value-added network (VAN)
- Connections to the Internet through an Internet service provider (ISP)

In both cases, the WAN traffic generated by devices at an organizational site is typically forwarded to a nearby facility operated by the NSP or ISP using telecommunications circuits. Once it is in the NSP or ISP network, the organization's traffic is grouped and processed with traffic from other sources. In the case of a WAN operated by an NSP, only the traffic from the NSP's clients is routed over the NSP network. Therefore, such networks can be described as semi-private WANs. Similarly, the Internet can be described as a public WAN, since it routes all traffic received from any ISP (see Figure 4.19). When considering an NSP WAN or the Internet as a means of linking organizational sites, issues such as security, reliability, and control must be considered in greater detail than with a private WAN.

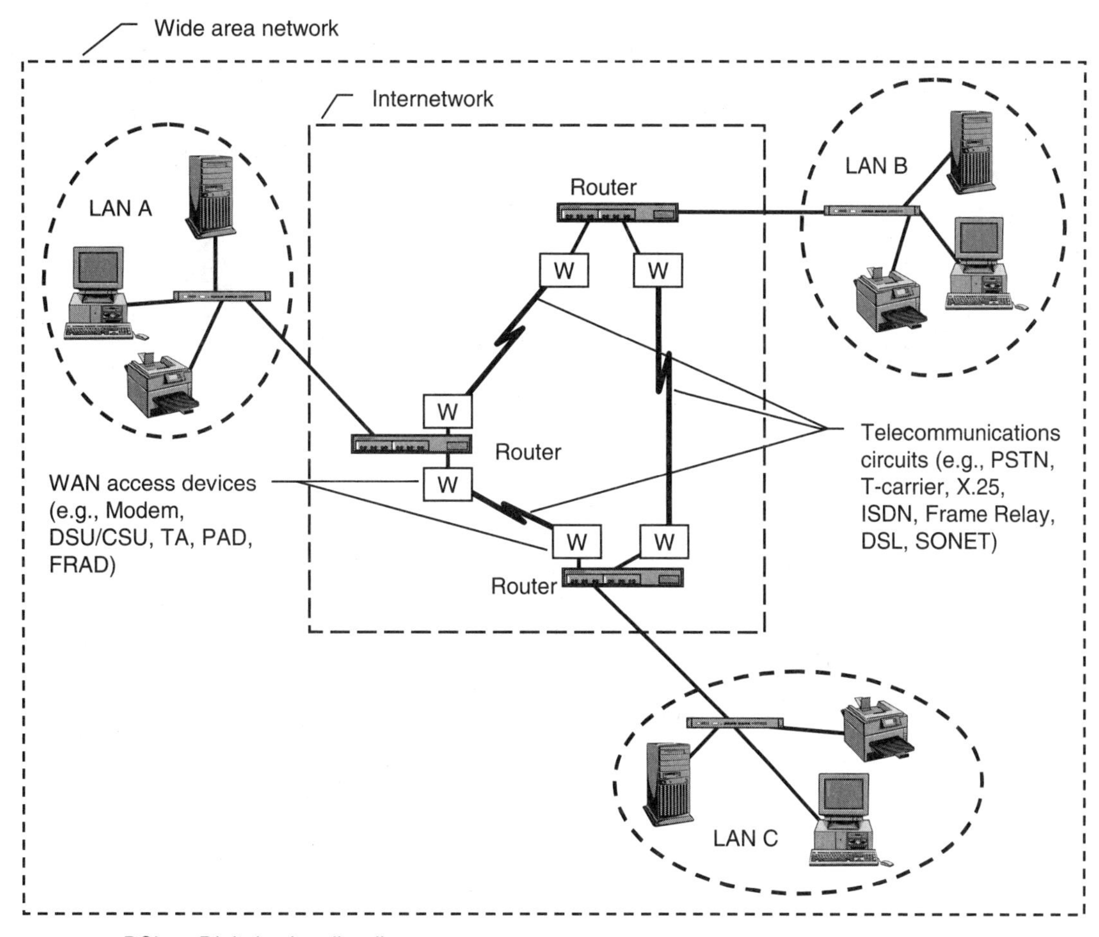

DSL = Digital subscriber line
DSU/CSU = Data service unit/channel service unit
FRAD = Frame Relay access device
ISDN = Integrated services digital network
LAN = Local area network
PAD = Packet assembler/disassembler
PSTN = Public switched telephone network
SONET = Synchronous optical network
TA = Terminal adapter
WAN = Wide area network
W = WAN access device

Figure 4.18 Example of a private wide area network.

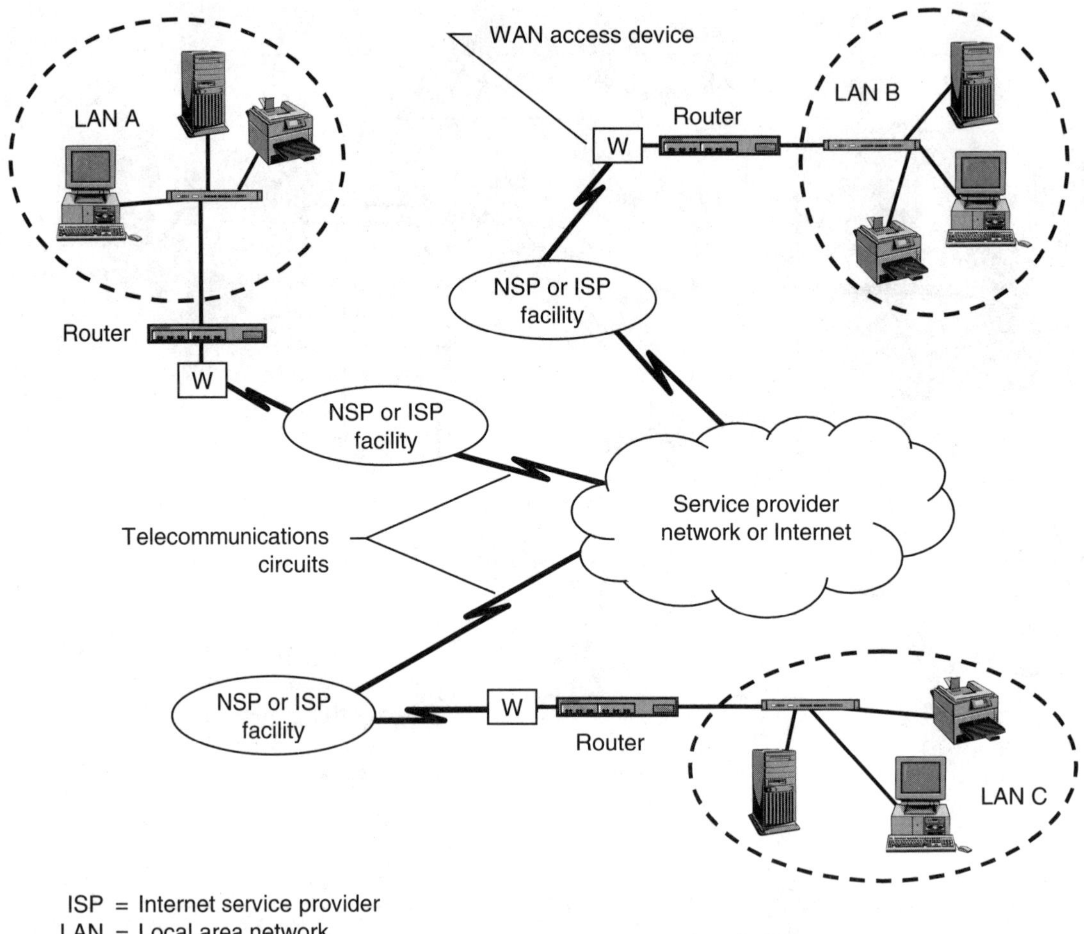

Figure 4.19 Example of a semi-private or public wide area network.

NOTE: A single WAN can use any combination of private point-to-point, semi-private NSP, or public Internet connections to link organizational sites.

Public Switched Telephone Network (PSTN)

PSTN telecommunications circuits are commonly referred to as dial-up lines, analog circuits, or generally as the telephone network. A PSTN circuit is the most common and least expensive type of link; however, it is optimized for voice communications, not data transmission. The telephone network is designed to transport continuous electrical signals in the 300 to 3300 hertz (Hz) frequency range, analogous to human speech. The 3000 Hz bandwidth is considered sufficient to recognize and understand the speaker at the other end of the connection.

NOTE: Frequency is measured in cycles per second, where one cycle per second equals one Hz.

The WAN access device used to connect to a PSTN circuit is called a modem. At the sending device, the modem converts digital signals to audible tones in the 300 to 3300 Hz range—the reverse process is performed by the modem connected to the receiving device.

Data transfer over all telecommunications circuits—including PSTN connections—is measured in digital bits per second (b/s), not analog cycles per second. Analog signal transfer is measured by counting the number of changes or transitions per second in the signal, defined as the baud. Using various coding techniques, multiple bits can be transmitted in a single transition. For example, using a certain code, one baud can transfer four bits. A modem using this code and operating at 2400 baud will transfer 9600 b/s (2400 transitions/second $\times$ 4 bits/transition).

NOTE: Modem standards and operating speeds are listed in Chapter 3: Network Components.

T-carrier

The T-carrier system is a digital communications service offered by circuit providers in many countries. An organization typically leases one or more T-carrier circuits, which are available in a range of operating speeds. T-carrier circuits can be used for the point-to-point connections in a private WAN or serve as the connections between an organizational site and an NSP or ISP facility. T-carrier circuit costs are usually distance-sensitive. Circuits spanning longer distances cost more to lease than shorter links.

In North America, the most common leased T-carrier circuits are T-1 and T-3, operating at 1.544 megabits per second (Mb/s) and 44.736 Mb/s respectively. Some circuit providers also provide lower cost fractional T-1 (FT-1) and fractional T-3 (FT-3) services to organizations that do not require the maximum rate available on the circuit. For example, a one-quarter FT-1 circuit provides 384 kilobits per second (kb/s) and a one-half FT-1 circuit provides 768 kb/s.

Technically, any FT-1 circuit is capable of operating at the full T-1 rate. Therefore, an organization can lease or upgrade circuits as needed to accommodate the growth in its WAN traffic.

T-1 was introduced in North America in 1962 as the first carrier system based on a digital communications infrastructure. In T-1 and other T-carrier systems, time division multiplexing (TDM) is used to group multiple digital communications channels into a single stream. For example, a single T-1 is a combination of twenty-four 64 kb/s digital channels, and a single T-3 consists of 28 T-1 channels. T-1 signaling is described in detail later in this section.

TDM combines multiple communications channels into a single stream by allocating a time slot to each channel. At the sending end of the circuit, a device called a TDM multiplexer (mux) divides the stream into time slots and allocates each slot to a specific channel. This enables each channel to remain distinct as it is grouped with others and transmitted over the link. The mux at the receiving end of the circuit separates (or demultiplexes) the incoming stream. This restores the individual channels, each of which is then forwarded by the mux to the appropriate device (see Figure 4.20).

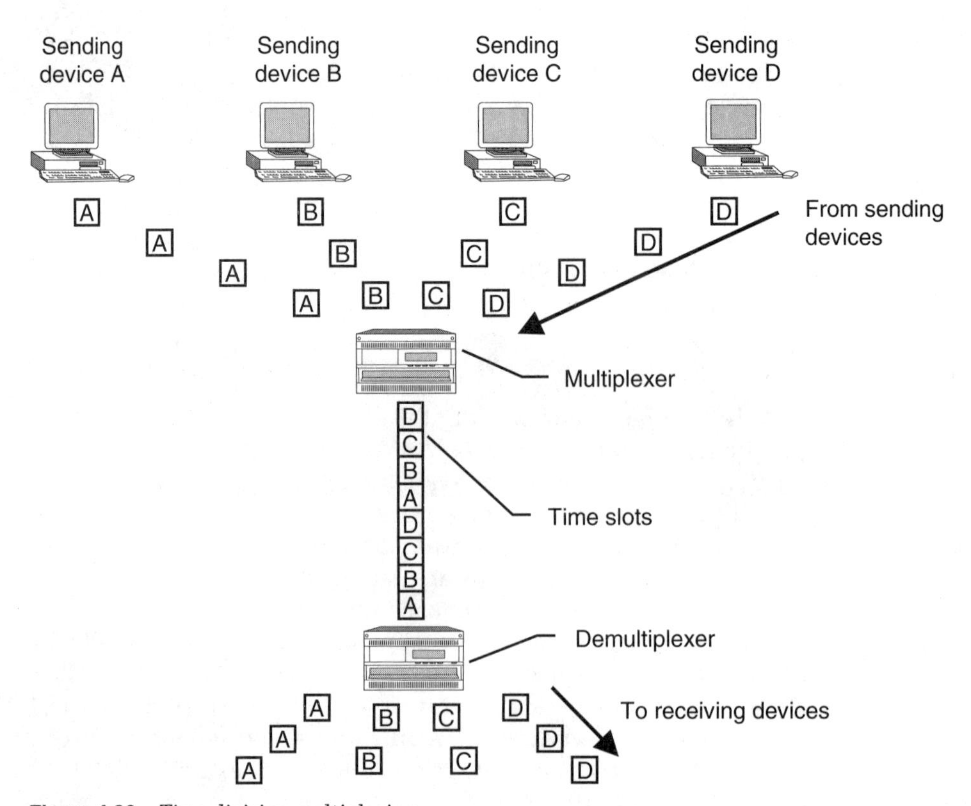

Figure 4.20 Time division multiplexing.

T-1 signaling. A T-1 circuit can use copper, optical fiber, or wireless media to transport signals. Regardless of medium used, the T-1 link transfers a stream of bits representing 24 channels using TDM on a frame-by-frame basis. Each frame consists of 192 bits—8 bits per channel multiplied by 24 channels plus one additional framing bit used for synchronization—for a total of 193 bits per frame (see Figure 4.21). On a T-1 circuit, 8000 frames are transmitted per second, resulting in an operating speed of 1.544 Mb/s, as follows:

193 bits/frame × 8000 frames/second = 1,544,000 bits/second = 1.544 Mb/s

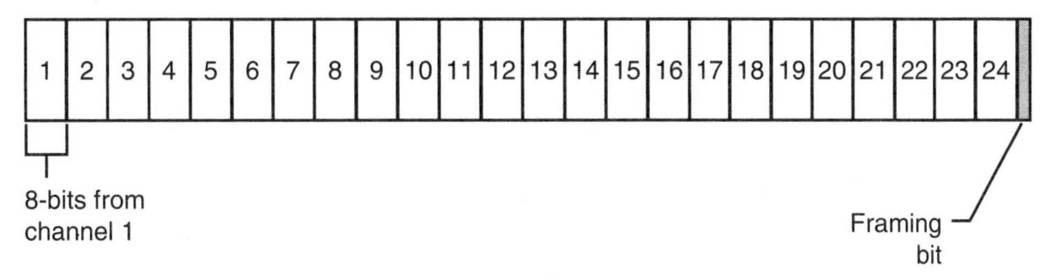

Figure 4.21 T-1 frame.

Signaling hierarchies. In T-carrier terminology, a 64 kb/s communications channel is referred to as digital signal level zero (DS-0), the lowest level in the signaling hierarchy. The next level is DS-1, consisting of 24 DS-0 channels and transported over a T-1 carrier system. Similarly, the DS-3 level used on T-3 carrier systems groups 28 DS-1 streams for a total of 672 DS-0 channels (28 streams × 24 channels/stream = 672 channels).

The signaling hierarchies used in Europe and Japan differ from the T-carrier hierarchy used in North America. To distinguish between them, the terms E-1/E-3 and J-1/J-3 are sometimes used to describe the European and Japanese counterparts to T-1/T-3. Table 4.4 provides a comparison of the three hierarchies.

TABLE 4.4 Comparison of signaling hierarchies.

Signal Level	North America	Europe	Japan
DS-0	64 kb/s	64 kb/s	64 kb/s
DS-1	1.544 Mb/s T-1 (contains 24 DS-0s)	2.048 Mb/s E-1 (contains 32 DS-0s)	1.544 Mb/s J-1 (contains 24 DS-0s)
DS-3	44.736 Mb/s T-3 (contains 672 DS-0s)	34.368 Mb/s E-3 (contains 512 DS-0s)	32.064 Mb/s J-3 (contains 480 DS-0s)

NOTE: Additional signal levels exist in the North American, European, and Japanese hierarchies. The levels listed here are those most commonly offered for lease by circuit providers.

Data Service Unit/Channel Service Unit (DSU/CSU). The WAN access device used to connect a network access device, such as a router, to a T-carrier circuit is called a DSU/CSU (see Figure 4.22). The DSU component is responsible for converting the outgoing bit stream generated by the router into a signaling format compatible with the T-carrier used—and the reverse for incoming signals. The CSU component serves as the termination point for the T-carrier circuit. It is responsible for maintaining the circuit, providing diagnostic information, and if necessary, preventing the entry of signals that may cause damage to the circuit.

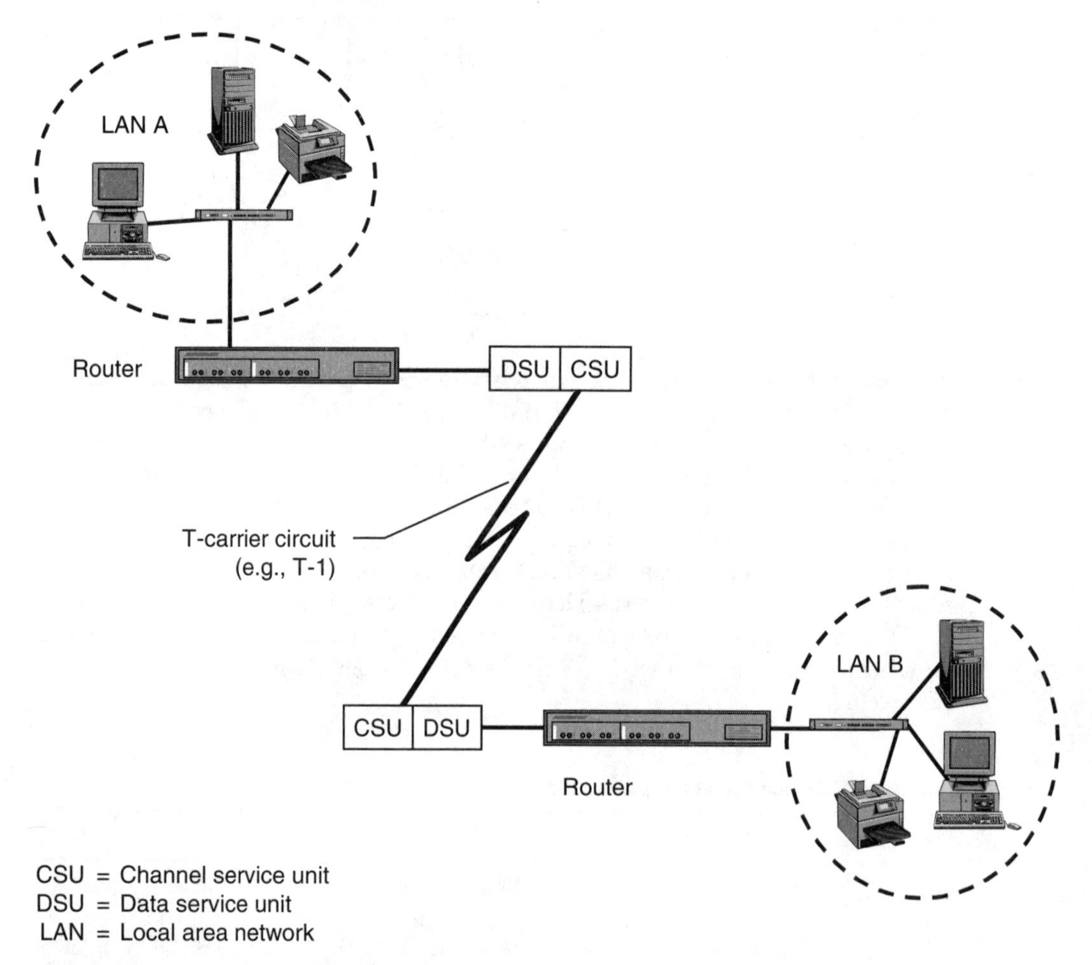

Figure 4.22 Example of a DSU/CSU configuration.

NOTE: The router shown in Figure 4.22 can also be equipped with an internal DSU/CSU.

X.25

X.25 is an established communications service for digital transmission over extended distances. It has been used to build private WANs and NSP networks in many countries. Developed in the 1960s, it is the first large-scale implementation of packet switching technologies.

NOTE: An X.25 network is sometimes referred to as a packet switched data network (PSDN).

In 1974, the International Telecommunication Union-Telecommunication (ITU-T) began development of a reliable mechanism for data transmission over telecommunications circuits. Recommendation X.25 was published in 1976, specifying the interface between data terminal equipment (DTE) and data circuit-terminating equipment (DCE) connected to a packet switching network. At the time, a maximum operating speed of 64 kb/s was specified. This value was increased to 2.048 Mb/s in 1992.

X.25 was developed to connect remote terminals to mainframes and minicomputers. Its flexibility and reliability encouraged international adoption by computer vendors and telecommunications circuit providers. Technologies such as Frame Relay and ATM—sometimes described as fast packet technologies—are based on concepts and techniques researched during the development of X.25. Currently, connections to X.25 networks are available from NSPs in nearly every country in the world. A related standard, X.75, specifies the interfaces for linking independent X.25 networks. Together, X.25 and X.75 make it possible to connect devices and networks on a global scale.

X.25 is used to transfer messages between devices over a packet switched network. The type of connection used in X.25 can be described as a virtual circuit, used to deliver messages called packets to their destination through a network consisting of switches. The virtual circuit forms a communications channel through which a remote device requests and receives services.

Data transfer using X.25 is considered very reliable but relatively inefficient when compared to alternatives. Reliability is assured since error-checking functions are performed by every switch in the path between the sending and receiving devices. Each switch reads every packet it receives to verify the address information before proceeding with error-detection operations on the contents of the packet. If the packet is valid, it is forwarded to the next switch in the path, which acknowledges receipt and performs the same operations. The sending switch retains a copy of the packet, which is used if a packet is corrupted and the receiving device requests a resend. These processes increase the reliability but limit the data transfer efficiency of the network.

X.25 service consists of three protocols that work together to enable communications (see Figure 4.23). The combination of the three is referred to as the X.25 protocol stack. These three layers of the X.25 protocol stack closely resemble the bottom three layers of the Open Systems Interconnection (OSI) reference model.

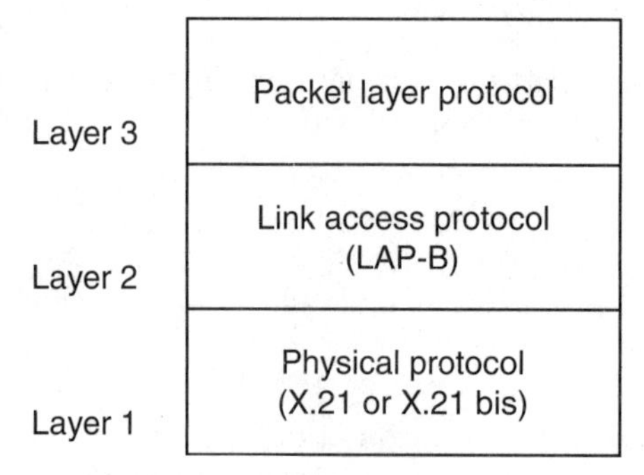

Figure 4.23 X.25 protocol stack.

NOTE: X.25 predates the OSI model, which is described in detail in Chapter 2: Networking Fundamentals.

Layer 1, or the physical protocol layer, is defined by the X.21 and X.21 bis protocols. These standards define the mechanical, electrical, functional, and procedural means for the transmission and reception of bits from a device connected to an X.25 network.

NOTE: The extension bis is used by the ITU to denote the second in a series of related standards.

Layer 2 specifies the use of the link access procedure-balanced (LAP-B) protocol for establishing virtual connections, handling flow control during a communications session, and terminating or tearing down the virtual circuit at the end of the session. The detection and correction of errors and the acknowledgement of receipt for each packet are also handled at this layer.

Layer 3 defines the procedures that allow two devices to control communications and reliably exchange data. Full X.25 connectivity is established at this layer. The packet layer protocol sets up reliable virtual connections through a packet switching network. Each connection creates two logical channels, one for the sending device and one for the receiving device. These channels represent a virtual circuit that guarantees that all packets will arrive in the same sequence as they were sent. Two types of virtual circuits are defined. One is a temporary circuit that is dismantled when data transfer is complete. The other is a permanent circuit that occupies a reserved path over the network.

Packet Assembler/Disassembler (PAD). The WAN access device used to connect to an X.25 packet switching network is called a PAD. At the sending end, the role of the PAD is to convert non-packetized data into packets, provide addressing information in each packet, and place the packets onto the X.25 network. At the receiving end, the PAD restores the data and delivers it to the

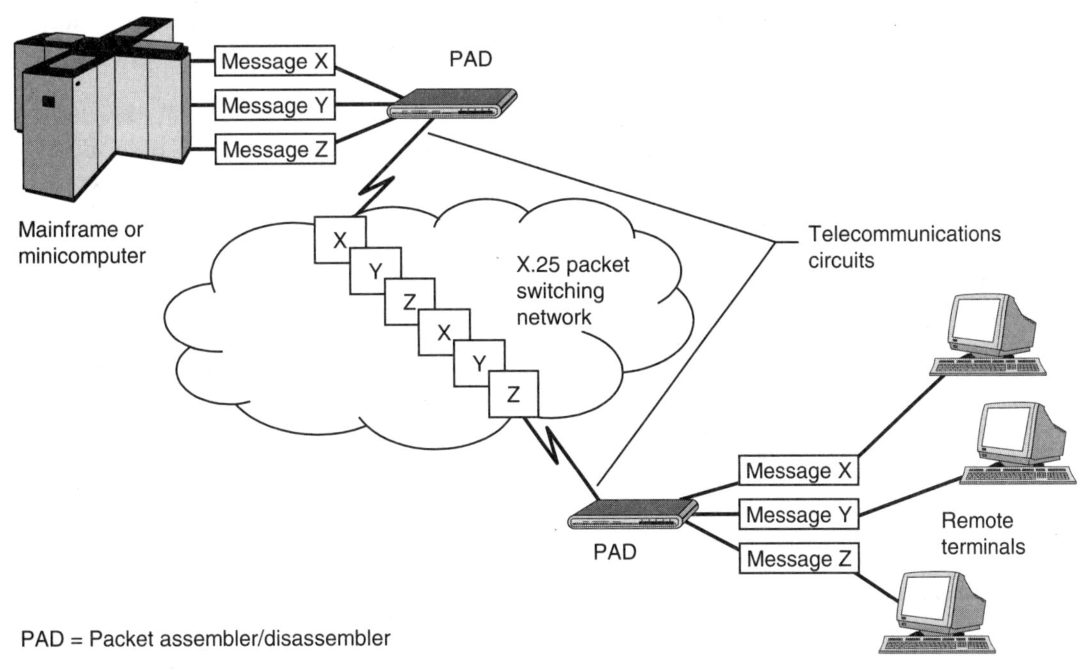

Figure 4.24 X.25 packet switching.

destination device (see Figure 4.24). A PAD can be equipped with multiple ports, each of which can establish a separate virtual circuit through the X.25 network.

Integrated Services Digital Network (ISDN)

ISDN is an international digital communications infrastructure intended as a replacement to the analog PSTN used for voice communications. As a switched end-to-end digital service, ISDN can transport all types of traffic—voice, data, facsimile, and video—over a common network.

The initial set of standards for ISDN were issued by the ITU-T in 1984. Also referred to as narrowband ISDN (N-ISDN), the specifications defined a maximum 128 kb/s operating speed over extended distances. At that time, analog modems had a maximum speed of 9.6 kb/s, which has since increased to the current 56 kb/s. Specifications for broadband ISDN (B-ISDN)—capable of operating at gigabit per second (Gb/s) rates—have been used to develop ATM technologies.

An ISDN circuit provides two types of channels—bearer channels, also called B channels and signaling channels, also called D channels. The B channels transport the data generated by the device connected to the ISDN circuit (e.g., a telephone, a fax machine, or a router). A single B channel can operate at a maximum speed of 64 kb/s. D channels are used by the ISDN network for administrative signaling and call control, using the link access procedure-D channel (LAP-D) protocol. It is also possible to use a D channel for X.25 packet data transfer.

The existence and functionality of the D channel is one of the distinguishing features of ISDN. In other types of circuits, administrative signals must be sent on the same channel as the data being transported—this is referred to as in-band signaling. The D channel enables out-of-band signaling for such activities as call setup, control, and termination. This makes it possible to use B channels entirely for data transfer, increasing the efficiency of the circuit. The D channel enables faster call setup—an ISDN connection can be established within two seconds or less, compared to the 30 seconds it can take for an analog modem to dial, connect, and begin transferring data.

ISDN provides two types of circuits, referred to as rate interfaces. A basic rate interface (BRI) circuit (see Figure 4.25) consists of two B channels, each with an operating speed of 64 kb/s, and one D channel with an operating speed of 16 kb/s. Each B channel can function like a PSTN circuit and have its own telephone number—called a directory number (DN) in ISDN terminology. Some WAN access devices can combine, or bond, the two B channels to form a single 128 kb/s channel.

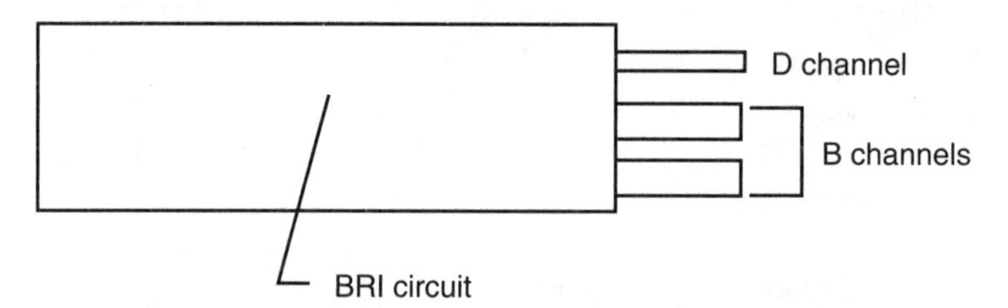

Figure 4.25 ISDN basic rate interface.

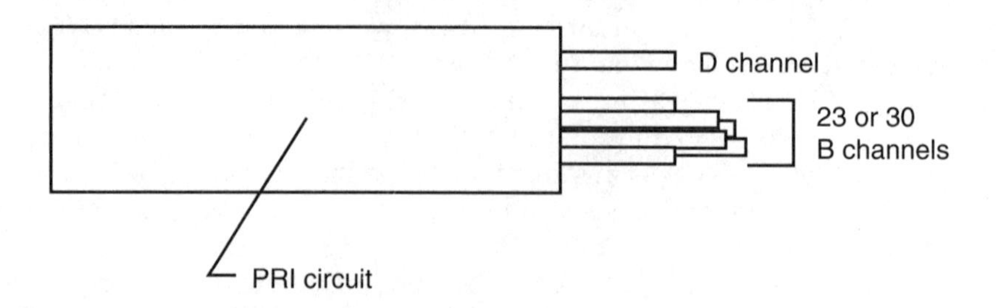

Figure 4.26 ISDN primary rate interface.

The second type of ISDN circuit is the primary rate interface (PRI). In North America and Japan, a PRI consists of 23 B channels and one D channel (23 B + D). Each of the 24 channels operates at 64 kb/s. This makes a North American or Japanese PRI the equivalent of a T-1 or J-1 circuit, each of which contains twenty-four 64 kb/s DS-0 channels. In Europe, an ISDN PRI consists of 30 B channels and one D channel (30 B + D), all operating at 64 kb/s—making the European PRI the equivalent of an E-1 circuit (see Figure 4.26).

It is possible to combine the B channels in a PRI into various groups, similar to bonding the two B channels in a BRI. Such groupings are referred to as H channels, some of which are the:

- H0 channel, formed by combining six B channels into a single 384 kb/s channel.
- H10 channel, formed by combining all 23 B channels in a North American or Japanese PRI into a single 1.472 Mb/s channel.
- H11 channel, formed by combining the 23 B + D channels in a North American or Japanese PRI into a single 1.536 Mb/s channel.
- H12 channel, formed by combining all 30 B channels in a European PRI into a single 1.920 Mb/s channel.

The ISDN BRI interface is intended for residential or small business use, where it can replace two separate PSTN circuits. The ISDN PRI interface can be dedicated to a single device for use as a high-speed circuit. Alternatively, multiple voice, data, facsimile, or video devices can share the channels in the PRI.

Network Terminal Adapter (NTA). The WAN access device used to connect a network access device such as a router to an ISDN circuit is called an NTA. Designations such as network termination 1 (NT1) and network termination 2 (NT2) are used to describe different types of NTAs. The NTA connects both terminal equipment (TE) and devices equipped with terminal adapters (TAs) to the ISDN circuit. TE devices compatible with ISDN are called termination equipment 1 (TE1) units. TE devices not compatible with ISDN are called termination equipment 2 (TE2) units. TA devices are used to connect TE2 devices to the NTA (see Figure 4.27).

An NTA device is equipped with two different types of interfaces, as follows:

- The U interface connects the NTA to the ISDN circuit.
- The S/T interface connects a TE1 or TA device to the NTA.

A TA device is also equipped with two different types of interfaces, as follows:

- The S/T interface connects an NTA device to the TA.
- The R interface connects a TE2 device (non-ISDN) to a TA.

NOTE: The S interface is a 4-wire bus that allows for the connecting of eight independent TE1s or TA devices.

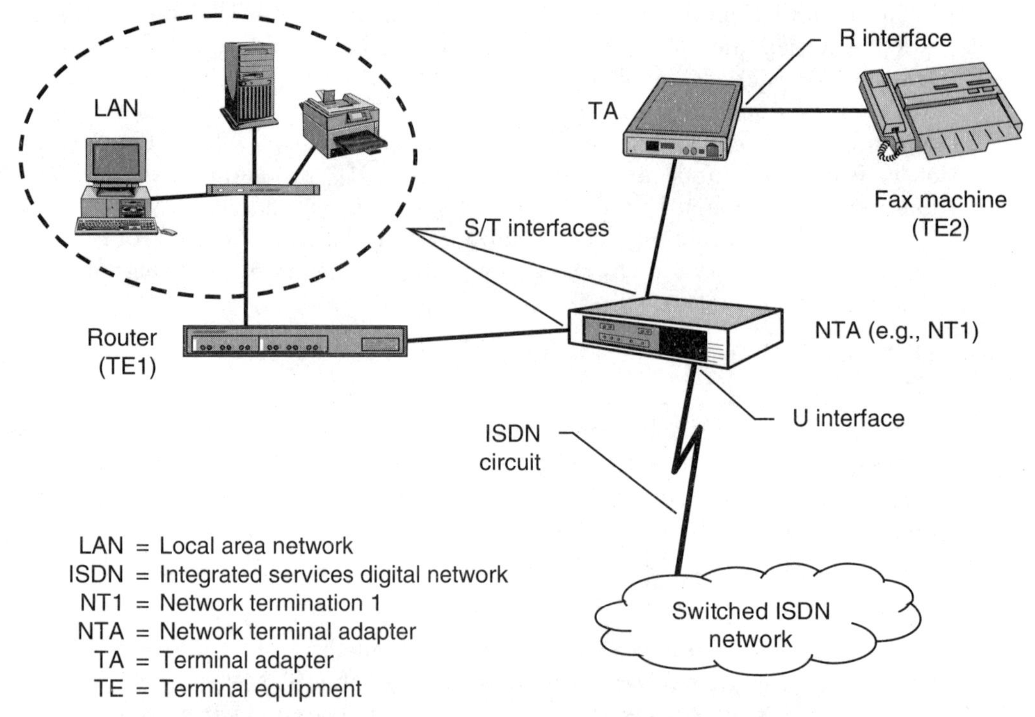

Figure 4.27 Example of a network terminal adapter configuration.

NOTE: The router shown in Figure 4.27 can also be equipped with an internal NTA.

Frame Relay

Frame Relay can be described as a high-performance version of X.25. It is a communications service designed to transport messages as rapidly as possible between networks over extended distances, using virtual circuits and switches. Messages are in the form of OSI Layer 2 frames rather than the OSI Layer 3 packets used in X.25. A user-to-network interface (UNI) is defined for connecting an organizational network to a Frame Relay switch owned by an NSP. To enable connectivity between different Frame Relay networks, a network-to-network interface (NNI) is defined for interconnecting switches owned by different NSPs.

Frame Relay operations are similar to X.25 and emerged from ISDN development. As a result, Frame Relay has elements in common with both X.25 and the D channel signaling technology used in ISDN—X.25 uses the LAP-B protocol, ISDN and Frame Relay use LAP-D. Due to the similarities, it is possible to establish Frame Relay connections over ISDN circuits, and devices can access a Frame Relay network over X.25 links.

NOTE: It is more common to use full or fractional T-carrier circuits and CSU/DSUs to connect to an NSP's Frame Relay network. In most cases, connection speeds range from 56 kb/s to 1.544 Mb/s, although a maximum of 44.736 Mb/s is possible.

A series of standards for Frame Relay has been issued by the ITU-T, many of them in the early 1990s. Initially, Frame Relay networks used permanent virtual circuits, where a predefined path through the Frame Relay switches must be in place between the sending and receiving devices prior to message transfer—the equivalent of leased telecommunications circuits. In 1993, switched virtual circuits were added to the Frame Relay standard, making it possible to set up, use, and tear down virtual circuits as needed—the equivalent of dial-up telecommunications circuits.

Frame Relay operates faster than X.25 because it is optimized for rapid message transport, not extensive error checking and correction. X.25 technologies were developed to connect simple terminals to remote computers over unreliable telephone circuits. The network was designed to provide a guaranteed, error-free packet delivery service. In current organizational networks, the devices connected to telecommunications circuits are capable of initiating error-handling procedures. The circuits currently available are of higher quality, capable of handling high-speed signals reliably and with fewer errors than in the past. Because of these developments, the decision was taken to eliminate most error handling operations and maximize data transfer speed on Frame Relay networks.

If a frame is corrupted or lost in a Frame Relay network, it is the responsibility of the receiving device to detect the problem and request a retransmission from the sending device. When a Frame Relay switch detects an error in a frame, the frame is discarded. If the input buffers in a switch are full, incoming frames are discarded until the congestion clears. Frame Relay networks do not have mechanisms to correct errors or maintain flow control.

In order to provide some assurance of message delivery, each Frame Relay virtual circuit is assigned a committed information rate (CIR). The CIR is the rate at which the Frame Relay network agrees to transport data on a given circuit. A second value, the committed burst size (B), measures the maximum number of bits that the Frame Relay network can transfer during a given interval. Dividing the committed burst size by the CIR provides the time interval (T) (see Example 4.1).

EXAMPLE 4.1 Frame Relay average operating speed.

$$T = \frac{B}{CIR}$$

If CIR = 32 kb/s and B = 256 kilobits (kb) for a given virtual circuit, then applying the above formula T equals:

$$T = \frac{256kb}{32\ kb/s} = 8\ \text{seconds}$$

This means that the Frame Relay network will attempt to transport 256 kb over an 8-second period on this circuit. The operating speed of the circuit can vary, but will average 32 kb/s when measured over any 8-second interval.

Frame Relay uses statistical multiplexing to combine or interleave data from several devices onto a single communications channel. Each device with data to transmit is assigned a slot in the channel. If a device has nothing to transmit, the slots it would have used are divided among the devices that have data to transmit. For example, if four devices share a channel and each has data to transmit, each device can transmit at one quarter of the operating rate of the channel. If only three devices have data to transmit, each can transmit at one third of the operating rate of the channel.

Frame Relay Access Device (FRAD). The WAN access device used to connect a network access device, such as a router, to a Frame Relay network is called a FRAD (see Figure 4.28). Data to be transferred between LANs connected by a Frame Relay network takes the following path:

1. The sending device on one of the LANs sends the data to its router.
2. The router forwards the data to a FRAD.
3. The FRAD divides the data into segments, encapsulates each segment in a Frame Relay frame, and directs the frames to a port connected to a permanent or switched virtual circuit.
4. The frames are transferred between switches in the Frame Relay network.
5. The FRAD attached to the router on the destination LAN removes the frame encapsulation and forwards the data to the router.
6. The router sends the data to the receiving device on the destination LAN.

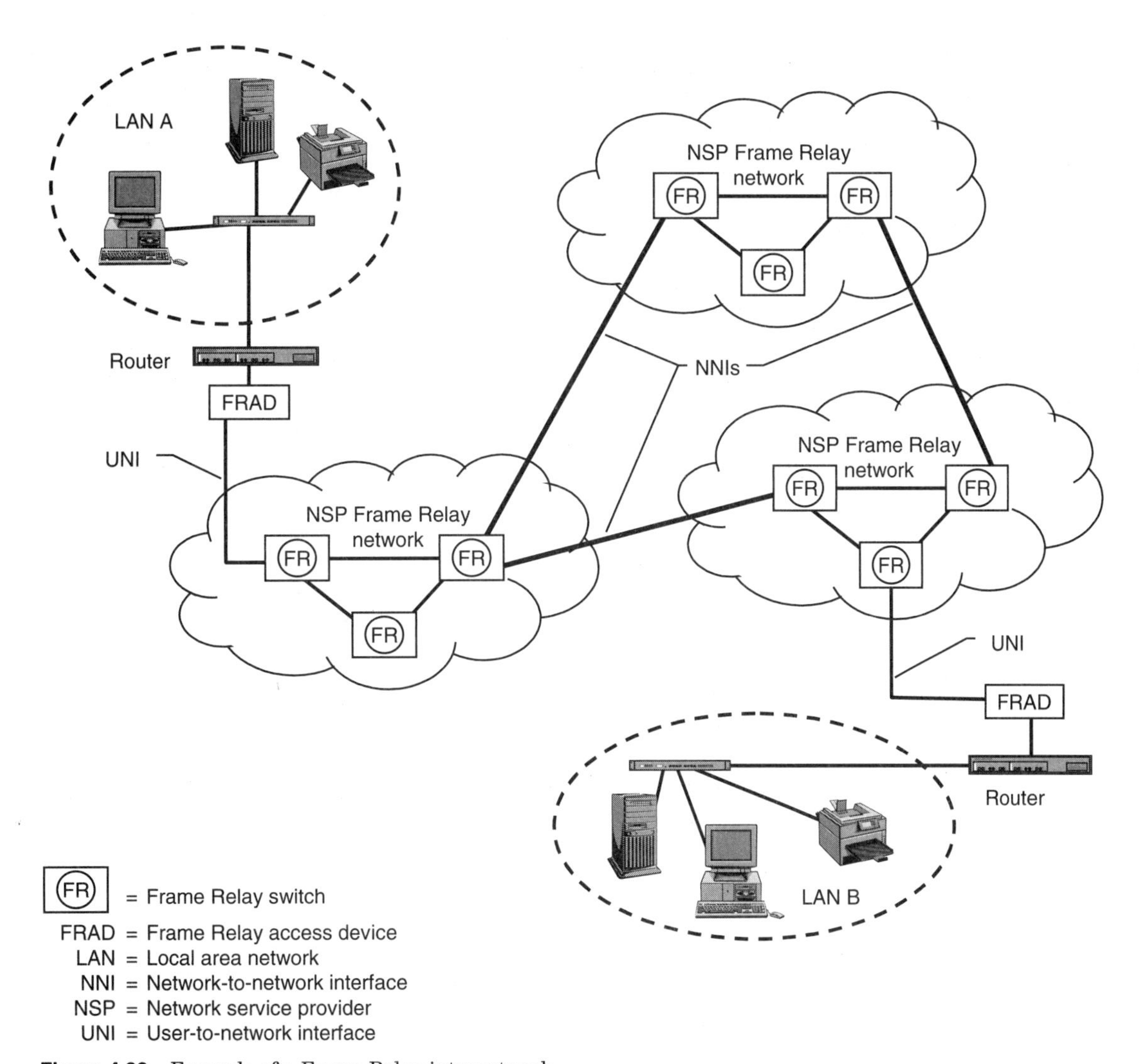

Figure 4.28 Example of a Frame Relay internetwork.

NOTE: The FRAD shown in Figure 4.28 can also be incorporated into a router.

Digital Subscriber Line (DSL)

DSL technology can be described as an updated version of ISDN, capable of providing high-speed data transfer services to residential and commercial subscribers over an infrastructure normally used for analog PSTN circuits. A DSL circuit can operate at various rates, ranging from 16 kb/s to 52 Mb/s. DSL circuits are divided into two channels—an upstream channel for data sent to the DSL network and a downstream channel for data received from the DSL network. The operating speeds of the two channels can be the same (symmetrical) or different (asymmetrical).

DSL makes it possible for a residential user to connect at high speed to an NSP or ISP network through the existing telephone network cabling infrastructure. Prior to the introduction of DSL, the most common choices available for data transfer over a residential telephone circuit were analog modem technologies (56 kb/s maximum operating speed) and ISDN BRI service (128 kb/s with bonded channels).

DSL uses a technique to split a telephone circuit into two parts. One part is used to provide the same analog telephone service that previously occupied the entire circuit. The second part is used to provide a continuous or always-on digital service for data transfer—the equivalent of a leased telecommunications circuit. The two parts can operate simultaneously, making it possible to carry on a voice conversation on the telephone while transferring data to and from a remote network.

The maximum operating speed of a DSL circuit is affected by the quality of the cabling infrastructure used to provide the circuit and the distance between the NSP site and the subscriber. In many cases, a maximum of 5.5 km (3.4 mi) can be attained without the need for additional equipment—the shorter the distance, the higher the maximum speed, all else being equal.

Due to the existence of a variety of DSL systems and the wide range of distances between potential subscribers and NSP sites, many types of DSL services have been introduced by vendors. The term xDSL is commonly used to describe all DSL services, some of which are the following:

Asymmetric Digital Subscriber Line (ADSL). A subscriber equipped with an ADSL circuit can receive data at a maximum rate of 6.1 Mb/s (downstream channel) and can transmit data at a maximum rate of 640 kb/s (upstream channel).

High Bit Rate Digital Subscriber Line (HDSL) and Symmetric Digital Subscriber Line (SDSL). Both HDSL and SDSL are symmetric circuits that can operate at the same rate as a T-1 or E-1 circuit—1.544 Mb/s or 2.048 Mb/s.

Rate-Adaptive Digital Subscriber Line (RADSL). An RADSL circuit can configure its operating speed—both upstream and downstream—to suit the quality of its cabling infrastructure.

Very High Bit Rate Digital Subscriber Line (VDSL). VDSL is intended to serve as a replacement for commercial high-speed telecommunications circuits that require optical fiber or coaxial cabling infrastructures. It can provide a downstream channel operating at 52 Mb/s and an upstream channel operating at 2.3 Mb/s. The range of a VDSL circuit is limited—it can be as short as 305 m (1000 ft) for a 52 Mb/s connection.

ADSL Transceiver Unit-Remote (ATU-R) and Digital Subscriber Line Access Multiplexer (DSLAM). The WAN access device used to connect a subscriber access device, such as a router, to a DSL network is called an ATU-R. At an NSP site, the connections to multiple subscriber ATU-Rs are managed by a DSLAM. The DSLAM can group the traffic from the ATU-Rs and transfer it to a common network. For example, the NSP can direct the Internet traffic from a group of residential subscribers to an ISP network (see Figure 4.29).

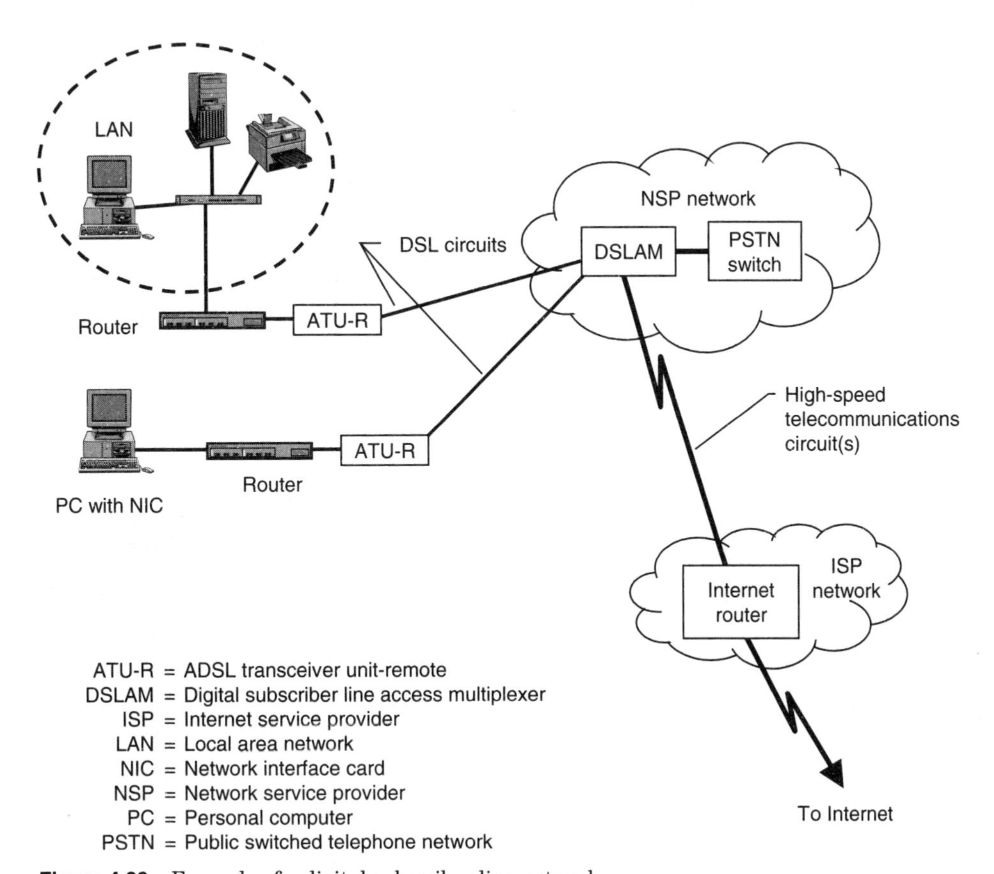

Figure 4.29 Example of a digital subscriber line network.

NOTE: The ATU-R shown in Figure 4.29 can also be incorporated into a router.

Synchronous Optical Network (SONET)

SONET is a transport technology designed to provide a uniform, consistent method of data transfer using fiber optic transmission systems. Prior to its introduction, vendors used incompatible and proprietary configurations and interfaces to link their products over optical fiber cabling. The need for a standard optical communications interface led to ANSI approval for SONET in 1988. After making minor modifications, the ITU-T adopted SONET as a global standard, changing its name to the synchronous digital hierarchy (SDH).

NOTE: It is common to refer to SONET as SONET/SDH, since the two standards are similar in scope and content.

The SONET standards define a high-speed digital transmission hierarchy and frame format for data transfer over singlemode optical fiber cabling. SONET was originally intended to be the OSI Physical layer service for B-ISDN, which has since evolved into ATM. Although all types of traffic—ATM cells, Internet protocol (IP) frames, T-carriers—can be transported over a SONET infrastructure, SONET has been optimized for time-sensitive traffic such as voice and video. It features extensive processes and designs intended to avoid noticeable delay, even when there is a failure in the path carrying the voice or video traffic between source and destination.

The basic unit of transport defined by SONET is the synchronous transport signal level one (STS-1) frame. This frame is organized as a cell matrix of nine rows and 90 columns. Each cell contains eight bits, for a total of 6480 bits per frame (see Figure 4.30).

Network control information is included in the frame and is referred to as the transport overhead. The transport overhead is placed in the first three

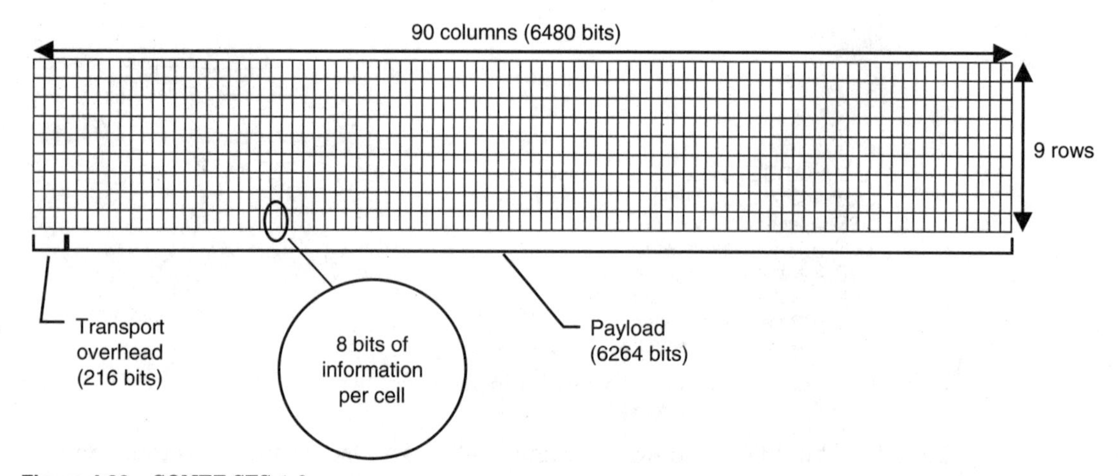

Figure 4.30 SONET STS-1 frame.

columns of the STS-1 frame, filling 27 cells with 216 bits of data. The remaining 87 columns, representing 783 cells or 6264 bits, is referred to as the synchronous payload envelope (SPE) or payload.

Cells are transmitted one at a time, from left to right, starting with the first row. One frame is transmitted every 125 microseconds (μs), resulting in a transmission of 8000 frames per second. The transmission rate is calculated as follows:

$$8000 \text{ frames/second} \times 6480 \text{ bits/frame} = 51{,}840{,}000 \text{ bits/second} = 51.84 \text{ Mb/s}.$$

The SONET transmission hierarchy is based on the STS-1 rate of 51.84 Mb/s. The optical equivalent of STS-1 is called optical carrier level one (OC-1).

NOTE: STS terminology is used when referring to the electrical equivalent of the optical signals.

SONET multiplexers are capable of merging frames from multiple STS-1 streams into a single high-speed, high-volume signal. The terms STS-n and OC-n—where n represents an integer—are used to indicate the number of streams multiplexed. Some of the rates in the SONET/SDH transmission hierarchy include those shown in Table 4.5.

TABLE 4.5 SONET/SDH transmission hierarchy.

Optical Level	Electrical Level	Transmission Rate
OC-1	STS-1	51.84 Mb/s
OC-3	STS-3/STM-1	155.52 Mb/s
OC-12	STS-12/STM-4	622.08 Mb/s
OC-48	STS-48/STM-16	2.488 Gb/s
OC-192	STS-192/STM-64	9.953 Gb/s
OC-768	STS-768/STM-256	39.813 Gb/s

NOTE: The ITU-T SDH uses the equivalent of STS-3 as its basic frame, referred to as the synchronous transport module level one (STM-1) frame.

SONET links can be implemented in a variety of configurations to provide different levels of fault tolerance. In a point-to-point configuration between two sites, two links can be established over different optical fiber cables—if the cables take different paths, it is called route diversity (see Figure 4.31). Both links carry the same traffic at all times and the SONET devices at each end continuously monitor the status and signal quality of each link. Any time one of the links is degraded or fails, the second is accessed, with no delay. This type of fault tolerance is called 1 + 1. In cases where a second link is available but does not carry duplicate traffic at all times, the term 1:1 is used. When 1:1

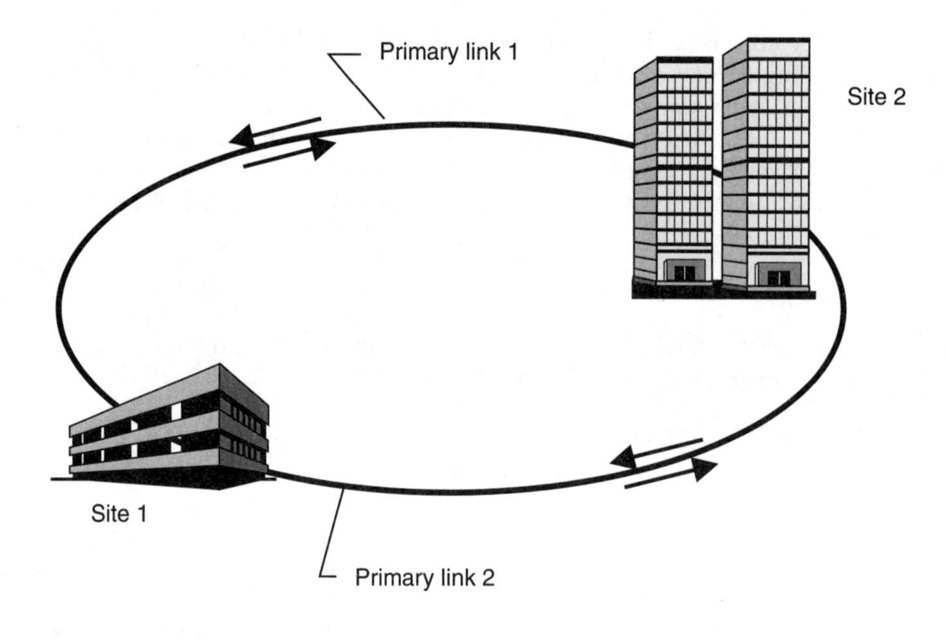

Figure 4.31 SONET point-to-point with route diversity.

links are used, there is a slight delay as the backup link is activated after failure of the primary link.

NOTE: The term 1:n—where n represents an integer—is used to describe cases where one backup is available for n links or systems.

In a SONET ring configuration between three or more sites, two links can be established between adjacent sites, creating a dual-ring cabling layout or topology. Under normal operating conditions, one ring is the active primary and the other is the unused secondary or backup (see Figure 4.32). If an active ring fiber is damaged, the sites at both ends of the damaged cable activate the secondary ring and use it to route traffic to and from the primary ring, bypassing the damaged segment of cable.

SONET technologies were developed to link sites operated by telecommunications circuit providers. As a result, SONET uses components and designs suitable for time-sensitive voice traffic. In cases where only non-time-sensitive data traffic is to be transported, SONET capabilities may be viewed as unnecessarily sophisticated and expensive. A technology called resilient packet ring (RPR) has been proposed as a substitute for SONET in such cases. The Institute of Electrical and Electronics Engineers, Inc.® (IEEE®) has established the 802.17 Resilient Packet Ring Working Group to develop a standard for this technology.

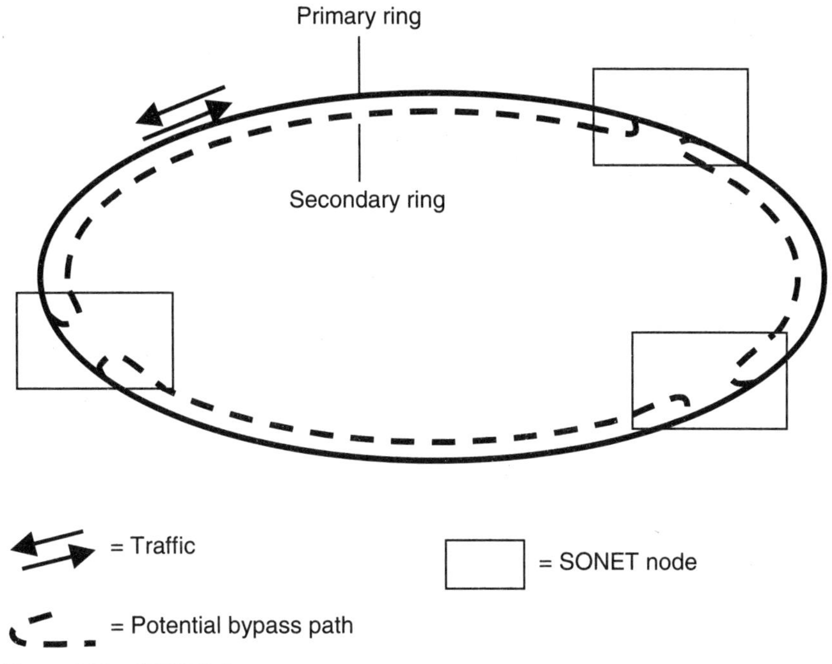

Figure 4.32 SONET ring.

Like SONET, RPR is implemented using a dual-ring topology. On an RPR network, however, both rings are used at all times. In the event of a fault on one ring, all traffic is placed on the second ring—the delay caused by the additional traffic is considered acceptable for a data network.

Chapter

5

Network Communications

Chapter 5 describes the fundamental characteristics of a network, which include topology and addressing conventions, as well as forms of access control, signaling, and routing.

Overview

General

The term communications is used to describe any process in which messages are exchanged. Whether the exchange is between devices or individuals, four conditions are necessary in order for the communications process to be successful.

- A communications channel must be in place between the devices or individuals that are to exchange messages. For example, two individuals can communicate using the telephone network or the postal system.
- For every type of communications channel, a common format—or translation services—must be used for the addressing, signaling, and content of all messages exchanged over the channel. When individuals use the telephone network:
  - A telephone is addressed using a sequence of numbers arranged in a specific format.
  - The human voice is the signal format transported by the network.
  - Each speaker must format the message content using a language familiar to the other to be understood.

- A common set of rules for communications—or protocol—must be used to exchange messages over a shared communications channel. In the case of a telephone conversation, individuals cannot talk and listen at the same time. If a segment of the conversation is not heard, the listener can ask the speaker to repeat the missing segment.
- The messages exchanged must have a common context. In the case of a telephone conversation, the topic of conversation must be familiar to both parties. In the case of data communications, the sending and receiving devices must be running compatible applications. For example, one computer can receive a spreadsheet file from another; however, unless the receiving computer is equipped with the same or a compatible type of spreadsheet software, it cannot make use of the file.

Network communications can be classified as follows:

- Shared media communications between devices in the same broadcast domain over a shared channel (or collision domain) (e.g., between a station and a server through a hub).
- Point-to-point communications between devices in the same broadcast domain over a dedicated channel (e.g., between a station and a server through a switch).
- Network-to-network communications between devices in different broadcast domains (e.g., between a station and a server through one or more routers).

The communications channels of a network are created using various types of media and telecommunications circuits, described in detail in Chapter 4: Network Connections. The shape of the cabling layout used to link devices is called the physical topology of the network. The path taken by messages to get from one device to another can be the same or different from the cabling layout—it is called the logical topology of the network. This chapter describes the topologies associated with local area networks (LANs) and internetworks as well as the following communications processes:

- Message addressing
- Network access
- Signaling
- Message transport

NOTE: Because of their widespread use, Ethernet LANs and Internet protocol (IP) internetworks are used to illustrate and describe many of the topics presented in this chapter.

Network Topologies

General

The physical topology of a network is determined by the capabilities of the network access devices and media, the level of control or fault tolerance desired, and the costs associated with cabling or telecommunications circuits. Networks can be classified according to their physical span as follows:

- LANs
- Building or campus internetworks
- Wide area internetworks

Point-to-point topology

A point-to-point topology is the most basic network or internetwork layout. It links two devices or networks in a building, on a campus, or over extended distances (see Figures 5.1 and 5.2).

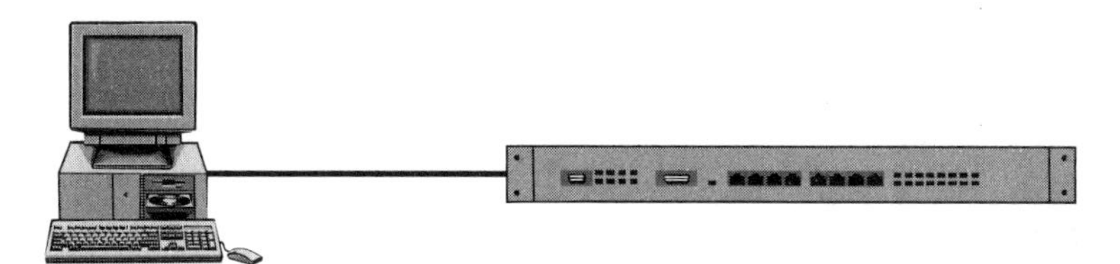

Figure 5.1 Point-to-point topology between devices.

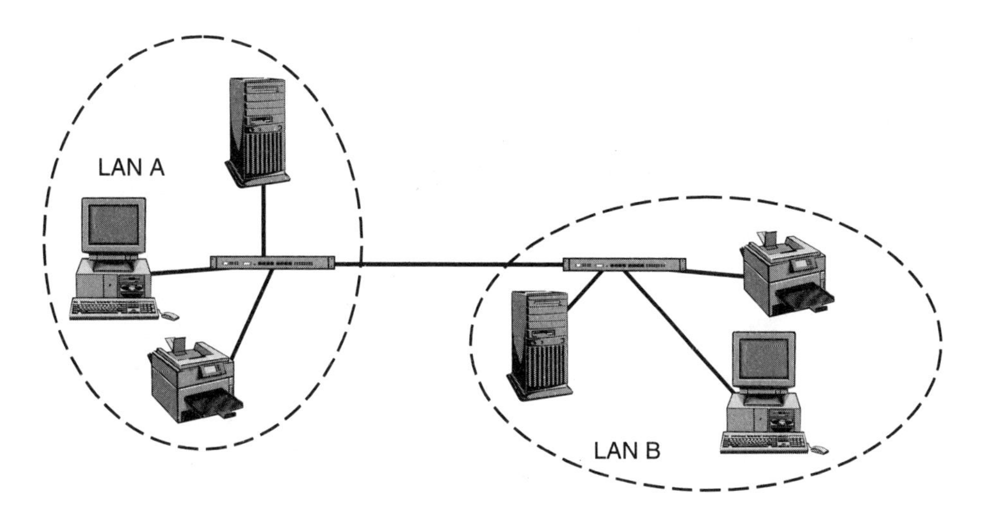

Figure 5.2 Point-to-point topology between networks.

Daisy-chain topology

A daisy-chain topology links devices or networks in sequential format (see Figures 5.3 and 5.4). Each device or network, other than the end units, has a separate connection to the previous and following device or network.

Figure 5.3 Daisy-chain topology between devices.

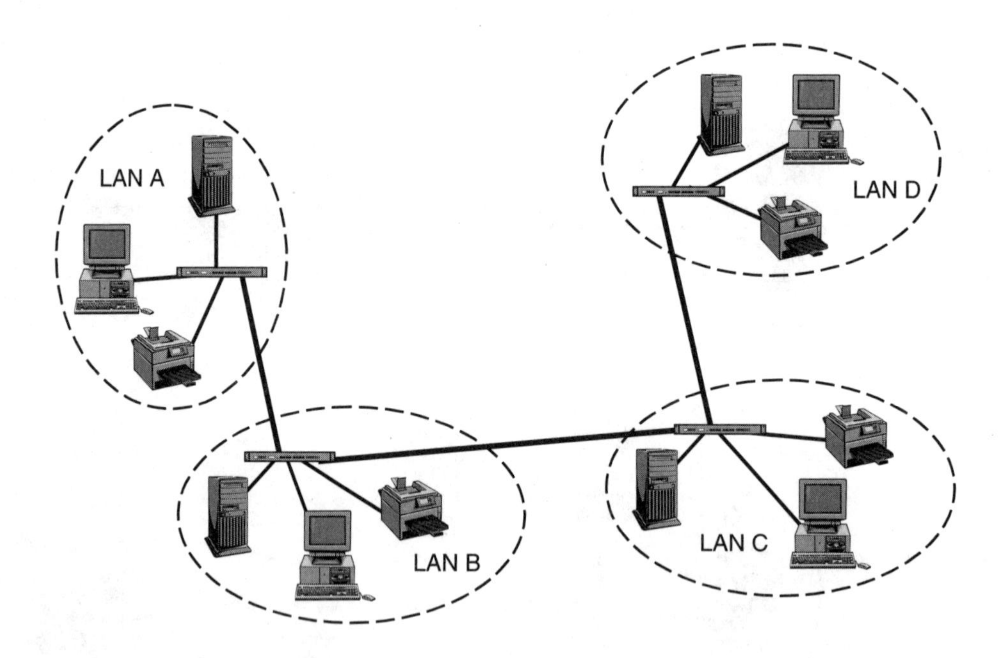

LAN = Local area network

Figure 5.4 Daisy-chain topology between networks.

Bus (point-to-multipoint) topology

A bus topology is sometimes referred to as a point-to-multipoint topology and is similar in appearance to a daisy-chain topology. Each device or network on a bus topology network has a single connection to a shared medium that serves as the communications channel (see Figures 5.5 and 5.6).

Figure 5.5 Bus topology between devices.

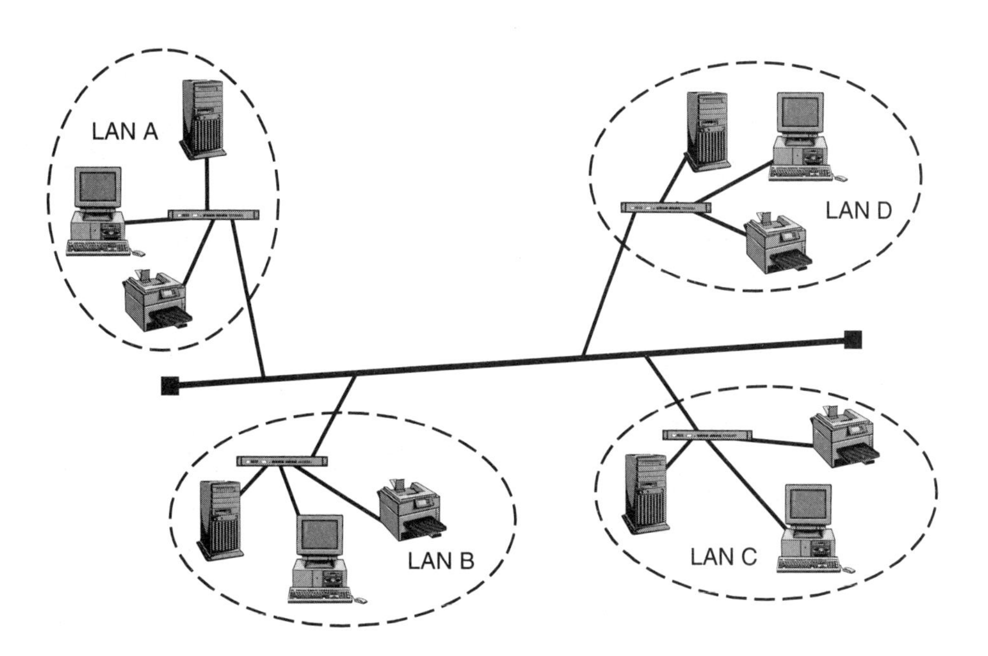

Figure 5.6 Bus topology between networks.

Ring topology

A ring topology can be described as a closed daisy-chain topology. Devices or networks are linked in sequential format, with the first device or network linked to the last (see Figures 5.7 and 5.8).

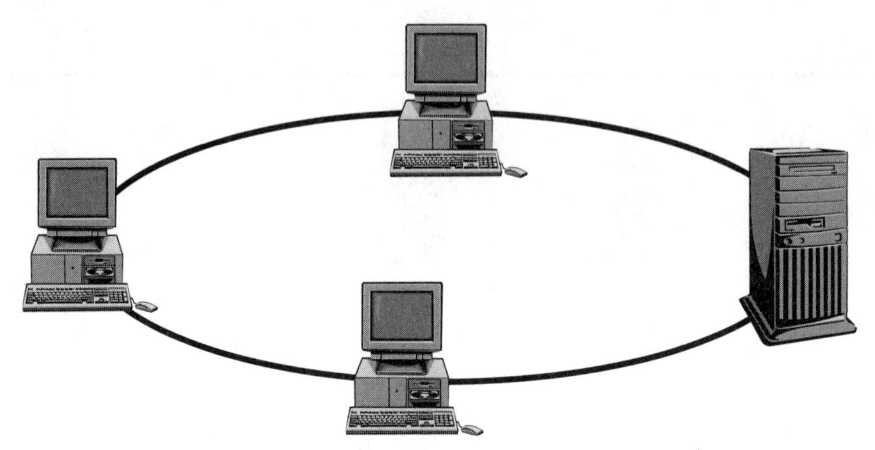

Figure 5.7 Ring topology between devices.

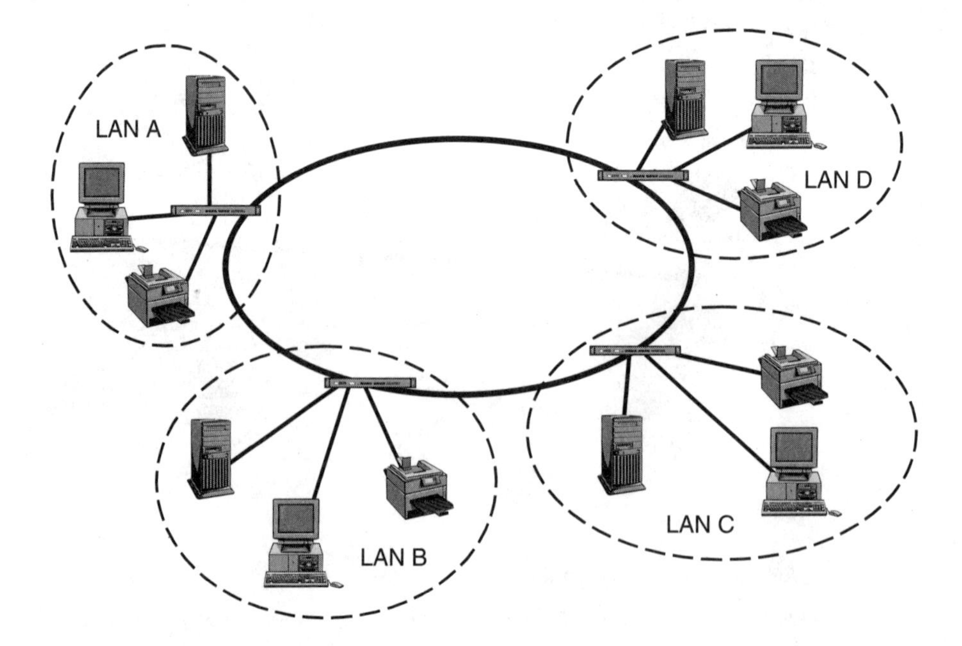

Figure 5.8 Ring topology between networks.

Dual-ring topology

A dual-ring topology is similar in configuration to the ring topology. On a dual-ring topology, each device or network has two connections to each adjacent device or network (see Figures 5.9 and 5.10). The second ring can be a backup path—to be used in the event of a failure of the primary ring—or both rings can be used simultaneously to double the traffic capacity of the network.

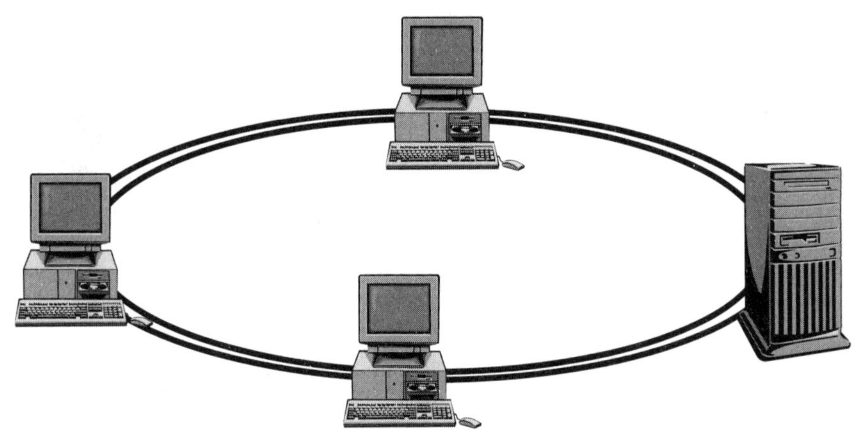

Figure 5.9 Dual-ring topology between devices.

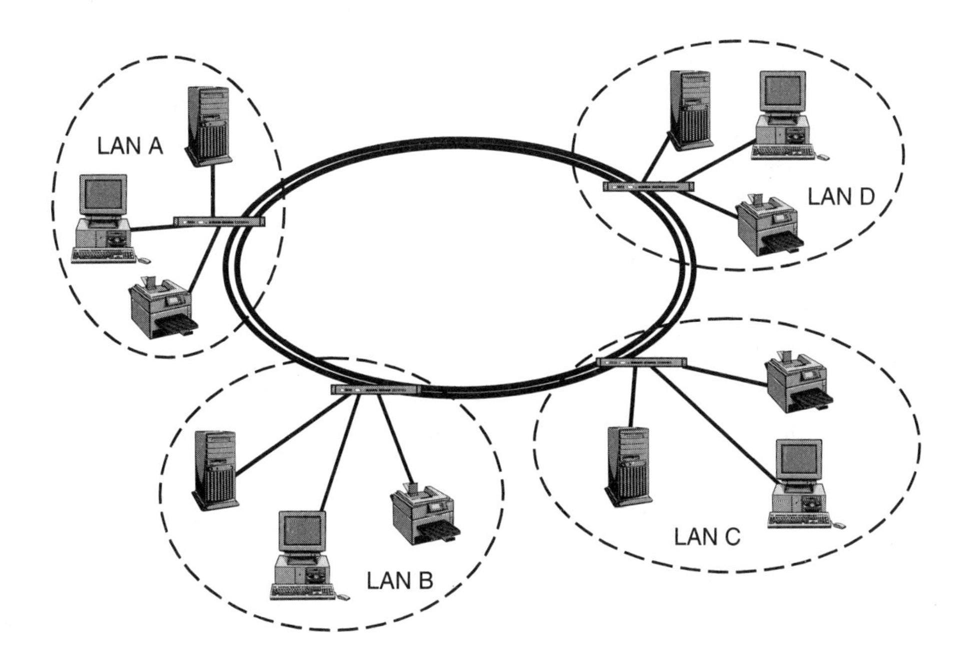

LAN = Local area network

Figure 5.10 Dual-ring topology between networks.

Star topology

A star topology is created when a single device or network is used as a central connection point for all other devices or networks (see Figures 5.11 and 5.12).

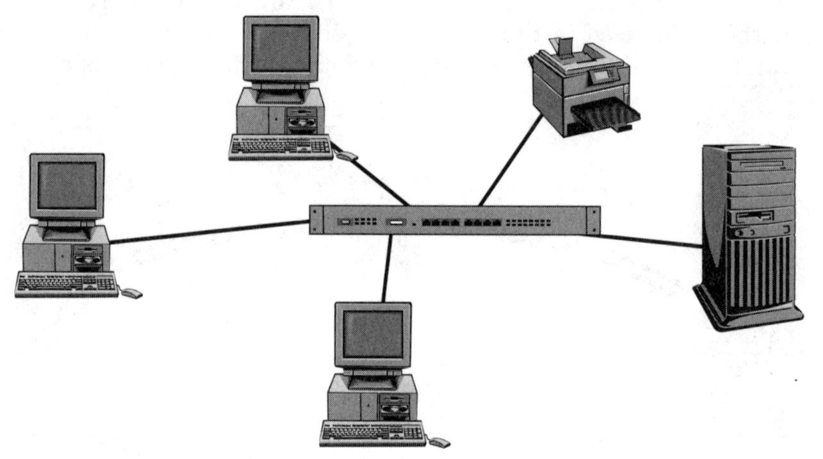

Figure 5.11 Star topology between devices.

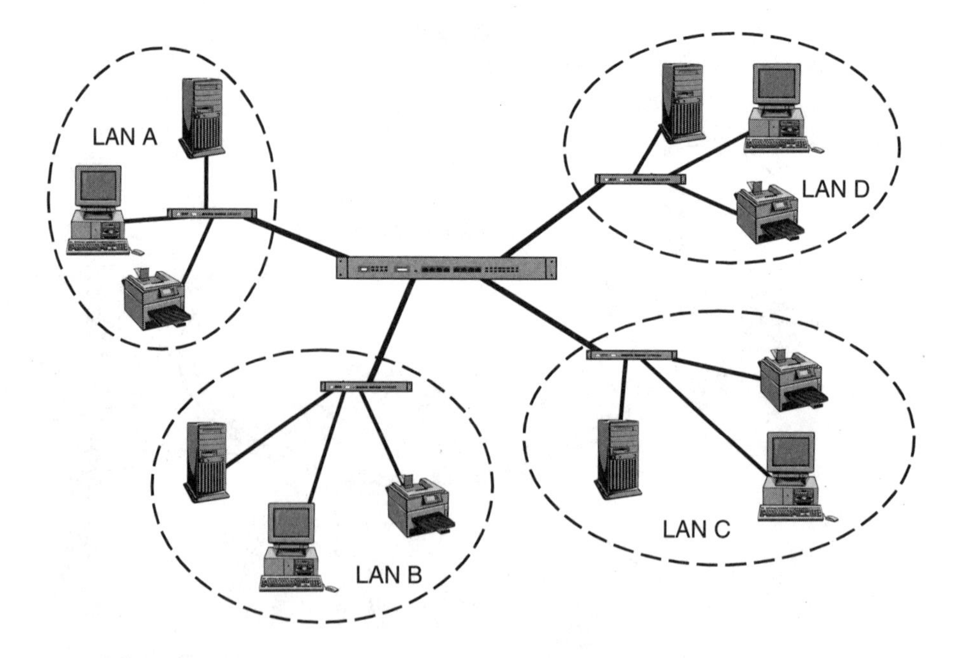

Figure 5.12 Star topology between networks.

NOTE: When used as an internetwork topology, the star topology is also referred to as a centralized or collapsed backbone.

Star-wired bus topology

A star-wired bus topology combines a physical star topology with a logical bus topology. Devices are linked to a central unit, forming the physical star. A single communications channel exists within the central unit, forming the logical bus. Any message sent by a device is forwarded to all devices, since they are all connected to the same bus (see Figure 5.13).

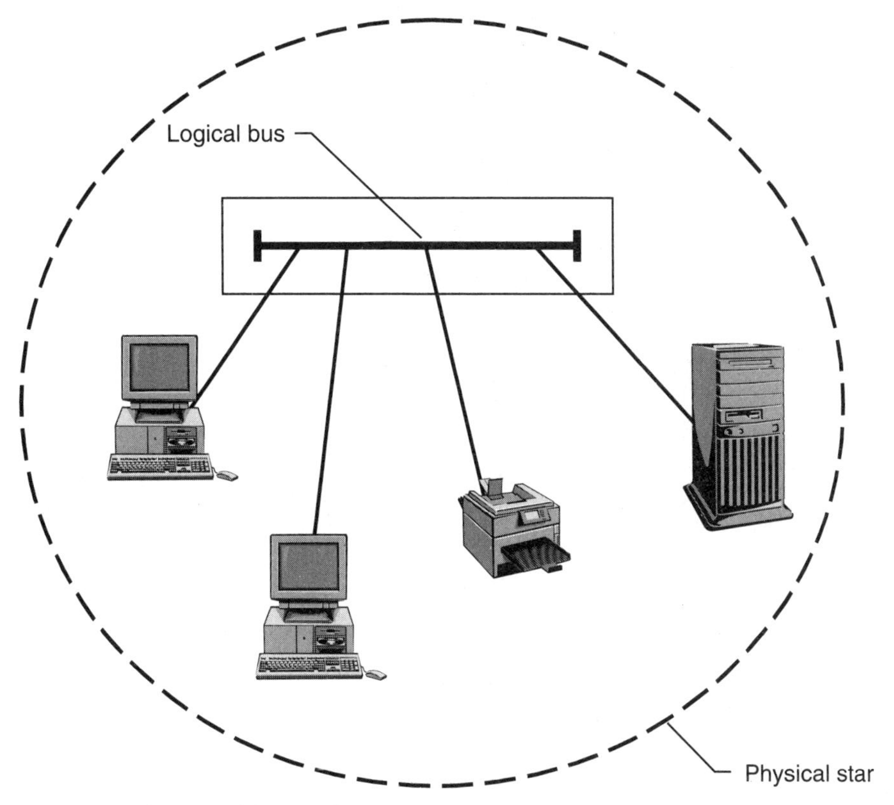

Figure 5.13 Star-wired bus topology.

Star-wired ring topology

A star-wired ring topology combines a physical star topology with a logical ring topology. Devices are linked to a central unit, forming the physical star. A single communications channel exists within the central unit, forming the logical ring. Any message sent by a device is forwarded to the next device on the ring. This process continues until the message reaches its destination (see Figure 5.14).

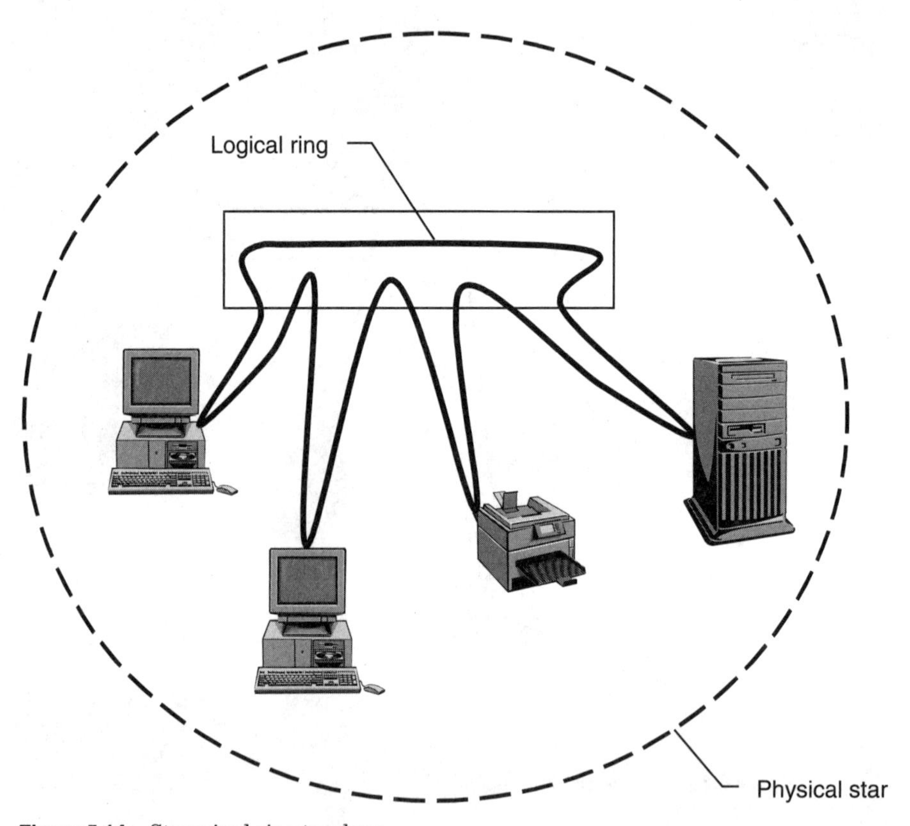

Figure 5.14 Star-wired ring topology.

Hierarchical topologies

A hierarchical topology links devices or networks using a series of levels, similar to an organizational chart. Each device or network is placed at a specific level in the hierarchy. The levels are linked using a series of connections (see Figures 5.15, 5.16, and 5.17).

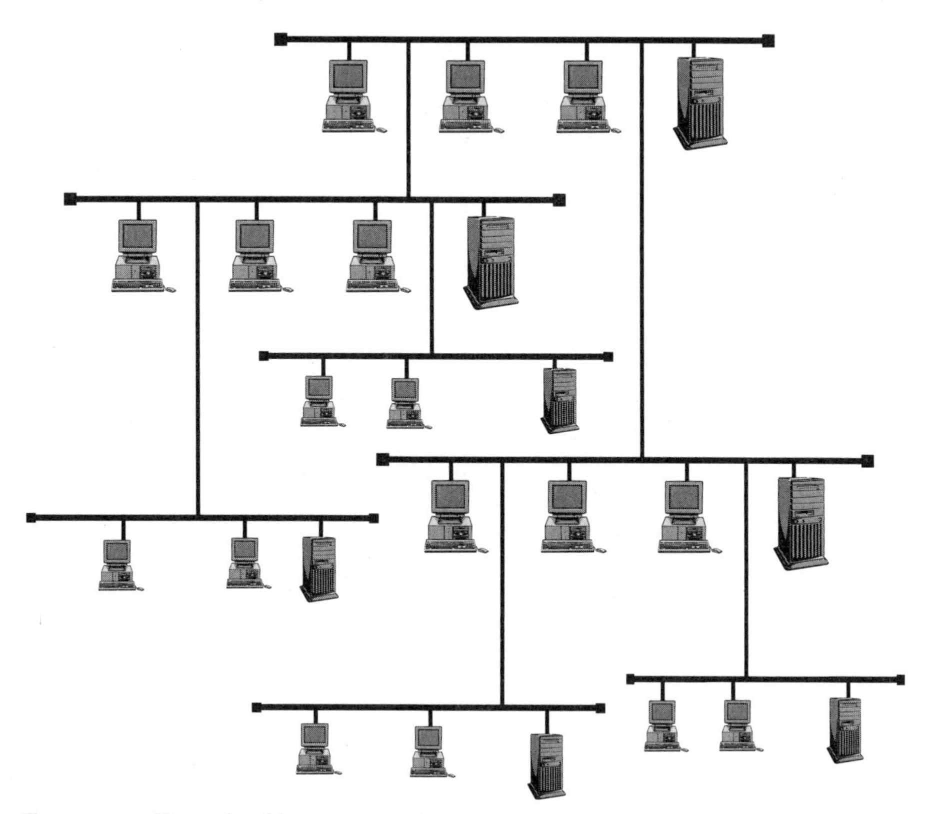

Figure 5.15 Hierarchical bus (tree) topology.

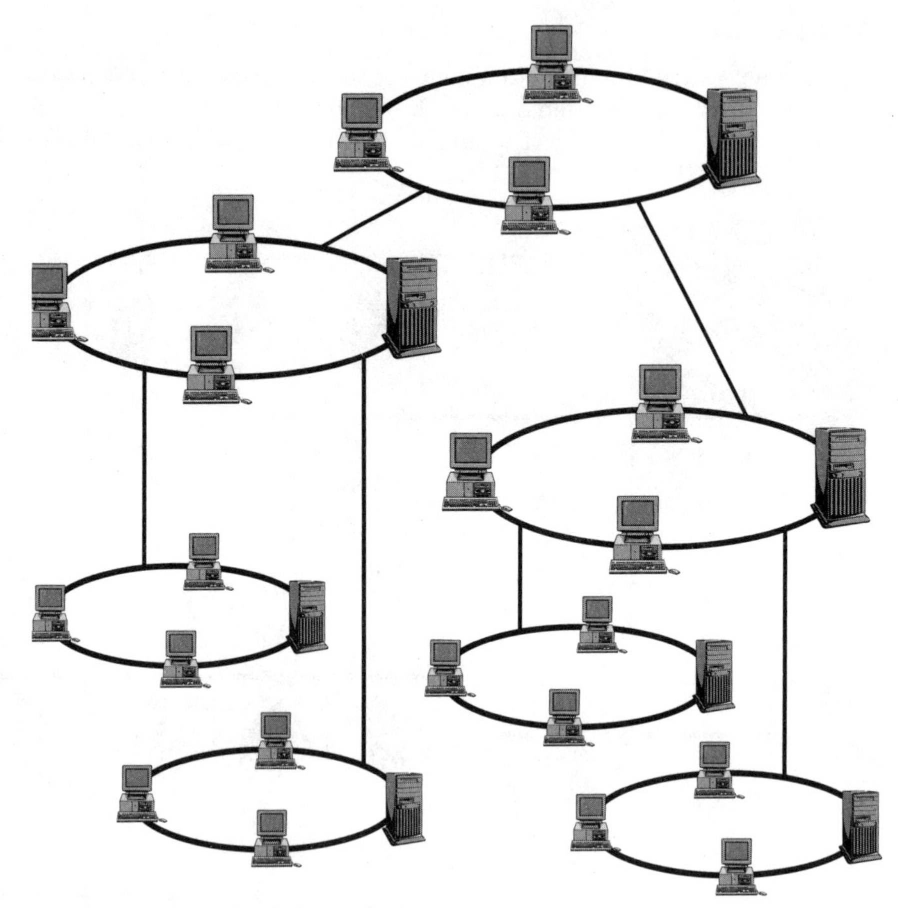

Figure 5.16 Hierarchical ring topology.

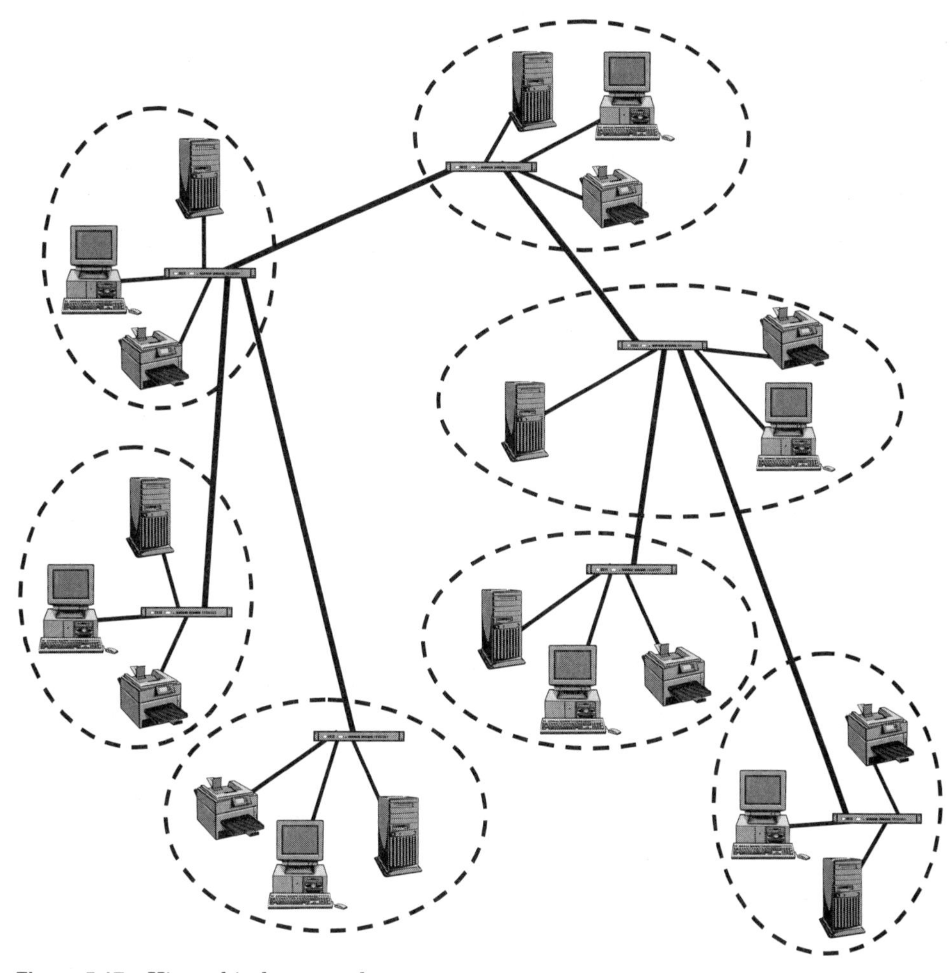

Figure 5.17 Hierarchical star topology.

NOTE: It is also possible to use a combination of bus, ring, and star topologies within a single hierarchy.

Dual homing topology

Each device or network on a dual homing topology is connected to two other devices or networks. This topology provides a high level of fault-tolerance, since a backup communications channel exists for each device or network (see Figure 5.18).

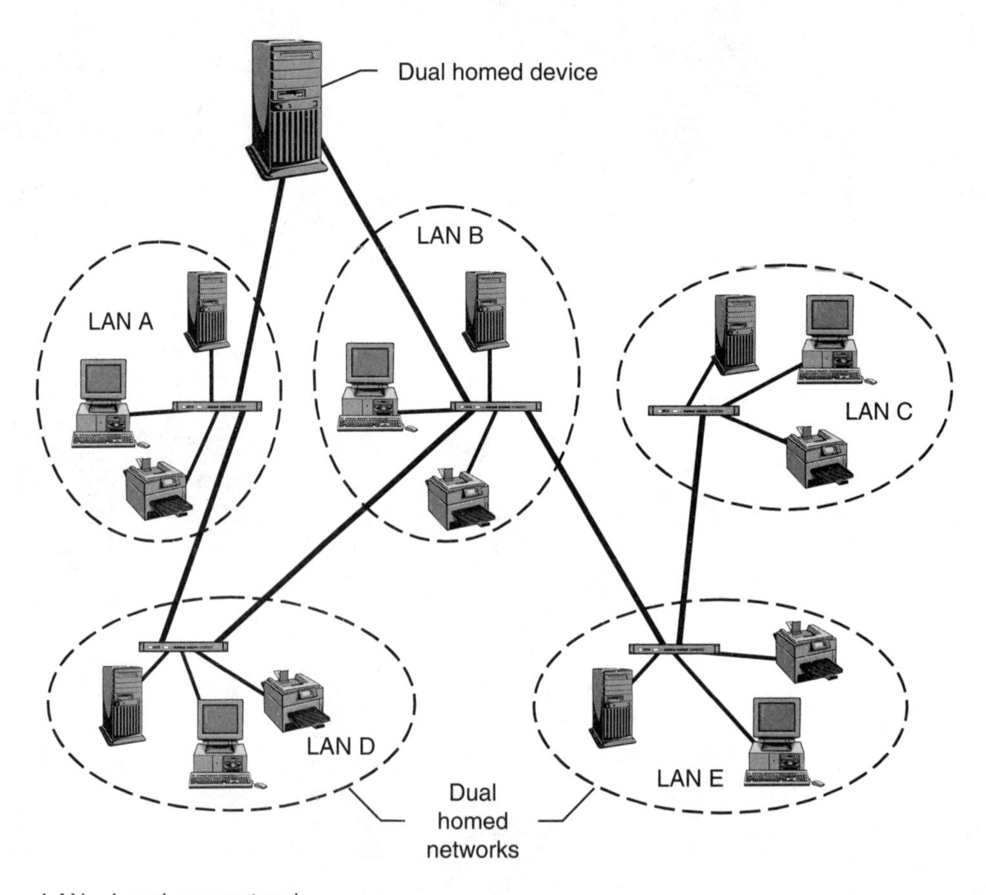

Figure 5.18 Dual homing topology.

Mesh topology

Each device or network on a mesh topology is connected to all other devices or networks. This topology provides the highest level of fault-tolerance, since multiple backup communications channels exist for each device or network (see Figure 5.19). This type of topology is typically associated with a wide area network (WAN).

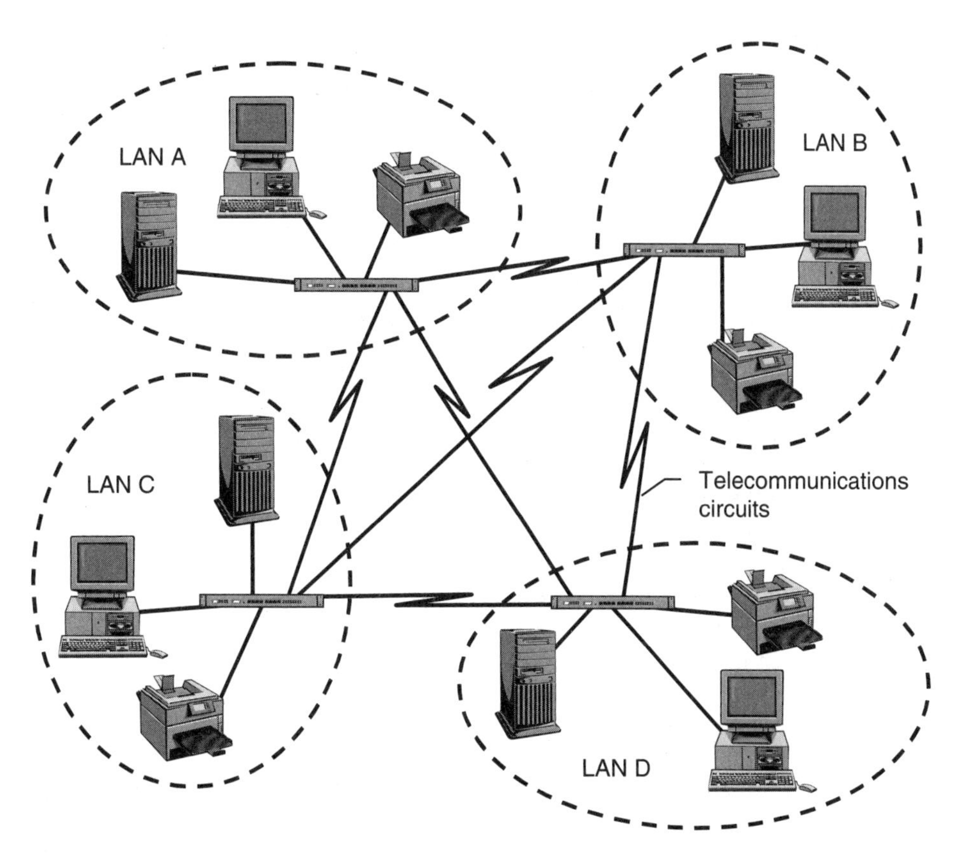

Figure 5.19 Mesh topology.

NOTE: If each device or network is connected to several (but not all) other devices or networks, the topology is called partial mesh.

Network Addressing

General

An organizational network uses multiple types of addresses to transfer messages.

- Every device on a network requires a unique device address.
- If the network is connected to others through an internetwork, unique network addresses are required in addition to unique device addresses.
- A single telecommunications circuit on an internetwork can contain multiple communications channels, in which case an address is required for each channel.
- It is also possible for a network device (e.g., a server or a station) to simultaneously run multiple applications and send messages using different communications protocols. In such cases, an incoming message must be delivered to the appropriate protocol and application, resulting in two additional types of addresses.

To summarize, it is possible to use five types of addressing information to deliver a message over a network, as follows:

- Application address
- Protocol address
- Network address
- Device address
- Communications channel address

The Open Systems Interconnection (OSI) model can be used to illustrate the relationships between the various types of addresses (see Figure 5.20).

Layer 7	Application layer	Application address
Layer 6	Presentation layer	Protocol address
Layer 5	Session layer	
Layer 4	Transport layer	
Layer 3	Network layer	Network address
Layer 2	Data Link layer	Device address
Layer 1	Physical layer	Communications channel address

Figure 5.20 Addressing and the OSI model.

LAN and internetwork addressing

The size and complexity of an organizational network determines the number of times a message is handled by a network device before reaching its destination. In cases where the organizational network consists of LANs linked by an internetwork, the endpoint of a message is a specific device on a LAN, which can be identified by an address assigned to its network interface card (NIC). Such device addresses, however, are not universal in format—different types of LANs use different forms of addressing.

NOTE: Additional terms used to describe device addresses, include:

- Hardware address.
- Physical address.
- Data Link layer address.
- Layer 2 address.
- Medium access control (MAC) address.
- NIC address.
- LAN address.
- Individual address.

An internetwork can link different types of LANs (e.g., Ethernet, token ring, and fiber distributed data interface [FDDI]). To uniquely identify each device on any LAN in a consistent manner, a second address called the network address is assigned to each device. The network address of a device identifies both the device and the LAN to which it is connected.

NOTE: The terms network identification (netid) and host identification (hostid) are commonly used to describe the two parts of the network address. In such cases, netid identifies the LAN broadcast domain and hostid identifies the device within the broadcast domain.

Using the same format for all network addresses makes it possible to connect different types of LANs to a common internetwork. One device can easily take the place of another (e.g., in the event of a breakdown or an upgrade) through a reassignment of the network address.

NOTE: Additional terms used to describe network addresses, include:

- Network layer address.
- Logical address.
- Layer 3 address.
- Internetwork address.
- Routing address.
- Group address.

On a LAN, device addresses are used to identify the source and destination of each message. The stations, servers, and shared peripherals in a common broadcast domain send and receive messages to each other using device addressing. When multiple LANs are connected to an internetwork using routers, the routers use both device and network addresses, as follows:

- All routers on an internetwork keep tables of the network addresses of the LANs connected to the internetwork.
- A router connected to a given LAN keeps a table of both the device address and the network address of each LAN device.
- When a router receives a message intended for a device on a LAN connected to itself, it uses the information in its tables to forward the message to the appropriate device, using the device address.
- When a router receives a message intended for a device on a LAN connected to another router, it uses the network address to place the message on a path to that router. On large internetworks, multiple router hops may be required to route a message from one LAN to another.

Message format. Different terms are used to distinguish between the message format used on a LAN and the message format processed by routers when directing traffic onto the internetwork. The term frame is used to describe LAN messaging units (e.g., there are Ethernet frames, token ring frames, and FDDI frames). On an internetwork, the term packet or datagram is used (e.g., IP packets or IP datagrams). Datagrams containing network addresses are encapsulated—placed in frames containing device addresses—comparable to placing an envelope addressed to an individual inside a larger envelope addressed to an organization. Figure 5.21 illustrates this process using an IP version 4 (IPv4) datagram and an Ethernet frame.

NOTE: IP datagrams and routing are described in detail later in this chapter, in the section titled Internetwork Routing.

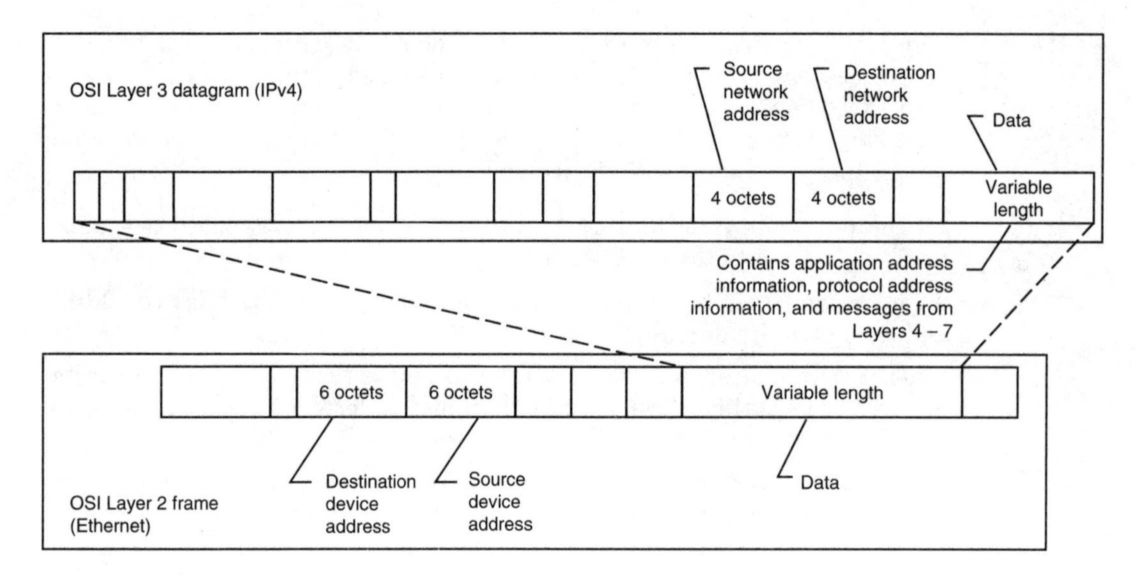

IPv4 = Internet protocol version 4
OSI = Open Systems Interconnection

Figure 5.21 Relationship between a datagram and a frame.

NOTE: An octet is a grouping of eight bits. It is frequently referred to as a byte.

Message transfer. Example 5.1 illustrates an example of internetwork message transfer using datagrams encapsulated in frames. In this figure, a router links two LAN broadcast domains. Each device on the network—including the router interfaces—has been assigned a device address, as shown.

NOTE: It is common to express device addresses using hexadecimal (hex) notation and IPv4 network addresses using dotted decimal notation instead of a sequence of zeros and ones. Each two-digit grouping in a hex address (e.g., A0) represents eight bits. Each decimal value in a dotted decimal address (e.g., 192) also represents eight bits. Device addresses are 48 bits in length and IPv4 network addresses are 32 bits in length. For more information, see Appendix A: Binary and Hexadecimal Numbering Fundamentals and Appendix B: Internet Protocol (IP) Addressing Fundamentals.

EXAMPLE 5.1 Internetwork message transfer

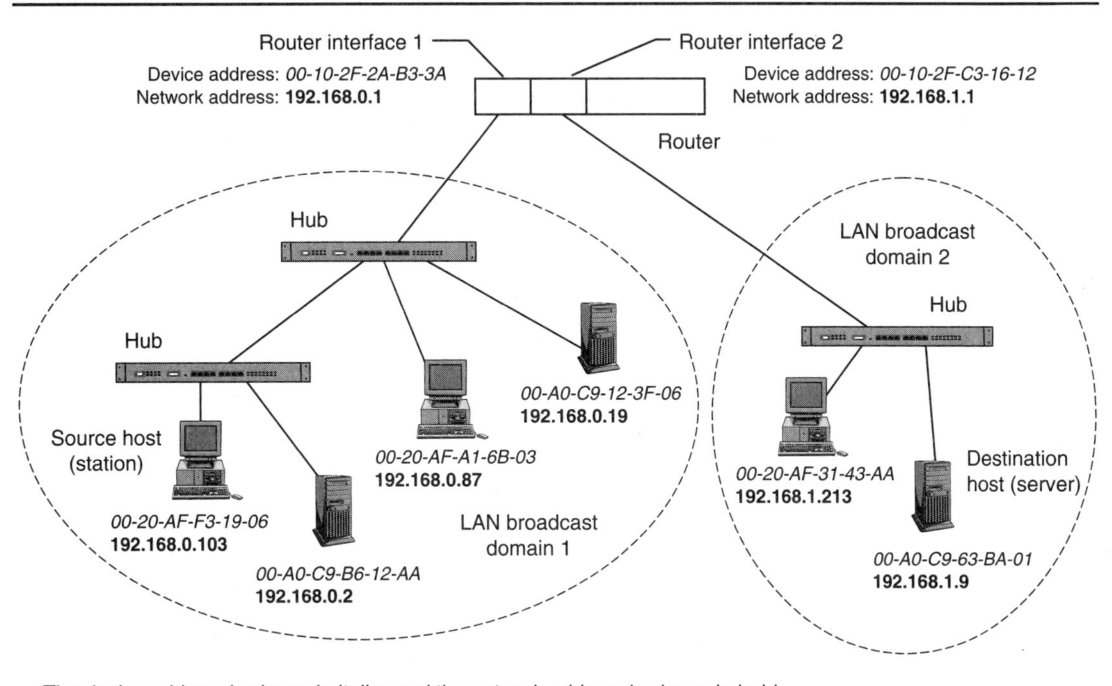

The device address is shown in italics and the network address is shown in bold.
The network address represents both the netid and hostid, as follows → **192.168.0.1** (netid = 192.168.0; hostid = 1)

hostid = Host identification
LAN = Local area network
netid = Network identification

The message transfer process occurs as follows:

1. The source host (station) in LAN broadcast domain 1 has an IP datagram to send to the destination host (server) in LAN broadcast domain 2.
2. Since the netid of the destination host (192.168.1) is different from the netid of the source host (192.168.0), the source host places the datagram in a LAN frame and sends the frame to the router, using the router's device address (00-10-2F-2A-B3-3A).
3. The router receives the frame on its interface 1, removes and discards the device addressing information, and inspects the network address.
4. Since the destination host is on a LAN connected to the same router, the router places the datagram in a LAN frame and sends the frame to the server, using the server's device address (00-A0-C9-63-BA-01).

This example illustrates how frames and device addresses are discarded and replaced as many times as necessary to get the datagram from source to destination. At all times, however, the datagram and the network addresses it contains remain unchanged.

Example 5.1 can also be used to describe the use of network addressing and routers to reduce broadcast traffic on a network. Routers confine broadcast traffic to the broadcast domain where they originate. In Example 5.1, any datagram addressed to all devices is sent to only those devices with the same netid as the source of the datagram. For example, a broadcast datagram issued by host 192.168.0.87 is prevented by the router from distribution to devices with netid equal to 192.168.1—only the hosts 192.168.0.19, 192.168.0.2, and 192.168.0.103 receive the broadcast.

If the level of broadcast traffic becomes excessive and results in delayed network response time, an existing broadcast domain can be split into two by changing the network address of some of the devices and using an additional router interface. Example 5.2 illustrates this modification—the LAN broadcast domain 1 shown in Example 5.1 has been converted into two broadcast domains. From an addressing point of view, all that is required is to change the network addresses of the devices transferred to the new broadcast domain. In this example, the netid for the new broadcast domain is 192.168.2, which is applied to all devices in this domain (see Example 5.2).

After this modification, a broadcast datagram issued by host 192.168.0.87 is prevented by the router from distribution to devices with netid equal to 192.168.1 or 192.168.2—only the host 192.168.0.19 receives the broadcast.

Network Access Control

General

Network message transfer requires data to be placed in frames and identified using device addresses before transfer over a LAN. In cases where a LAN communications channel is shared by two or more devices (e.g., when hubs are used), a process for regulating access to the channel is required to ensure fair access to each device that has frames to transmit. This process is called medium access control (MAC). A related process called logical link control (LLC) is used to provide compatibility between different types of MAC processes and services to the network layer.

NOTE: In cases where each LAN device has a dedicated communications channel (e.g., when switches are used), access control is not required. Switch operations are described in detail in Chapter 6: Switching and Virtual LANs (VLANs).

EXAMPLE 5.2 Creating a new broadcast domain.

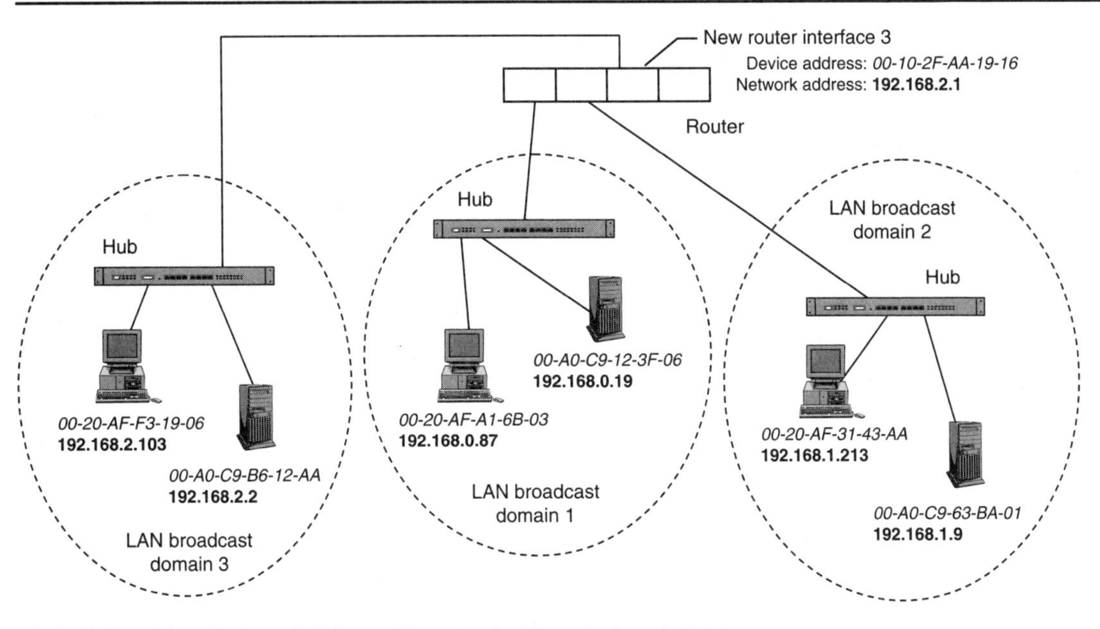

The device address is shown in italics and the network address is shown in bold.
The network address represents both the netid and hostid, as follows → **192.168.0.1** (netid: 192.168.0; hostid: 1)

hostid = Host identification
LAN = Local area network
netid = Network identification

The OSI model can be used to illustrate network access and communications on a LAN. As shown in Figure 5.22, the bottom two layers of the model provide the services needed to transfer frames from one LAN device to another. Higher layers provide the Data Link and the Physical layers with the contents to be sent over the LAN, as follows:

- A Network layer protocol prepares and issues datagrams to the Data Link layer.
- A Data Link layer protocol places the datagrams into frames and obtains access to the medium.
- The frames are transmitted by a Physical layer protocol in the form of a bit sequence, using a signal suitable to the medium.

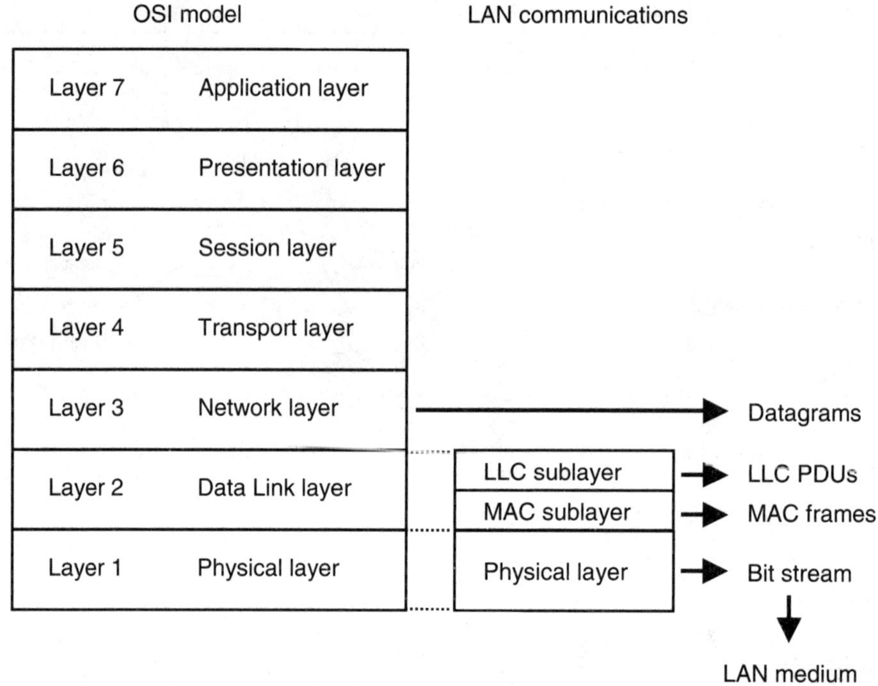

LAN = Local area network
LLC = Logical link control
MAC = Medium access control
OSI = Open Systems Interconnection
PDU = Protocol data unit

Figure 5.22 OSI model and LAN communications.

Logical Link Control (LLC) sublayer functions

The LLC sublayer defines a common set of services used by all LAN technologies when transferring a datagram from the Network layer to the MAC sublayer of the Data Link layer. Since many protocols exist for both the Network layer and the MAC sublayer, the LLC provides a means to identify the Network layer protocol that generated a given datagram. LAN devices running multiple Network layer protocols can send and receive different types of datagrams using a common frame format.

The LLC sublayer provides a consistent interface to all Network layer protocols—the MAC process used on the LAN is not visible to the Network layer protocol. This simplifies the development of new Network layer protocols and new MAC processes. As long as both are compatible with the LLC sublayer, any LAN technology can be used to identify and transport any type of datagram.

The LLC sublayer provides three types of communication services between devices on a LAN, as follows:

- LLC type 1 service provides an unacknowledged connectionless delivery of frames. No mechanism exists at the Data Link layer to confirm to the sending device that a given frame was received successfully. Type 1 service is requested in cases where higher layer protocols are responsible for flow control, error control, and error recovery processes.
- LLC type 2 is a connection-oriented service that provides flow control, error control, and error recovery processes at the Data Link layer. A connection is established between devices before the transfer of frames, similar to the call setup process used over the telephone network. Type 2 service is requested in cases where higher layer protocols do not have the capability to ensure the successful transfer of messages.
- LLC type 3 service provides an acknowledged connectionless delivery of frames. As with LLC type 1, there is no need to establish a connection before the transfer of frames. Unlike LLC type 1, however, there is a confirmation of receipt issued by the receiving device. LLC type 3 services have not been widely adopted.

The output of the LLC sublayer is the LLC protocol data unit (PDU).

Medium Access Control (MAC) sublayer functions

The MAC sublayer specifies the protocols and processes used by devices to send and receive messages over a LAN communications channel. This sublayer defines how each device can obtain access to a shared medium and manages the communications process used to transfer frames between devices. Multiple MAC sublayers are available, each corresponding to a different LAN technology (e.g., Ethernet and token ring LANs have different MAC sublayers).

The processes available at the MAC sublayer include:

- Medium access management, used to enable sharing of the LAN medium.
- Framing, used to address a datagram to the appropriate device on the LAN.
- Error detection, used to verify the content integrity of a received frame.

Three types of access control are commonly used to share a LAN communications channel. They are:

- Carrier sense multiple access with collision detection (CSMA/CD).
- Carrier sense multiple access with collision avoidance (CSMA/CA).
- Token passing.

The output of the MAC sublayer to the Physical layer is the MAC PDU, more commonly referred to as the MAC frame.

Physical layer functions

The Physical layer is responsible for establishing, maintaining, and releasing physical connections between network devices. In the case of LAN communications, the combination of the Physical layer and MAC layer defines the LAN technology. For example, Ethernet is a set of Physical and MAC layer specifications. It is possible to use multiple media types for Physical layer operations—twisted-pair, coaxial, optical fiber, and wireless media can all be used to transport Ethernet frames.

The processes available at the Physical layer include:

- Encoding, used to transform the frame to be transmitted into a format compatible with the medium.
- Signal generation.
- Timing and synchronization between the devices at each end of a connection.

The output of the Physical layer is a signal representing a stream of bits.

Network Signaling

General

The term network signaling refers to the process used to transfer a sequence of bits over a communications medium. Normally, digital signaling is used since the bits represent messages generated by digital devices. An exception is data transfer over the public switched telephone network (PSTN), which requires analog signaling. In such cases, modems are used to transform digital pulses into analog tones at the sending device and the reverse at the receiving device.

NOTES: An analog signal can be described as a wave that varies continuously over its duration. The amplitude, frequency, and the phase of the wave can vary in different combinations to represent a bit stream.

A digital signal can be described as a sequence of discrete pulses separated by intervals. The pulses and intervals can be grouped in various combinations to represent control and message information.

Baseband and broadband signaling

The terms baseband and broadband are also associated with signaling. Historically, baseband was used to describe a form of digital signaling where the medium provided a single communications channel. In comparison, broadband referred to a form of analog signaling where the medium was divided into multiple communications channels, as in the case of cable television signals.

Currently, the term broadband is more commonly used to refer to high-speed digital signaling, where the communications channel is capable of transporting the data streams simultaneously generated by many devices. For example, a telecommunications circuit capable of operating at megabit per second (Mb/s) rates is often described as a broadband connection.

The following terms are also used to describe different types of signaling:

- Simplex
- Half-duplex
- Full-duplex
- Dual-duplex
- Asynchronous
- Synchronous
- Isochronous

These terms are described below.

Simplex signaling

When simplex signaling is used, the message flow is in one direction only. The receiver cannot send a message back to the sender.

Half-duplex signaling

When half-duplex signaling is used, messages can flow in both directions, but not at the same time. The single communications channel linking the sender and the receiver can transport messages in one direction at a time.

Full-duplex signaling

When full-duplex signaling is used, messages can flow in both directions at the same time—two communications channels link the sender and the receiver.

Dual-duplex signaling

When dual-duplex signaling is used, messages can flow in both directions at the same time over a single communications channel. The single channel linking the sender and the receiver can simultaneously transport messages in both directions.

Asynchronous signaling

In asynchronous signaling, control bits precede and follow each character in a message. These control bits are referred to as start, stop, and parity bits. Since each character can be distinguished from the others using the control bits, it is possible to send characters at various intervals.

Synchronous signaling

In synchronous signaling, the communicating devices are synchronized using a common timing mechanism, or clock. The precise intervals defined by the clock are used to distinguish each character from the next; therefore, control bits are not required.

Isochronous signaling

In isochronous signaling, no timing mechanism exists between the communicating devices. However, isochronous data is associated with time-sensitive applications such as voice and video messaging, where delays are undesirable. Isochronous signaling places timing information into the message stream, which is then used to transport the stream over a network without excessive delay.

Signal encoding

Signal encoding refers to the conversion of data to be transmitted into a form compatible with the communications medium. The objective is to ensure that the binary zeros and ones transmitted by the device at one end of a link arrive at the other end in a recognizable form. Signal encoding is also used to maintain accurate clocking and, in cases where the signals are electrical, to minimize electromagnetic interference (EMI).

In digital signaling, data is transmitted as a sequence of electrical or optical pulses. Variations in the pulses are used to represent binary zeros and ones. The following are two examples of digital signal encoding, using variations in voltage.

NOTE: The same techniques can also be used for signaling over optical fiber media, using variations in light pulses instead of voltage levels.

Manchester encoding. Manchester encoding, which is used on 10 Mb/s Ethernet networks, is a type of biphase encoding. In all types of biphase encoding, there is at least one transition in voltage level per bit transmitted. Since there is a predictable transition for each bit transmitted, the receiver can synchronize with the sender using this transition. Therefore, biphase-encoded signals can also be described as self-clocking signals. Any absence of an expected transition indicates that an error has occurred. For an error to pass undetected, both the signal before and the signal after the expected transition would have to be altered.

When Manchester encoding is used, there is a voltage transition in the middle of each bit sequence, or bit time. The bit value of zero is represented by a high-to-low transition—the high voltage level is used during the first half of the bit time and the low voltage level is used during the second half of the bit time. The bit value of one is similarly represented by a low-to-high transition (see Figure 5.23).

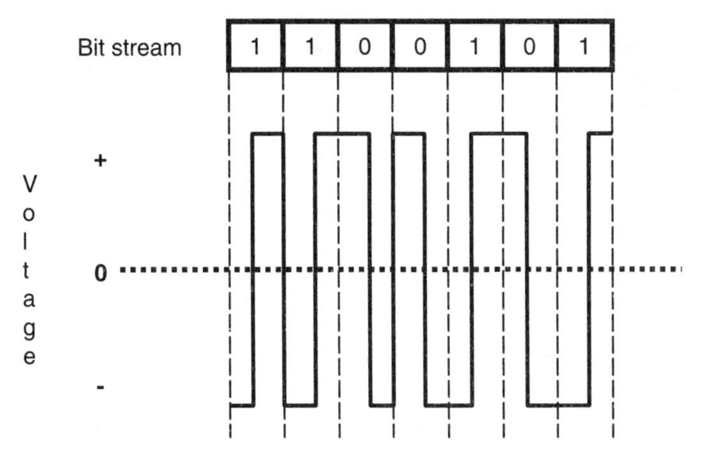

Figure 5.23 Manchester encoding.

Differential Manchester encoding. Differential Manchester encoding, which is used on 4 and 16 Mb/s token ring networks, is another type of biphase encoding. When differential Manchester encoding is used, there is a voltage transition in the middle of each bit time, as with Manchester encoding. However, the bit value of zero is represented by an additional transition at the beginning of the bit time. The bit value of one does not have this additional transition (see Figure 5.24).

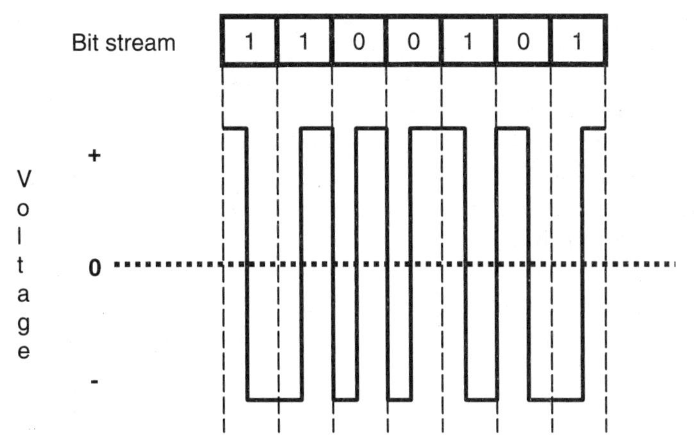

Figure 5.24 Differential Manchester encoding.

Internetwork Routing

General

Routers are used to interconnect various types of LANs at the Network layer of the OSI model. Routers make it possible to segment an organizational network into multiple broadcast domains, also referred to as subnetworks or subnets, using network address assignments. In cases where multiple LANs are interconnected using routers, the role of each router is to direct datagrams to their destination LANs—and ultimately, their destination devices—over the most efficient path.

A router is a combination of hardware and software, typically in the form of a specialized device. It is also possible to install routing software on a general-purpose server to convert it to a route server, or router. In both cases, the software used to implement routing operations can be described as the internetwork operating system (IOS). The functions of an IOS include:

- Enabling communications between a variety of LAN technologies and subnets.
- Choosing the most appropriate path through an internetwork for any given datagram.
- Providing tools to identify and manage internetwork traffic.

A router can process traffic selectively, using the values found in various fields of a datagram. For example, broadcast traffic generated on one subnet is prevented from reaching other subnets, a process called filtering. By discarding broadcast datagrams instead of forwarding them, the filtering capability of a router eliminates a common source of unnecessary traffic on an organizational network. The filtering capabilities of a router can also be used to grant or deny access to routing resources on the basis of network address information. Each interface on a router can be programmed with a set of rules listing access privileges for that interface.

Routers connected to the Internet must use IP as their Network layer protocol. Since the Internet is the only public internetwork spanning the globe, its popularity and influence has made IP the Network layer protocol of choice for many organizational networks. The worldwide demand for Internet connections has resulted in IP address shortages, making it necessary to develop various short-term solutions to manage a limited pool of unassigned addresses. To resolve this and other IP-related issues, an updated version of IP, initially called IP next generation (IPng) and later renamed IP version 6 (IPv6), has been introduced. IPv6 is intended to coexist with, and eventually replace, the current version of IP, which is IPv4.

Internet Protocol Version 4 (IPv4) datagrams

The specifications for IPv4 were published in January 1980 as request for comment number 760 (RFC 760) and updated in September 1981 as RFC 791. The format and fields of an IPv4 datagram are shown in Figure 5.25.

	Field	Size
Header	Version	4 bits
Header	Internet header length	4 bits
Header	Type of service	8 bits
Header	Total length	16 bits
Header	Identification	16 bits
Header	Flags	3 bits
Header	Fragment offset	13 bits
Header	Time to live	8 bits
Header	Protocol	8 bits
Header	Header checksum	16 bits
Header	Source address	32 bits
Header	Destination address	32 bits
Header	Options (optional)	Variable length
Header	Padding (optional)	Variable length
	Data	Variable length; maximum 524,120 bits (65,515 octets)

Figure 5.25 IPv4 datagram fields and field sizes.

NOTE: The maximum size of an IPv4 datagram is 524,280 bits (65,535 octets), of which a maximum of 480 bits (60 octets) can represent header information. If the options and padding fields are not used, the header length is 160 bits (20 octets).

IPv4 datagram components

Version. The version field indicates the version number of the protocol. The version described here is four.

Internet Header Length (IHL). The IHL field indicates the length of the header. This value represents a multiple of 32 bits and is in the range 5 to 15, indicating a header length of 160 to 480 bits.

Type of Service (TOS). The TOS field indicates the class of service desired for the datagram. The bits in this field allow a given datagram to have a higher precedence than others have when a router handles it. In cases where multiple paths exist to the destination, the TOS field can be used to choose a path with low delay, high throughput, low cost, or high reliability.

Total length. The total length field indicates the total length of the datagram in octets, including the header.

Identification. A datagram may need to be fragmented to fit into the frames used on a network. For example, an Ethernet frame can contain a maximum of 12,000 bits (1500 octets). If a datagram requires fragmentation, the identification field is used to associate a fragment with a specific datagram.

Flags. The flags field indicates whether a given fragment is the last one in a series. It can also be used to instruct routers not to fragment a given datagram.

Fragment offset. The fragment offset field indicates the position a given fragment should occupy when the datagram it is associated with is reassembled.

Time to Live (TTL). The TTL field is used to ensure that any datagram will eventually be removed from the internetwork whether or not it has reached its destination. This prevents datagrams from circulating endlessly as a result of routing loops. The value in the TTL field serves as a counter, and is decremented each time the datagram is processed by a router. When the TTL value reaches zero, the datagram is discarded.

Protocol. The protocol field identifies the higher layer protocol that generated the information found in the data field of the datagram. For example, the transmission control protocol (TCP) is commonly used as the Transport layer companion to the Network layer IP. The combination is referred to as TCP/IP.

Header checksum. The header checksum field is used to validate the contents of the header.

NOTE: This value is changed each time the datagram is processed by a router, since the TTL field in the header is decremented.

Source address. The source address is the network address of the device that originated the datagram, expressed in dotted decimal notation (e.g., 192.168.0.1).

NOTE: Each decimal value represents eight bits. Conversions between decimal, hexadecimal, and binary values are described in Appendix A: Binary and Hexadecimal Numbering Fundamentals.

Destination address. The destination address is the network address of the device(s) that must receive the datagram, expressed in dotted decimal notation.

NOTE: IPv4 addressing is described in detail in Appendix B: Internet Protocol (IP) Addressing Fundamentals.

Options. The options field is optional and is typically used for internetwork testing and measurement purposes. Multiple options exist and it is possible to indicate one or more options in a datagram, which makes this field variable in length.

Padding. The padding field is optional and consists of a sequence of zeros. It is used to make the header length equal to a multiple of 32.

Data. The data field contains information generated by higher layer protocols and processes. If the options and padding fields are not used, a maximum of 524,120 bits (65,515 octets) can be placed in this field.

Internet Protocol Version 6 (IPv6) datagrams

The specifications for IPv6 were published in December 1995 as RFC 1883 and updated in December 1998 as RFC 2460. The format and fields of an IPv6 datagram are shown in Figure 5.26.

	Field	Size
Header	Version	4 bits
	Traffic class	8 bits
	Flow label	20 bits
	Payload length	16 bit
	Next header	8 bits
	Hop limit	8 bits
	Source address	128 bits
	Destination address	128 bits
	Data	Variable length; maximum 524,280 bits (65,535 octets)

Figure 5.26 IPv6 datagram fields and field sizes.

NOTE: The maximum size of an IPv6 datagram is 526,840 bits (65,855 octets), of which 320 bits (40 octets) represent header information. Under certain conditions, provisions in IPv6 allow for the use of:

- Additional headers called extension headers.
- Larger datagrams called jumbo payload datagrams or jumbograms.

Extension headers and jumbograms are not discussed in this document.

IPv6 datagram components

Version. The version field indicates the version number of the protocol. The version described here is six.

Traffic class. The traffic class field can be used to indicate the class of service desired for the datagram. The bits in this field allow a given datagram to have a higher precedence than others have when a router handles it. This field is intended to be similar in function to the TOS field in an IPv4 datagram.

Flow label. Like the traffic class field, the flow label field can be used to indicate special handling. Flow labeling makes it possible to group and identify a series of datagrams that originate from a common application (e.g., voice or videoconferencing traffic).

Payload length. The payload length field indicates the length—in octets—of the data field in the datagram.

Next header. The next header field is used to identify the higher layer protocol that generated the information found in the data field of the datagram. This field is intended to be similar in function to the protocol field in an IPv4 datagram.

Hop limit. Like the TTL field in an IPv4 datagram, the hop limit field is used to ensure that any datagram will eventually be removed from the internetwork—whether or not it has reached its destination. This prevents datagrams from circulating endlessly as a result of routing loops. The value in the hop limit field serves as a counter, and is decremented each time the datagram is processed by a router. When the hop limit value reaches zero, the datagram is discarded.

Source address. The source address is the 128-bit network address of the device that originated the datagram. Due to its length, this value is often expressed in colon hexadecimal notation (e.g., FF03:FEDB:1234:10A8:36EA:BC82:A3D1:3070).

NOTE: Each group of four hexadecimal values represents 16 bits. Conversions between decimal, hexadecimal, and binary values are described in Appendix A: Binary and Hexadecimal Numbering Fundamentals.

Destination address. The destination address is the network address of the device(s) that must receive the datagram.

Data. The data field contains information generated by higher layer protocols and processes. A maximum of 524,280 bits (65,535 octets) can be placed in this field.

Routing operations

Routers forward datagrams by examining address information at the Network layer and then directing the datagrams over a specific route. When a router begins to process a Network layer datagram, it must first determine the datagram's destination network. The process is as follows:

1. The checksum value found in the frame is used to perform an error check on the frame.
2. Any information added by Physical and Data Link layer protocols of the sending device is removed from the incoming frame to extract the internetwork datagram.
3. The router then evaluates the information added by the Network layer protocol of the sending device. This information typically includes the Network layer destination address (and in some cases, a predetermined path through the network).

This is illustrated in Figure 5.27.

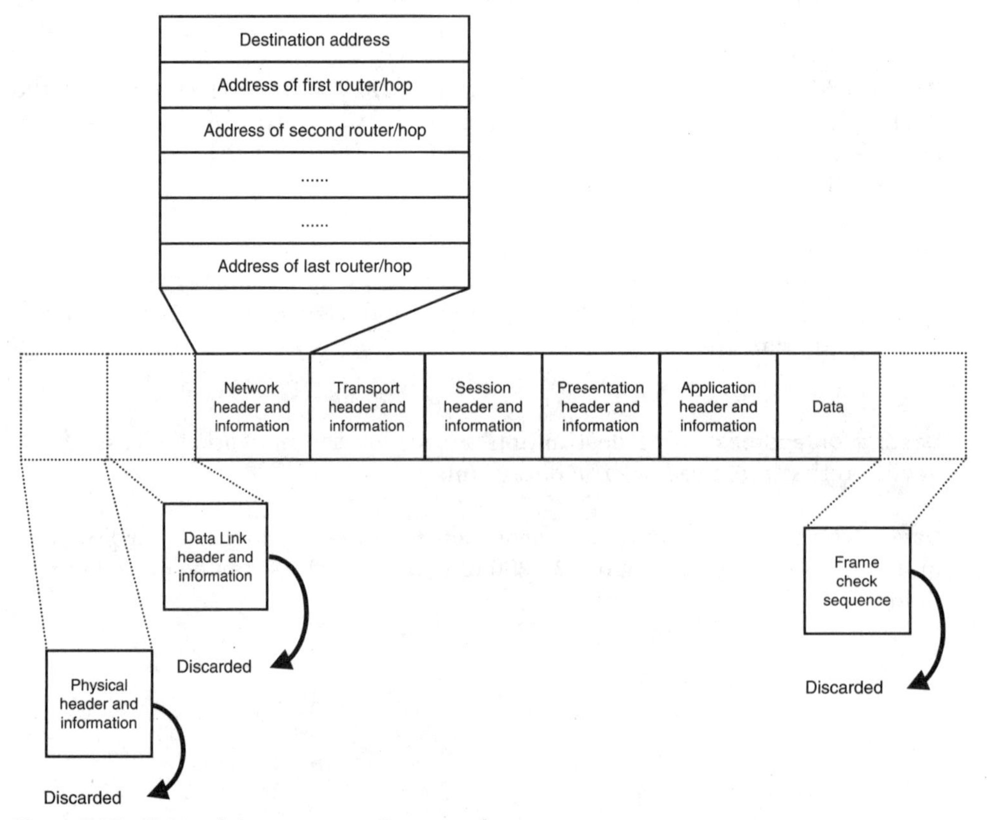

Figure 5.27 Router datagram processing example.

Based on the information found, the router can perform a variety of operations. If the:

- Datagram is addressed to the router itself, the router will evaluate the remaining information in the datagram and act accordingly.
- Router has created a filtering list, the source and destination addresses are checked to see if the datagram should be sent to the destination network or discarded. A datagram may not be allowed into or out of a network for security reasons.
- Datagram is addressed to a device on an attached network, it is placed in a frame and forwarded to the destination device.
- Datagram contains the address of its next hop—the next router to which it should be sent—the datagram is placed in a frame and forwarded to the indicated router.
- Router is unable to find the destination address or cannot determine a path to the destination network, the datagram is discarded. In some cases, an error message is also sent to the sending device.
- Datagram contains control fields used to make routing decisions, the router examines, and possibly modifies, these fields. For example, some datagrams contain a TTL field. This field contains data that indicates how long the datagram has been on the internetwork or the number of hops it has taken. Each router that the datagram passes through modifies this data. If the datagram has not been delivered to its destination in a given period or has reached a preset number of hops, it is assumed that the datagram is in a loop. When this happens, the router discards the datagram.

Once the datagram reaches its destination network, the last router creates the appropriate Data Link layer frame. This enables the delivery of the message to the correct device on that network.

Routing tables

Overview. Routing tables are used to forward datagrams to their destination. Each router maintains tables, in part containing information about any adjacent or neighboring router(s), as well as about the LAN(s) to which it is directly connected. These tables are essential for determining the best path for a datagram through an internetwork.

A router table is a database made up of information on the routes a datagram can take through the internetwork. It is much more comprehensive than the simple address table used by bridges and switches.

Rarely is geographic distance the best measure for choosing a path. Such a measure does not account for the economic costs associated with various types of telecommunications circuits.

The best path through the internetwork depends on many factors, most of them defined by the internetwork designer. The criteria used to choose routes are known as metrics, and can include the:

- Amount of transfer delay.
- Available capacity of the communications channel.
- Average traffic load on the internetwork links and the routers.
- Total number of hops the datagram will encounter over the path.
- Cost associated with a given path.

Usually, the route with the lowest overall path cost is selected for the datagram.

The internetwork designer assigns a value to each metric in order to introduce bias into the route calculation. For example, the designer can influence whether a given datagram is directed along the fastest route or the lowest-cost route—a necessary decision in pay-per-datagram communications.

In cases where no path metric has been established, the router uses a default metric. For example, the default metric can be the duration of a single bit in a 10 nanosecond (ns) interval on the path. This acts as a rough approximation of the path delay. With this metric, a 10 Mb/s Ethernet connection is assigned a default value of 10 and a 56 kb/s link is assigned a default value of 1785—the faster the data transfer rate, the lower the bit duration and the corresponding default metric value. The router selects the path with the lowest overall value (the fastest path to the destination).

Static and dynamic tables. A router typically has multiple interfaces that are used to receive and forward datagrams. The router must use a table to find out on which interface a particular address is located.

Early-generation routers did not exchange path information with other routers. Such units would forward datagrams on every connecting path, routinely flooding the network and placing some datagrams on endless path loops. Subsequent technologies made it possible for routing path information to be shared between routers and maintained in routing tables.

There are two types of routing tables, static and dynamic.

- Static routing tables—As the name implies, static routing tables do not change by themselves. Such tables are established and populated manually by the internetwork administrator. A database entry must be made for each segment of every possible path through the internetwork. All changes or additions to this database are also made manually. Although time-consuming to maintain, static routing tables may be considered an advantage in internetwork environments requiring a high level of security. With static routing, no path information messages are exchanged over the internetwork.

- Dynamic routing tables—Dynamic routing tables automatically adapt to a changing internetwork environment without manual intervention. Such tables are automatically constructed by the routers during initialization and are maintained and updated at regular intervals. Routers broadcast special datagrams containing path information to each other to synchronize and maintain their tables.

Routing protocols

Routing protocols, which are also referred to as routing algorithms, are used to dynamically configure and update routing tables. Such protocols are critical to the efficient operations of a router. However, as an internetwork grows, routing table exchange traffic among the routers can increase network response time.

Routing protocols perform the calculations necessary to determine the best path to a given destination. Given that an internetwork can consist of multiple routers—each connected to multiple LANs—it is necessary to rapidly calculate how a given datagram can be transferred reliably and efficiently through the internetwork. At a minimum, the protocol must incorporate the following factors in its design:

- End-to-end reliability
- Calculation of appropriate route metrics
- Traffic monitoring
- Data Link layer processes
- Network management requirements

Autonomous domains. The concept of autonomous (or administrative) domains is used by a variety of routing protocols. Usually referred to simply as domains, these are defined as a collection of routing devices that use the same routing protocols and are administered by a single authority. Routing within a domain is referred to as interior (or intradomain) routing. Routing between domains is called exterior (or interdomain) routing (see Figure 5.28).

For routing purposes, the Internet can be described as a collection of interlinked autonomous domains, each operated by a different organization. Examples include educational institutions, government agencies, and Internet service providers (ISPs). Each of these organizations typically operates its own internetwork, using interior routing. Such private internetworks are connected to the Internet using exterior routing.

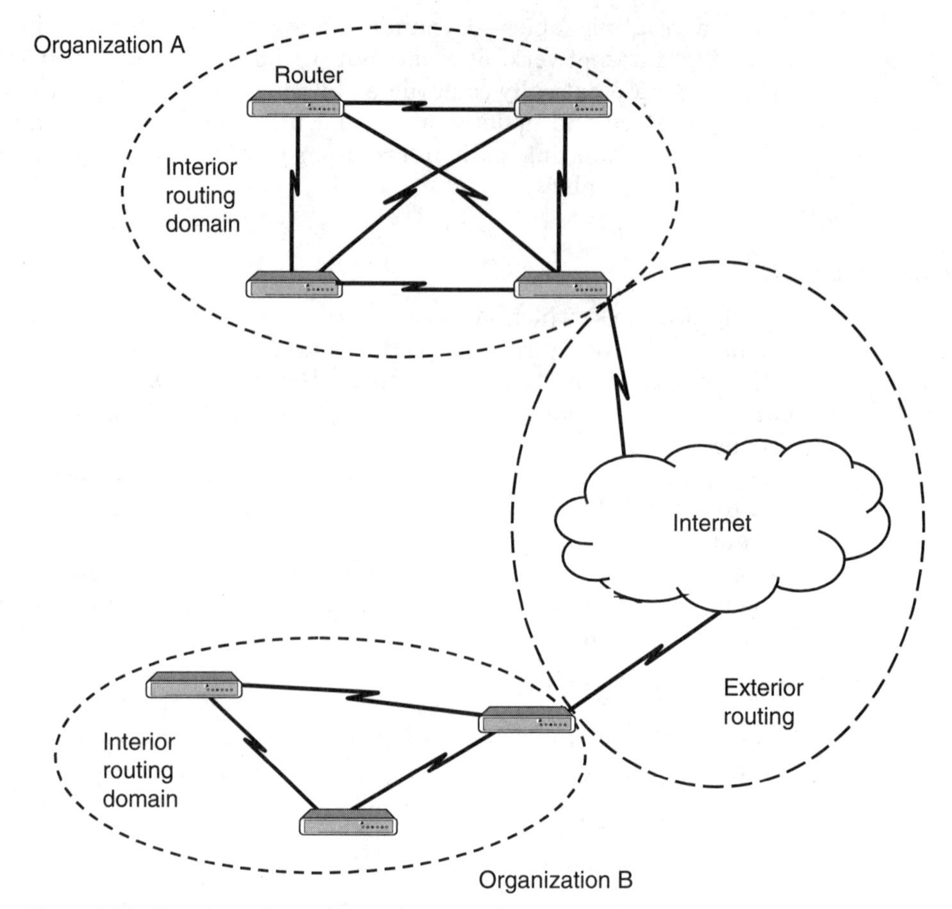

Figure 5.28 Interior and exterior routing example.

Interior routing protocols. Routing protocols that are designed to be used within an autonomous domain can be divided into two types—distance-vector protocols and link state protocols. The main difference between the two is the manner in which each calculates the path cost.

Distance-vector routing. Distance-vector protocols are also called Bellman-Ford algorithms. A distance-vector protocol makes routing decisions based on the number of router hops required to arrive at the destination or the total cost of the path. Costs are assigned by the internetwork administrator as a means of representing how much it costs to use a given router-to-router link or as a way to indicate a preference for one link over another.

Additional path information is gathered from neighboring routers. As information passes from router to router, each router identifies internetwork paths.

This type of algorithm makes it necessary for each router to send its entire routing table to its nearest neighboring routers. Every time a router receives an updated table from a neighboring router, it compares the values in this table with those in its own table. It then uses any new information to update its table with new or deleted routes.

Information found in these tables includes:

- Network number.
- Port number.
- Cost metric—the value that allows the router to choose a path when forwarding datagrams.
- Address of any neighboring router.

Distance-vector routing is not appropriate for large internetworks with hundreds of routers or internetworks in which paths are constantly changing. On such internetworks, the table update process can take so long that path information may remain inconsistent between routers for extended periods.

Routing information protocol (RIP) is a widely used distance-vector routing protocol. The specifications for RIP were published in 1988 as RFC 1058. The current version is RIP version 2, most recently updated in November 1998 as RFC 2453.

RIP is used in a variety of internetwork environments for the exchange of routing information between routers. RIP creates paths through an internetwork by using the number of hops a datagram must take to reach its final destination. It defines the best path as the one requiring the fewest hops between the source and the destination networks. The maximum number of hops allowed in a single path is 15, which makes RIP unsuitable for large internetworks.

RIP-based routers perform the following tasks to ensure that their internal routing tables contain the latest internetwork information:

- Routing information is requested from other routers and is used to update internal routing tables.
- Routers respond to route requests that are issued by other routers.
- Periodically, a router advertises its presence to ensure that other routers are aware of its continued presence on the internetwork.
- When a router detects a change in the internetwork configuration, it broadcasts this change.

When using RIP, each router transmits its entire routing table every 30 seconds, which can use up a significant amount of the available internetwork communications channel. Large internetworks can experience a reduction in performance and reliability as a result of those updates. RIP-based routers also routinely broadcast their presence, as described above. If a router does not advertise its status within 180 seconds, other routers may assume that the router or the paths leading to it have failed.

Link state routing. Link state protocols are also called Dijkstra algorithms. A link state protocol is able to adjust to changes more quickly than a distance-vector protocol and is considered to be more robust. With link state routing, a complete map of the internetwork is contained in each router. Since each router has a complete view of the internetwork, it can independently compute the shortest path to the destination network.

NOTE: Routers using link state protocols require more processing power and more memory for table storage than routers using distance-vector protocols.

The shortest path can be calculated using a number of different criteria, including the:

- Number of routers the datagram must pass through to get to its destination. Fewer hops result in a more efficient path.
- Data transfer rates of the various inter-router links. Links between routers can range from slow dial-up connections to dedicated high-speed digital connections.
- Need to avoid congested areas, defined as paths known to have high traffic volumes.
- Cost of a route as estimated by the internetwork administrator. The least expensive route might not be the fastest but may be preferable for some types of traffic.

In link state protocols, table updates are communicated only when required and not at regular intervals, to reduce the internetwork traffic created by the router.

Open shortest path first (OSPF) is a widely used link state routing protocol. The specifications for OSPF were published in October 1989 as RFC 1131. The current version is OSPF version 2, most recently updated in April 1998 as RFC 2328.

OSPF is designed for use on large internetworks, where it overcomes many of the deficiencies of RIP. OSPF determines the best path based on factors such as delay, usage, communications channel capacity, and transmission costs. It also supports multi-path routing, which allows for load balancing between routers. OSPF also provides a service feature that allows a router to select the path best suited for the type of information contained in the datagram. For example, a path that has more hops may be preferable for an urgent message if the connections in the path have higher data transfer rates. Alternatively, a different path may be selected for less urgent traffic from the same source because the connections are low speed and have a lower cost.

With OSPF, routing table exchanges are hierarchical. Routers transmit their link changes dynamically only when needed and only to specific routers. This enables updates to be processed more quickly and efficiently than with RIP.

Exterior routing protocols. Exterior routing protocols are used by the routers located at the borders of organizational routing domains and are used to exchange information between autonomous domains.

Two examples of exterior routing protocols are:

- Exterior gateway protocol (EGP).
- Border gateway protocol (BGP).

Exterior Gateway Protocol (EGP). The specifications for EGP were published in October 1982 as RFC 827 and updated in April 1984 as RFC 904.

EGP was the original interdomain routing protocol used on the Internet. It provides a mechanism for neighboring routers located at the edges of their respective domains to exchange messages and information about the networks in their domains. Each domain has one or more routers that are selected to function as EGP routers.

The primary administrative functions of EGP are that it:

- Performs a neighbor connection procedure, in which two exterior routers connect and agree to exchange information.
- Periodically checks the links to neighboring routers by signaling them and waiting for a response.
- Periodically exchanges routing information with neighboring routers.

EGP was designed when the Internet consisted of a single backbone. It is inefficient in the current mesh environment. EGP routers are maintained using static routing tables, which are not feasible on an internetwork the size of the current Internet.

Border Gateway Protocol (BGP). The specifications for BGP were published in June 1989 as RFC 1105. The current version is BGP version 4, most recently updated in March 1995 as RFC 1771.

As the replacement for EGP, BGP is designed to minimize communications channel usage by exchanging routing information between routers incrementally, rather than by sending entire routing tables. There is also greater control over the routing function, such as the ability to prioritize some types of traffic. Additional route attributes, such as the security of a path, have also been incorporated into BGP.

Institute of Electrical and Electronics Engineers (IEEE®) Standards

General

The Institute of Electrical and Electronics Engineers, Inc.® (IEEE®) is the organization responsible for the standardization of most LAN technologies. Specifically, the IEEE 802 Local and Metropolitan Area Network Standards Committee (LMSC) guides the development and standardization of networking technologies that correspond to the Physical and Data Link layers of the OSI model. This makes it possible for higher layer protocols and services to operate over a variety of network types. An example is the IP datagram, which can be transported over any combination of Ethernet, token ring, and wireless LANs.

NOTE: In the context of the IEEE 802 LMSC standards, the term local refers to an area large enough to include a multi-building campus, and the term metropolitan indicates an area large enough to include a city.

The objective of the IEEE 802 LMSC is to publish standards that are supported by both the vendors and the buyers of networking products. The greater the level of acceptance of a given standard, the higher the level of assurance of interoperability among products described as standards-compliant. Many of the standards published by the IEEE have subsequently been adopted by the:

- American National Standards Institute (ANSI) as United States national standards.
- International Organization for Standardization/International Electrotechnical Commission (ISO/IEC) as global standards.

The IEEE 802 LMSC has a number of working groups (WGs) and technical advisory groups (TAGs). Each focuses on a specific set of topics and technologies. Since its inception in February 1980, the IEEE 802 LMSC has formed a total of 16 WGs and TAGs, some of which are no longer active. The following is a listing and description of the standards issued by these groups.

IEEE 802.1—Higher Layer LAN Protocols Working Group

IEEE 802.1 is responsible for developing standards for:

- LAN and MAN architectures.
- Communications between different 802 LAN and MAN technologies.
- Network management.
- The OSI layers above the Data Link layer.

Some of the standards developed by this group include:

- ANSI/IEEE 802.1B – 1995 (ISO/IEC 15802-2 – 1995).
Information Technology—Telecommunications and Information Exchange Between Systems—Local and Metropolitan Area Networks—Common Specifications—Part 2: LAN/MAN Management
- ANSI/IEEE 802.1D – 1998 (ISO/IEC 15802-3 – 1998).
Information Technology—Telecommunications and Information Exchange Between Systems—Local and Metropolitan Area Networks—Common Specifications—MAC Bridges
- ANSI/IEEE 802.1G – 1998 (ISO/IEC 15802-5 – 1998).
Information Technology—Telecommunications and Information Exchange Between Systems—Local and Metropolitan Area Networks—Common Specifications—Part 5: Remote MAC Bridging
- IEEE 802.1Q – 1998.
IEEE Standards for Local and Metropolitan Area Networks: Virtual Bridged Local Area Networks
- IEEE 802.1v – 2001.
Amendment to IEEE 802.1Q: IEEE Standards for Local and Metropolitan Area Networks: Virtual Bridged Local Area Networks: VLAN Classification by Protocol and Port

IEEE 802.2—Logical Link Control (LLC) Working Group

IEEE 802.2 is no longer active.

The standard developed by this group is:

- ANSI/IEEE 802.2 – 1998 (ISO/IEC 8802-2 – 1998).
Information Technology—Telecommunications and Information Exchange Between Systems—Local and Metropolitan Area Networks—Specific Requirements—Part 2: Logical Link Control

IEEE 802.3—Carrier Sense Multiple Access with Collision Detection (CSMA/CD) Working Group

IEEE 802.3 is responsible for developing standards for the Ethernet family of LAN technologies, which includes:

- 10 Mb/s Ethernet.
- 100 Mb/s Fast Ethernet.
- 1000 Mb/s (Gigabit) Ethernet.
- 10,000 Mb/s (10 Gigabit) Ethernet.

One of the standards developed by this group is:

- ANSI/IEEE 802.3 – 2000 (ISO/IEC 8802-3 – 2000).
  Information Technology—Telecommunications and Information Exchange Between Systems—Local and Metropolitan Area Networks—Specific Requirements—Part 3: CSMA/CD Access Method and Physical Layer Specifications

IEEE 802.4—Token Bus Working Group

IEEE 802.4 is no longer active.

The standard developed by this group is:

- ANSI/IEEE 802.4 – 1990 (ISO/IEC 8802-4 – 1990).
  Information Processing Systems—Local Area Networks—Part 4: Token-Passing Bus Access Method and Physical Layer Specifications

IEEE 802.5—Token Ring Working Group

IEEE 802.5 is responsible for developing standards for the token ring family of LAN technologies, which includes:

- 4 and 16 Mb/s token ring.
- 100 Mb/s high-speed token ring (HSTR).
- 1000 Mb/s gigabit token ring (GTR).

One of the standards developed by this group is:

- ANSI/IEEE 802.5 – 1998 (ISO/IEC 8802-5 – 1998).
  Information Technology—Telecommunications and Information Exchange Between Systems—Local and Metropolitan Area Networks—Specific Requirements—Part 5: Token Ring Access Method and Physical Layer Specifications

IEEE 802.6—Metropolitan Area Network (MAN) Working Group

IEEE 802.6 is no longer active.

One of the standards developed by this group is:

- ANSI/IEEE 802.6 – 1994 (ISO/IEC 8802-6 – 1994).
  Information Technology—Telecommunications and Information Exchange Between Systems—Local and Metropolitan Area Networks—Specific Requirements—Part 6: Distributed Queue Dual Bus (DQDB) Access Method and Physical Layer Specifications

IEEE 802.7—Broadband Technical Advisory Group

IEEE 802.7 is no longer active.
The standard developed by this group is:

- IEEE 802.7 – 1989.
IEEE Recommended Practices for Broadband LANs

IEEE 802.8—Fiber Optic Technical Advisory Group

IEEE 802.8 is no longer active.
This group has developed the following draft standard:

- IEEE 802.8 – 1998.
IEEE Recommended Practice for Fiber Optic Local and Metropolitan Area Networks

IEEE 802.9—Integrated Services LAN (IS-LAN) Working Group

IEEE 802.9 is no longer active.
This group has developed the following standard:

- IEEE 802.9 – 1996 (ISO/IEC 8802-9 – 1996).
Information Technology—Telecommunications and Information Exchange Between Systems—Local and Metropolitan Area Networks—Specific Requirements—Part 9: Integrated Services (IS)-LAN Interface at the MAC and Physical Layers

IEEE 802.10—Standards for Interoperable LAN/MAN Security (SILS) Working Group

IEEE 802.10 is no longer active.
One of the standards developed by this group is:

- IEEE 802.10 – 1998.
IEEE Standards for Local and Metropolitan Area Networks: Interoperable LAN/MAN Security

IEEE 802.11—Wireless LAN (WLAN) Working Group

IEEE 802.11 is responsible for developing standards for WLAN technologies.
One of the standards developed by this group is:

- ANSI/IEEE 802.11 – 1999 (ISO/IEC 8802-11 – 1999).
Information Technology—Telecommunications and Information Exchange Between Systems—Local and Metropolitan Area Networks—Specific Requirements—Part 11: Wireless LAN MAC and Physical Layer Specifications

IEEE 802.12—Demand Priority Working Group

IEEE 802.12 is no longer active.

This group has developed the following standard:

- ANSI/IEEE 802.12 – 1998 (ISO/IEC 8802-12 – 1998).
  Information Technology—Telecommunications and Information Exchange Between Systems—Local and Metropolitan Area Networks—Specific Requirements—Part 12: Demand Priority Access Method (DPAM), Physical Layer and Repeater Specification for 100 Mb/s Operation

NOTE: There is no IEEE 802.13 working group.

IEEE 802.14—Cable Modem Working Group

IEEE 802.14 is no longer active. It was formed to develop standards for digital communication services over cable television networks.

No standards were finalized by this group.

IEEE 802.15—Wireless Personal Area Network (WPAN) Working Group

IEEE 802.15 is responsible for developing standards for WPAN technologies.

To date, no standards have been finalized by this group.

IEEE 802.16—Broadband Wireless Access Working Group

IEEE 802.16 is responsible for developing standards for fixed point to multipoint wireless systems, also described as wireless MAN (WMAN) technologies.

To date, no standards have been finalized by this group.

IEEE 802.17—Resilient Packet Ring (RPR) Working Group

IEEE 802.17 is responsible for developing standards for a high-speed MAN technology, described in Chapter 4: Network Connections.

To date, no standards have been finalized by this group.

Internet Engineering Task Force (IETF) Standards

General

The IETF is one of a group of organizations responsible for the overall development of the Internet and the standardization of internetworking technologies. The influence of the IETF on internetworking can be compared to the influence of the IEEE on LAN technologies.

In addition to the IETF, a number of organizations contribute to the overall development of the Internet, including the:

- Internet Society (ISOC), which oversees the overall development of the Internet.
- Internet Architecture Board (IAB), whose members are appointed by the ISOC, which serves as the technical advisory group of ISOC. It is responsible for the overall development of the protocols and architecture associated with the Internet.
- Internet Engineering Steering Group (IESG), whose members are approved by the IAB, which oversees the activities of the IETF and manages the process used to introduce or update Internet standards.
- Internet Research Task Force (IRTF), which is the research counterpart of the IETF. Technical topics considered theoretical or experimental in nature are explored by IRTF. When a technology is judged suitable for practical use on the Internet, it is forwarded to the IETF for consideration.
- Internet Research Steering Group (IRSG), which oversees the activities of the IRTF in the same way the IESG manages the IETF.
- Internet Corporation for Assigned Names and Numbers (ICANN), which replaced the Internet Assigned Numbers Authority (IANA) in October 1998. ICANN oversees Internet naming and addressing.

All Internet standards are published and updated as RFCs. A protocol under consideration for eventual adoption as an Internet standard is first published as a proposed standard. After additional study, the protocol may be promoted to the draft standard stage. The final step in the standardization process is the promotion of a draft standard to standard status.

NOTE: Additional classifications include informational standard, experimental standard, and historic standard. Such standards are excluded from consideration as operational Internet standards.

TCP/IP protocol stack

The Internet is the best example of an internetwork that enables message transfer between all types of networks—LANs, mainframe and minicomputer environments, and others.

The Network layer protocol common to all internetwork devices connected to the Internet is IP. Two Transport layer protocols use IP to deliver messages—TCP and user datagram protocol (UDP). Higher layer service and application protocols use TCP/IP or UDP/IP to exchange messages with their counterparts over the internetwork. The combined group of Internet protocols is often referred to as the TCP/IP protocol suite or protocol stack (see Figure 5.29).

<table>
<tr><td>Layer 7</td><td>Application layer</td><td rowspan="3">Hypertext markup language (HTML)
Hypertext transfer protocol (HTTP)
Network news transfer protocol (NNTP)
Simple mail transfer protocol (SMTP)
File transfer protocol (FTP)
Domain name system (DNS)
Dynamic host configuration protocol (DHCP)
Simple network management protocol (SNMP)
Telnet</td></tr>
<tr><td>Layer 6</td><td>Presentation layer</td></tr>
<tr><td>Layer 5</td><td>Session layer</td></tr>
<tr><td>Layer 4</td><td>Transport layer</td><td>Transmission control protocol (TCP)
User datagram protocol (UDP)</td></tr>
<tr><td>Layer 3</td><td>Network layer</td><td>Internet protocol (IP)</td></tr>
<tr><td>Layer 2</td><td>Data Link layer</td><td></td></tr>
<tr><td>Layer 1</td><td>Physical layer</td><td></td></tr>
</table>

Figure 5.29 OSI model and the TCP/IP protocol stack.

NOTE: To simplify Figure 5.29, other protocols in the TCP/IP protocol stack are not shown.

The following is a brief description of the protocols listed in Figure 5.29.

Layer 3—Network Layer Protocol

Internet Protocol (IP). IP is responsible for delivering datagrams from source to destination without any guarantees. Higher layer protocols are expected to provide guaranteed delivery services.

Layer 4—Transport Layer Protocols

User Datagram Protocol (UDP). UDP serves as a connection point between IP and upper layer services and applications. UDP makes it possible to assign unique port numbers to individual service and application protocols. This enables a higher layer protocol in the sending device to address messages to a port number corresponding to the same protocol in the receiving device.

Transmission Control Protocol (TCP). Like UDP, TCP also associates higher layer protocols with specific port numbers. However, TCP also provides the means to guarantee the delivery of datagrams from source to destination over a network, in order and without error.

Layers 5, 6, and 7—Higher Layer Protocols

Telnet. Telnet makes it possible to use IP to connect to a remote device and issue commands in the same manner as a terminal directly attached to the remote device.

Simple Network Management Protocol (SNMP). SNMP is used to collect operational and status information from network devices, using criteria defined in a management information base (MIB).

Dynamic Host Configuration Protocol (DHCP). A device can use DHCP to obtain a network address and other configuration information from a DHCP server, eliminating the need to manually configure the device.

Domain Name System (DNS). DNS is used to associate names with numeric IP addresses, making it easier to access devices and services (e.g., www.bicsi.org versus 206.96.226.20).

File Transfer Protocol (FTP). FTP is used to reliably transfer files between devices over a network.

Simple Mail Transfer Protocol (SMTP). SMTP makes it possible to exchange messages created using any type of SMTP-compliant e-mail software. It enables organizations using different messaging applications to communicate without changing their software.

Network News Transfer Protocol (NNTP). NNTP is used to read and reply to messages posted on any NNTP server.

Hypertext Transfer Protocol (HTTP). HTTP is used to access hypermedia documents—commonly referred to as Web pages—stored on any HTTP server.

Hypertext Markup Language (HTML). HTML is used to create and interlink the Web pages that are accessed using HTTP.

Chapter

6

Switching and Virtual LANs (VLANs)

Chapter 6 describes technologies developed to increase the number of available communications channels on a network. This makes it possible to increase the number of connected devices without a corresponding increase in network response times.

Overview

General

The term switching is used when describing many types of network technologies. In each case, a unique set of hardware, software, processes, and rules guide the switching operations on the network. Examples of switching include:

- Local area network (LAN) switching.
- Asynchronous transfer mode (ATM) switching.
- Frame Relay switching.
- Switching at Layers 3 through 7 of the Open Systems Interconnection (OSI) model.
- Telephone network switching.

Since the term switch is used to describe the switching device used on many types of networks, its definition is broad. For the purposes of this chapter, which focuses on LAN switching, a switch is defined as a device consisting of two basic components:

- Two or more ports, each of which is used to connect other LAN devices to the switch
- Forwarding logic, which represents the set of rules used to process the frames received from attached LAN devices

The LAN switching process operates at the OSI Data Link layer, applying forwarding logic historically used in bridging technologies to Data Link layer frames. For this reason, LAN switching is also referred to as:

- Multiport bridging.
- Data Link layer switching.
- Medium Access Control (MAC) layer switching.
- Layer 2 switching.

LAN switches are multiport devices that transfer frames between their ports based on information contained in the frames, as follows:

1. A frame arrives at a port on the switch, referred to as the incoming or inbound port.
2. The switch examines the destination address contained in the frame.
3. The frame is transmitted out of the port connected to the destination device, referred to as the outgoing or outbound port. If the destination address is unknown, the frame is transmitted out of all ports other than the inbound port.

In most cases, devices connected to a LAN switch use the same Data Link layer networking technology (e.g., Ethernet, token ring, or fiber distributed data interface [FDDI]). By comparison, when similar forwarding logic is applied to OSI Network layer packets or datagrams, the device is called a router.

NOTE: Routing is described in detail in Chapter 5: Network Communications, in the section titled Internetwork Routing. Hybrid devices that combine frame switching and datagram routing in a single unit are sometimes described as Layer 3 switches. Such devices are described later in this chapter, in the section titled Higher Layer Switching.

Message types

Switching is used to divide a LAN into multiple collision domains, which has the effect of improving network response time by segregating traffic. Without switching, all devices on a LAN share a single collision domain where the frames generated by any device are delivered to all devices, not simply the destination device (see Figure 6.1). This creates unnecessary traffic on the network and makes it necessary to limit network access to one device at a time, since all devices receive each transmission. If two or more devices are permitted to transmit at the same time—as is the case with Ethernet technologies—the frames collide, forcing each device to try again after waiting a random period of time.

In a switched LAN environment, each switch port is a separate collision domain. Devices connected to different switch ports can transmit simultaneously without causing a collision between the frames they generate. For this reason, switches are also described as segmentation devices. In many cases, both switches and hubs are used on a LAN, creating a hybrid environment (see Figure 6.2).

NOTE: The term microsegmentation is used to describe cases where no hubs are used—each station, server, and shared peripheral device on the LAN is connected to a port on a switch (see Figure 6.3).

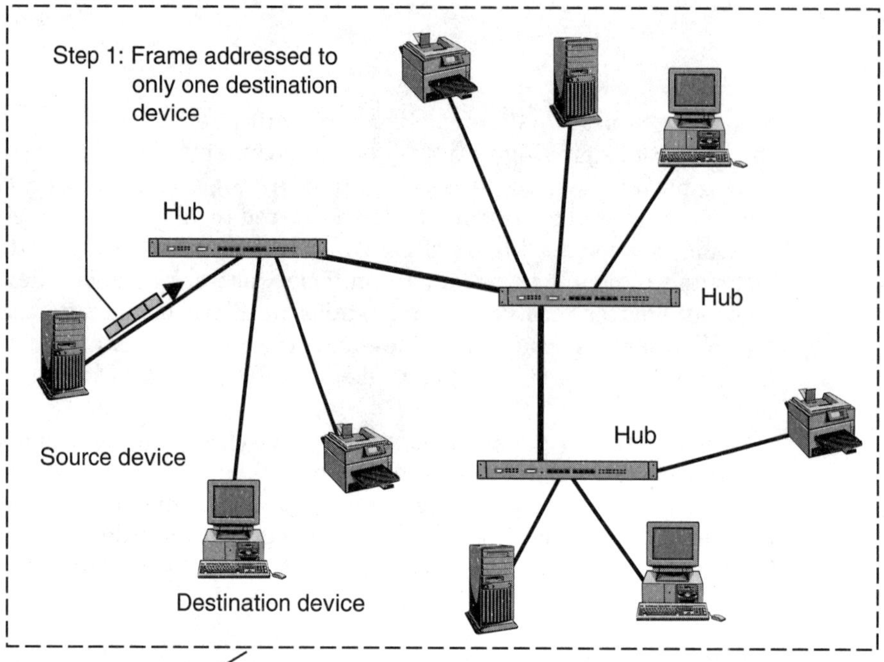

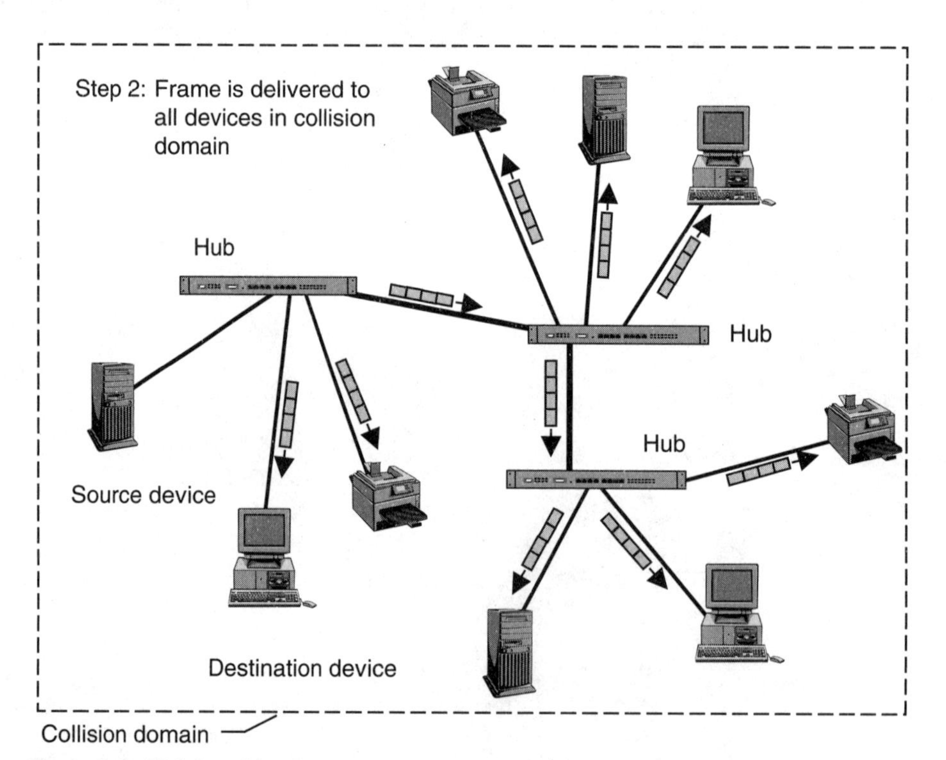

Figure 6.1 Hub-based local area network.

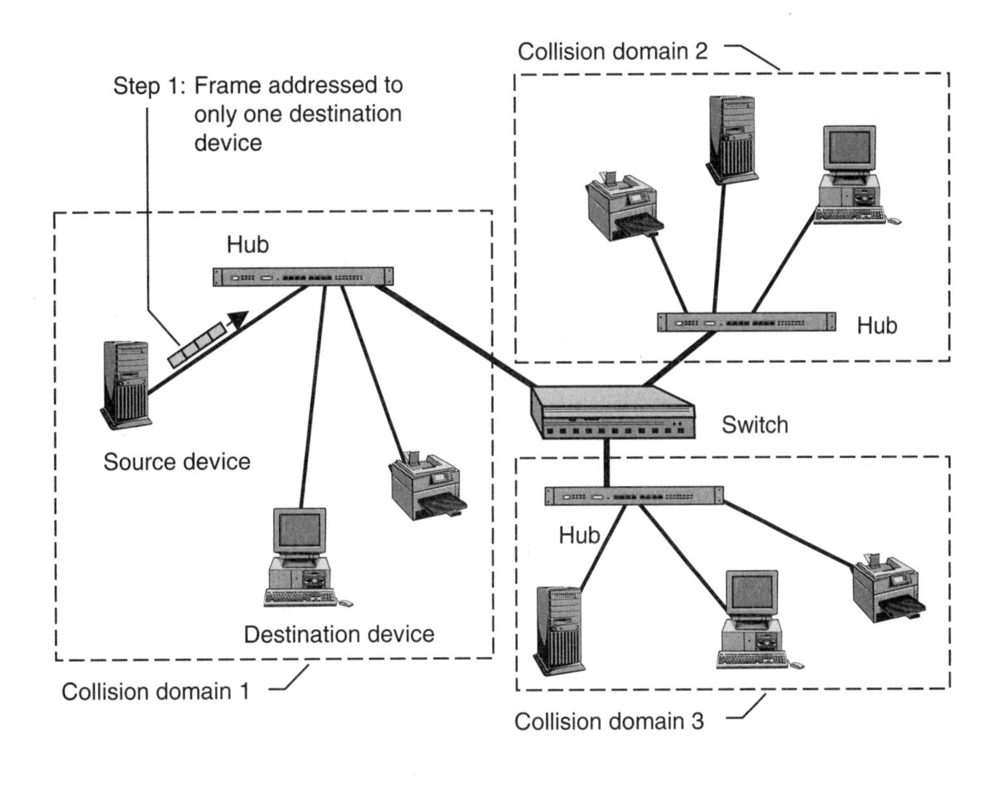

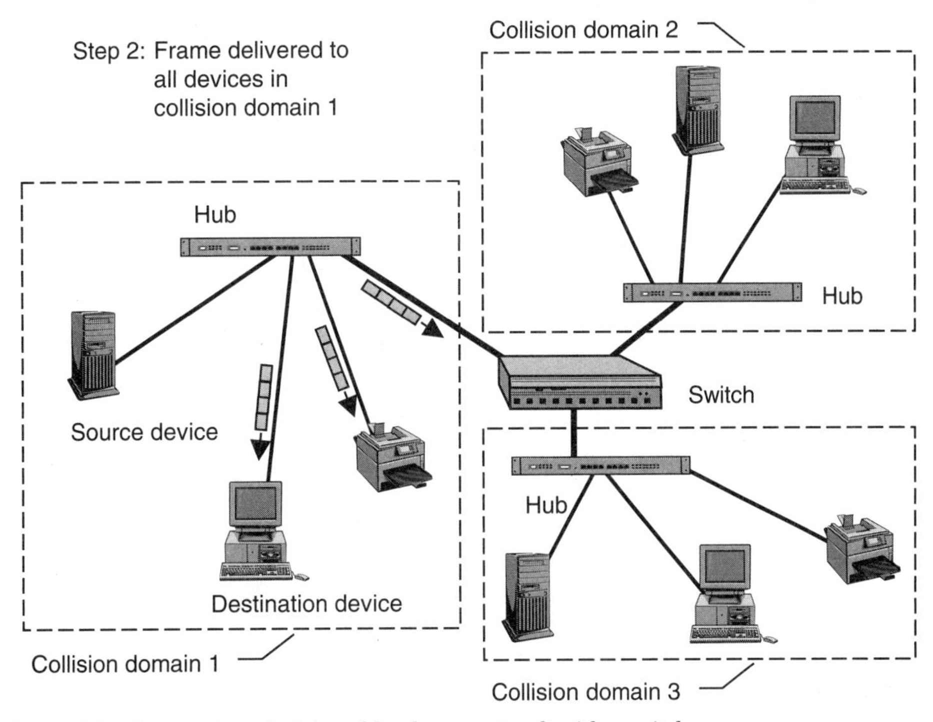

Figure 6.2 Segmenting a hub-based local area network with a switch.

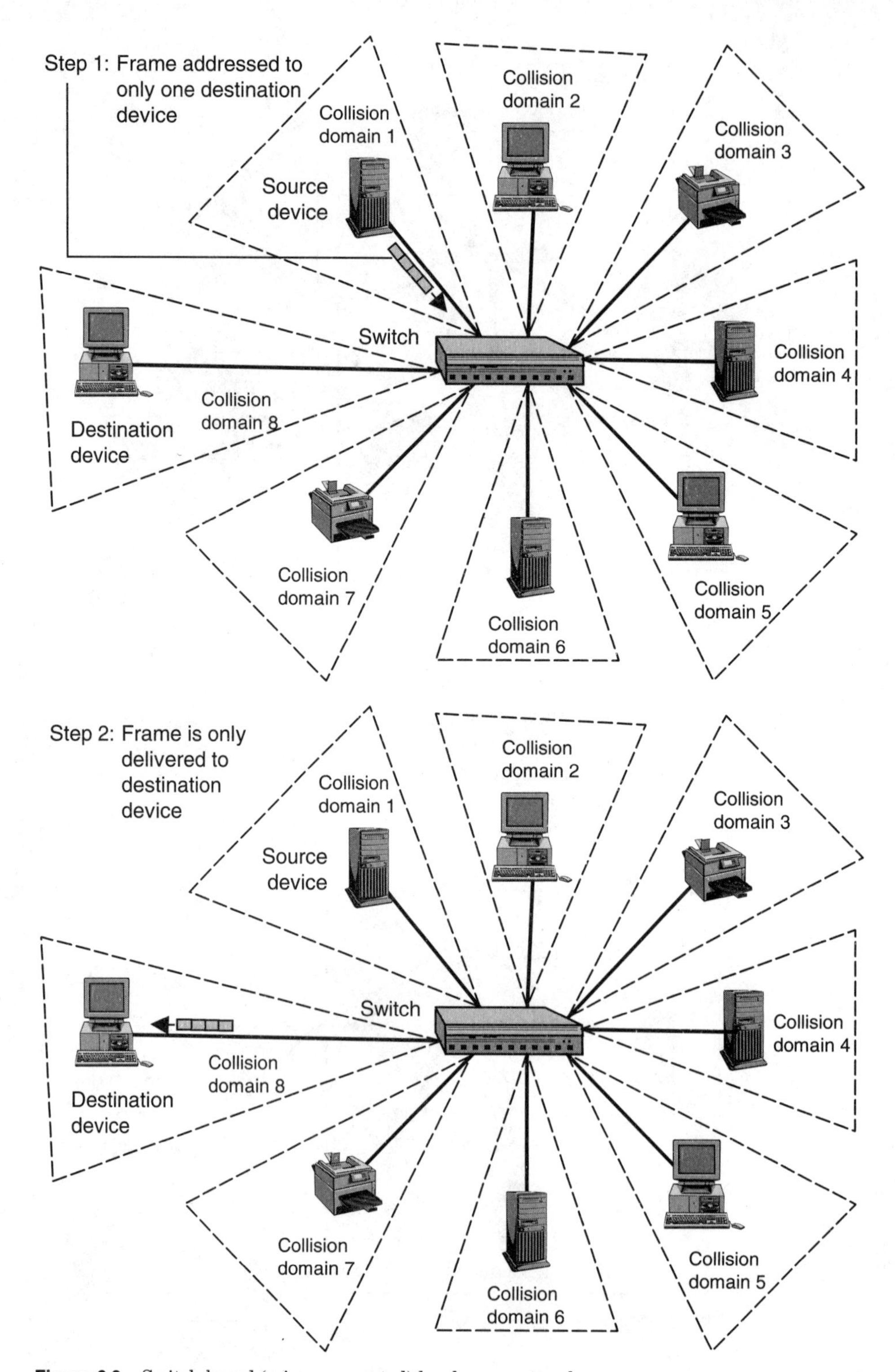

Figure 6.3 Switch-based (microsegmented) local area network.

The introduction of LAN switching technologies has made it possible to connect more devices to a LAN without increasing overall network response time. Switches conserve network resources by inspecting and directing frames to their destination devices rather than all devices. In some cases, however, a message must be directed to a group of devices or all devices on the LAN. When this occurs, a switch-based LAN operates in a similar manner to a hub-based LAN, eliminating many of the advantages of switching. A sending device can generate three types of messages:

- Unicast
- Broadcast
- Multicast

Unicast messaging. In unicast messaging, or unicasting, each message is addressed to one recipient (see Figure 6.4). If a device needs to send the same message to multiple destinations, it must perform a replicated unicast, where the same transmission is repeated as many times as there are destinations (see Figure 6.5).

With unicasting, there is no risk of sending a message to an unintended recipient, since the switch directs each frame to the device with the given destination address. This process is also referred to as a point-to-point transfer. However, generating multiple frames containing identical data is an inefficient use of network resources and requires additional processing to take place in the sending device.

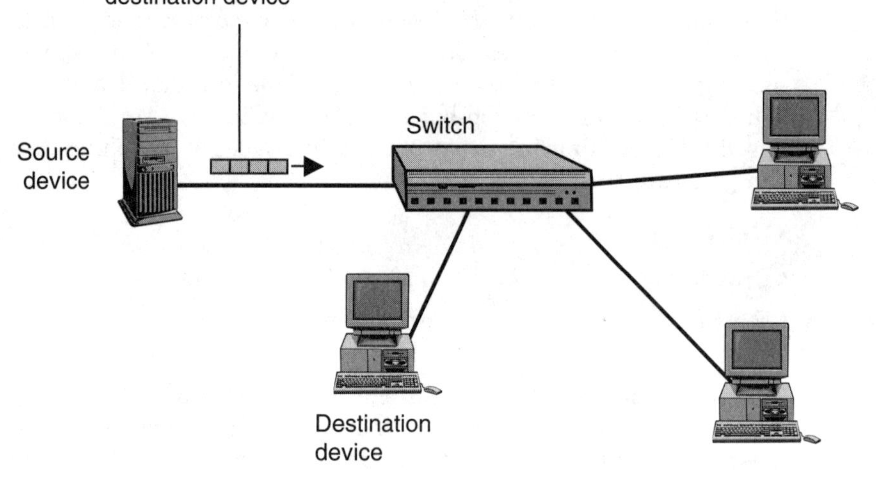

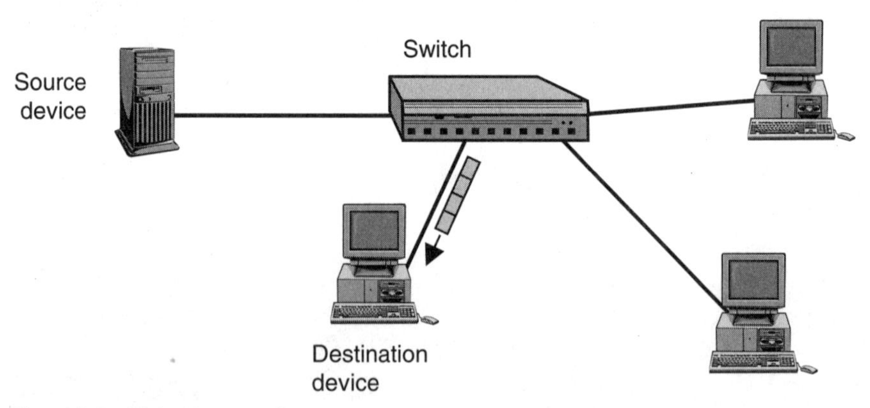

Figure 6.4 Unicast messaging.

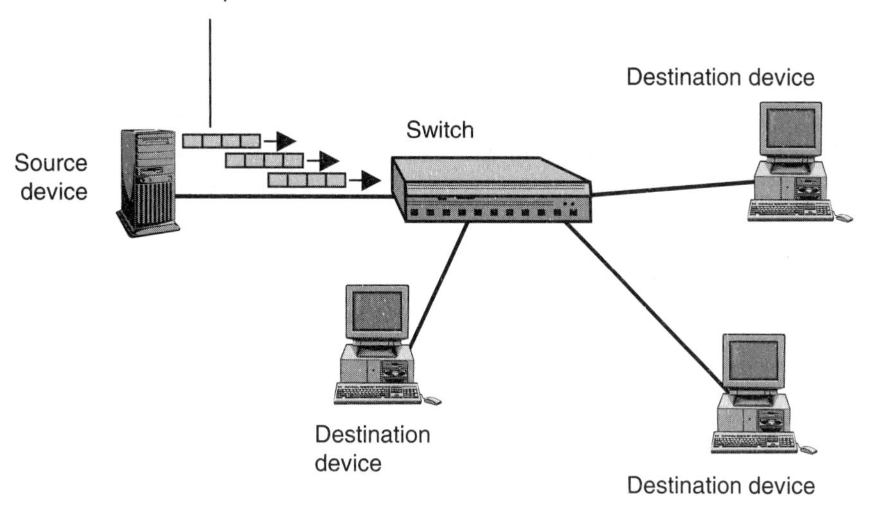

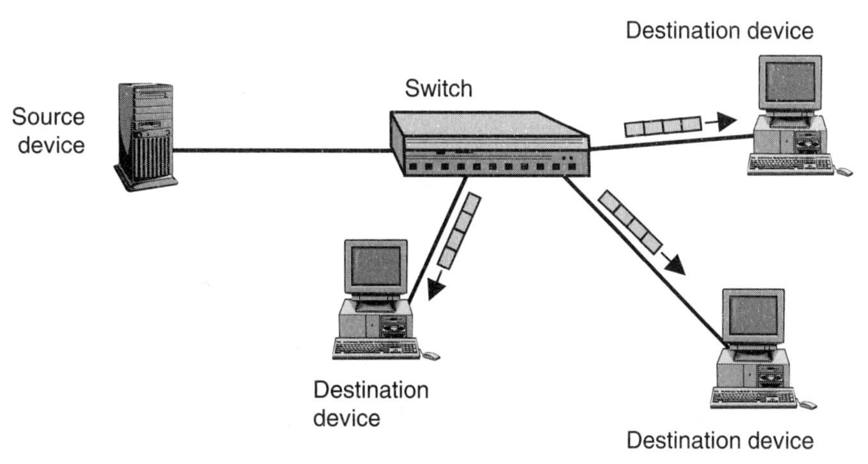

Figure 6.5 Replicated unicast messaging.

Broadcast messaging. In broadcast messaging, or broadcasting, each frame contains a distinct address to indicate that the destination is all devices in the broadcast domain. Such transfers are also referred to as point-to-multipoint—the sending device transmits a broadcast frame once and the switch directs the frame to all other network devices (see Figure 6.6).

Broadcasting is efficient in cases when all network devices can make use of the message being broadcast. If this is not the case, a recipient that does not require the data wastes processing resources examining and discarding the incoming frame. When the number of discarded frames exceeds the number processed, the broadcast can be classified as an inefficient use of network resources.

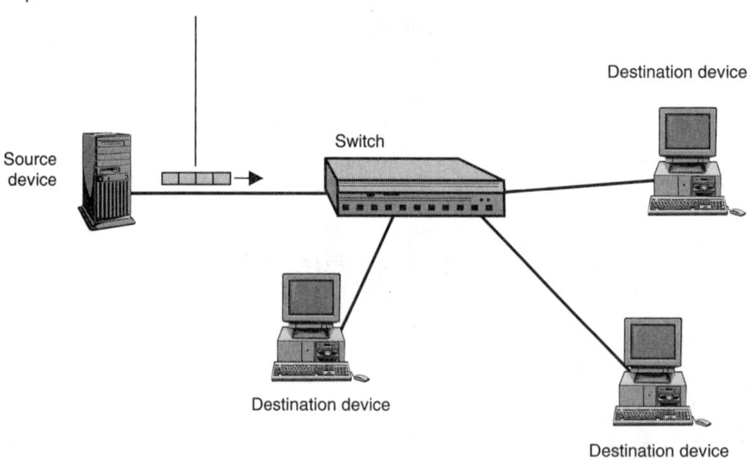

Step 2: Each destination device receives a copy of the frame

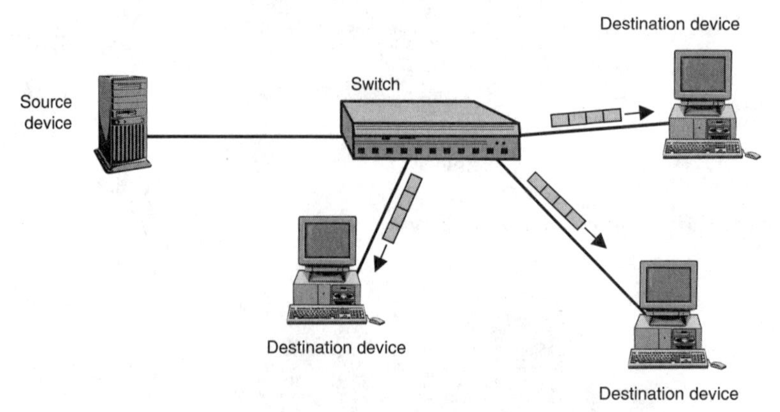

Figure 6.6 Broadcast messaging.

Multicast messaging. In multicast messaging, or multicasting, the network delivers a transmitted message to a select number of devices—not all devices as in the case of a broadcast. The sending device transmits the message once to a special multicast group address and the switch directs the message only to those devices that are listed as members of the group (see Figure 6.7). Multicasting can be described as selective or directed broadcasting—it is the intelligent form of point-to-multipoint message transfer.

Historically, routers have been used to limit the propagation of broadcast and multicast messages. Another solution is to use virtual LAN (VLAN) technologies to contain broadcasts and multicasts in switched LAN environments. VLANs are described in detail later in this chapter, in the section titled Virtual LANs.

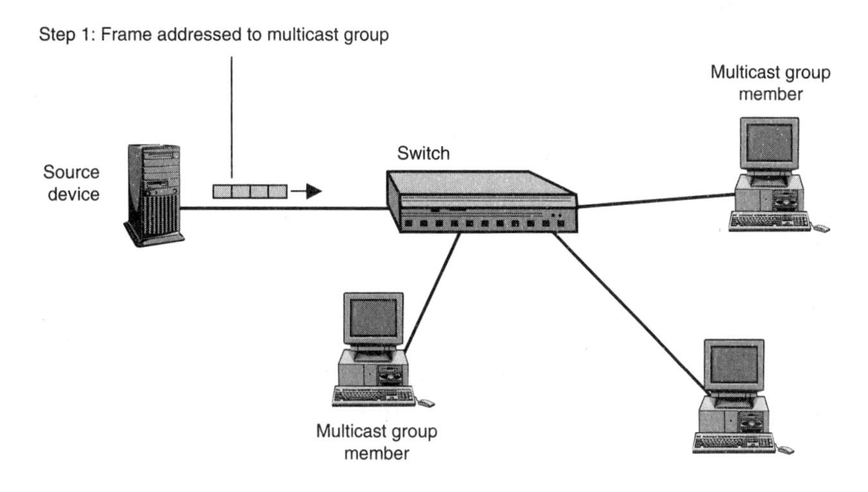

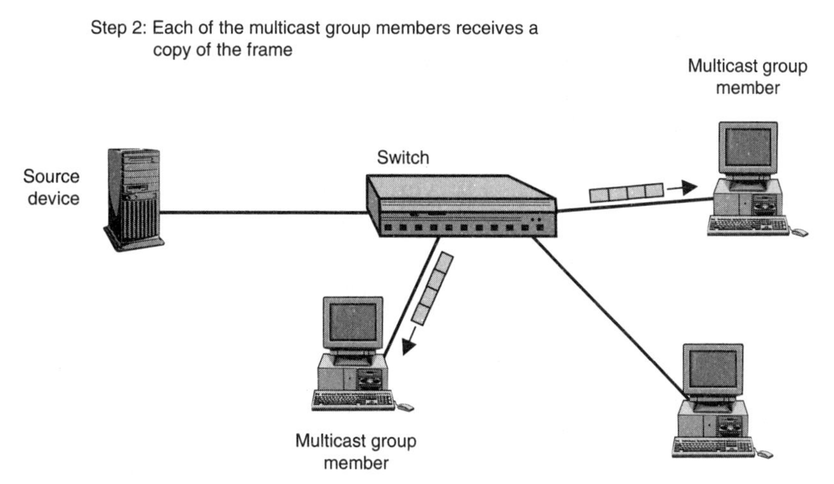

Figure 6.7 Multicast messaging.

Bridging, Switching, and Virtual Local Area Network (VLAN) Standards

This section describes the major Institute of Electrical and Electronics Engineers, Inc.® (IEEE®) standards associated with bridging, switching, and VLANs. They are presented in chronological order.

IEEE 802.1D (1998)

IEEE 802.1D, *IEEE Standard for Information Technology—Telecommunications and Information Exchange Between Systems—Local and Metropolitan Area Networks—Common Specifications—MAC Bridges*, represents the specifications for bridging in a LAN environment. The first version of this specification was approved in 1986.

IEEE 802.1G (1998)

IEEE 802.1G, *IEEE Standard for Information Technology—Telecommunications and Information Exchange Between Systems—Local and Metropolitan Area Networks—Common Specifications—Part 5: Remote MAC Bridging*, represents the specifications for remote bridging, where two bridging units are linked using telecommunications circuits.

IEEE 802.1p (forms part of 802.1D)

IEEE 802.1p was initially a separate document. It represents the specifications for traffic classification and management in a bridged or switched LAN environment. The intent of this standard is to provide a way to identify and prioritize the handling of frames containing time-sensitive data or multicast traffic.

IEEE 802.1Q (1998)

IEEE 802.1Q, *IEEE Standards for Local and Metropolitan Area Networks—Virtual Bridged Local Area Networks*, represents the specifications for VLAN operations in a switched LAN environment.

IEEE 802.3ac (1998)

IEEE 802.3ac, *IEEE Standard for Information Technology—Telecommunications and Information Exchange Between Systems—Local and Metropolitan Area Networks—Common Specifications—Specific Requirements Part 3: CSMA/CD Frame Extensions for VLAN Tagging*, represents the specifications for new VLAN fields in Ethernet frames, which are optional but necessary if VLANs are to be implemented in a switched Ethernet environment.

IEEE 802.3ad (2000)

IEEE 802.3ad, *IEEE Standard for Information Technology—Local and Metropolitan Area Networks—Part 3: CSMA / CD Access Method and Physical Layer Specifications—Aggregation of Multiple Link Segments*, represents the specifications for link aggregation, also referred to as port aggregation or trunking. Link aggregation permits multiple ports on a switch to operate as a single communications channel, providing additional network resources to connected devices (e.g., servers equipped with multiple network interface cards [NICs]).

IEEE 802.1v (2001)

IEEE 802.1v, *IEEE Standards for Local and Metropolitan Area Networks: Virtual Bridged Local Area Networks—Amendment 2: VLAN Classification by Protocol and Port*, is an amendment to IEEE 802.1Q. It represents additional specifications for the VLAN technologies standardized in 802.1Q.

Local Area Network (LAN) Bridging

General

Before the introduction of LAN switching, bridging was used to expand the number of devices on a LAN without excessively degrading network response time. The term multiport bridge continues to be used to describe a switch, illustrating the origins of switching technologies.

This section describes how bridges operate and the standard technique—called the spanning tree algorithm (STA)—used by both bridges and switches to prevent excessive broadcast traffic on the bridged or switched LAN.

NOTES: This document defines a LAN as a single broadcast domain containing any number of bridges or switches, each of which is used to create two or more collision domains.

VLAN technologies, described later in this chapter, make it possible to create multiple broadcast domains—each of which is called a virtual network—on the same physical LAN. In such cases, each of the virtual LANs is considered the functional equivalent of a separate physical LAN.

Bridge operations

In most cases, the introduction of a bridge does not require any changes to be made to other devices on the network. All stations, servers, and shared peripheral devices continue to operate without awareness of the existence of the bridge. In some cases, users may notice the improvement in network response time as the bridge reduces the number of devices in the shared collision domain. For this reason, the term transparent bridging is used to describe this type of bridge operations—the bridge is transparent to other network devices.

A bridge uses an internal database called a source address table (SAT) to keep track of all devices connected to each of its ports. At a minimum, each record in the SAT consists of the device address, port number, and time of entry of the record. The bridge builds the SAT by recording the values in the source address fields in the frames visible on each port. Once the SAT is in place, the bridge can use the records in the SAT to process an incoming frame, as follows:

1. When a frame enters a port, its contents are copied to an internal buffer prior to processing. This is called store-and-forward operations.
2. The bridge examines the source address field in the frame and checks if the device has a record in the SAT. If this is the first frame sent by the device, an entry is created in the SAT. This process is referred to as learning.
3. Next, the destination address field in the frame is examined and used to process the frame. There are three processing possibilities, as follows:

   - The destination address exists in the SAT and is associated with a port other than the one the frame was received on. In this case, the frame is transmitted (forwarded) out of the port associated with the destination address.
   - The destination address exists in the SAT and is associated with the port the frame was received on. In this case, the frame is discarded (filtered) since both the sending and receiving devices are in the same collision domain and can communicate without the services of the bridge.
   - The destination address does not exist in the SAT or the destination address is the broadcast address, indicating that the frame is to be processed by all devices. In both cases, the frame is forwarded out of all ports other than the one it was received on. This process is referred to as flooding.

The time of entry of each record is used to keep track of the amount of time that has passed since a given device has transmitted a frame. When this value reaches a defined threshold, the record is deleted from the SAT. This process, called aging, keeps the entries in the SAT current. Removing outdated entries improves bridge performance by keeping the SAT as small as possible, making searches faster.

Bridge operations are described in Example 6.1. This example shows how a single shared collision domain Ethernet LAN is divided into four collision domains using a four-port bridge.

EXAMPLE 6.1 Bridge operations.

Figure 6.8 shows an Ethernet LAN prior to the introduction of a four-port bridge. In this configuration, all devices share a single collision domain.

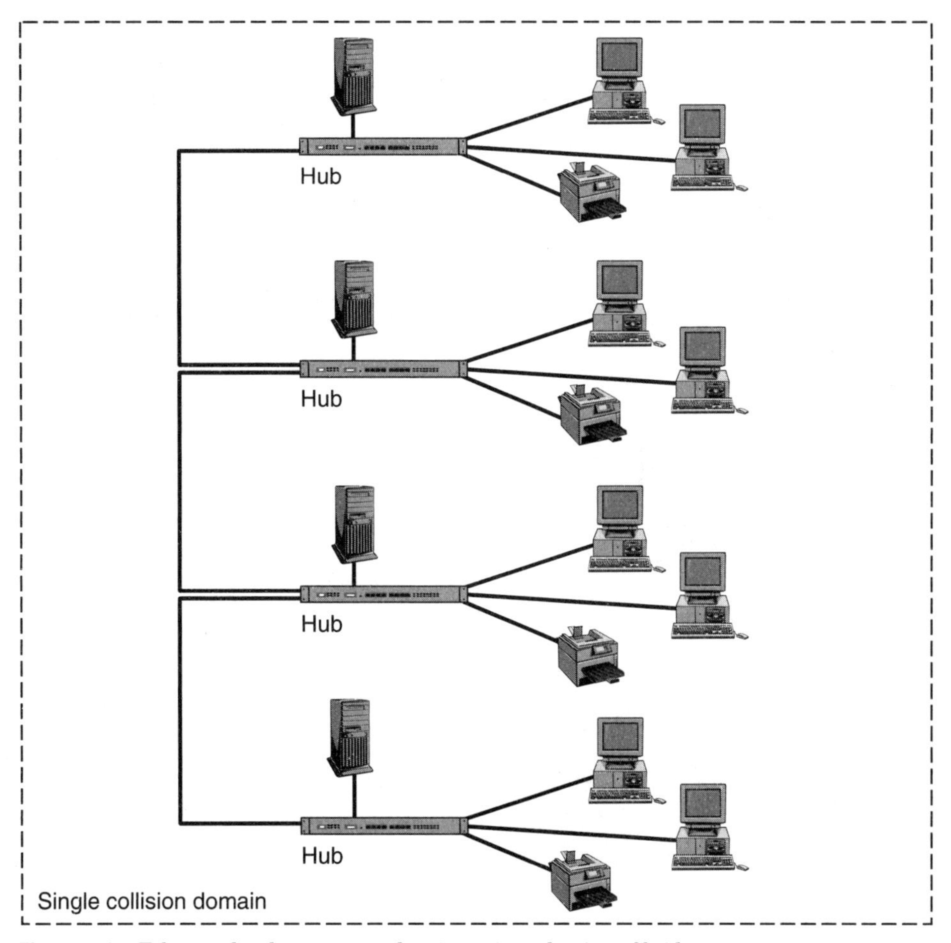

Figure 6.8 Ethernet local area network prior to introduction of bridge.

After the four-port bridge is installed, this Ethernet LAN is divided into four collision domains (see Figure 6.9).

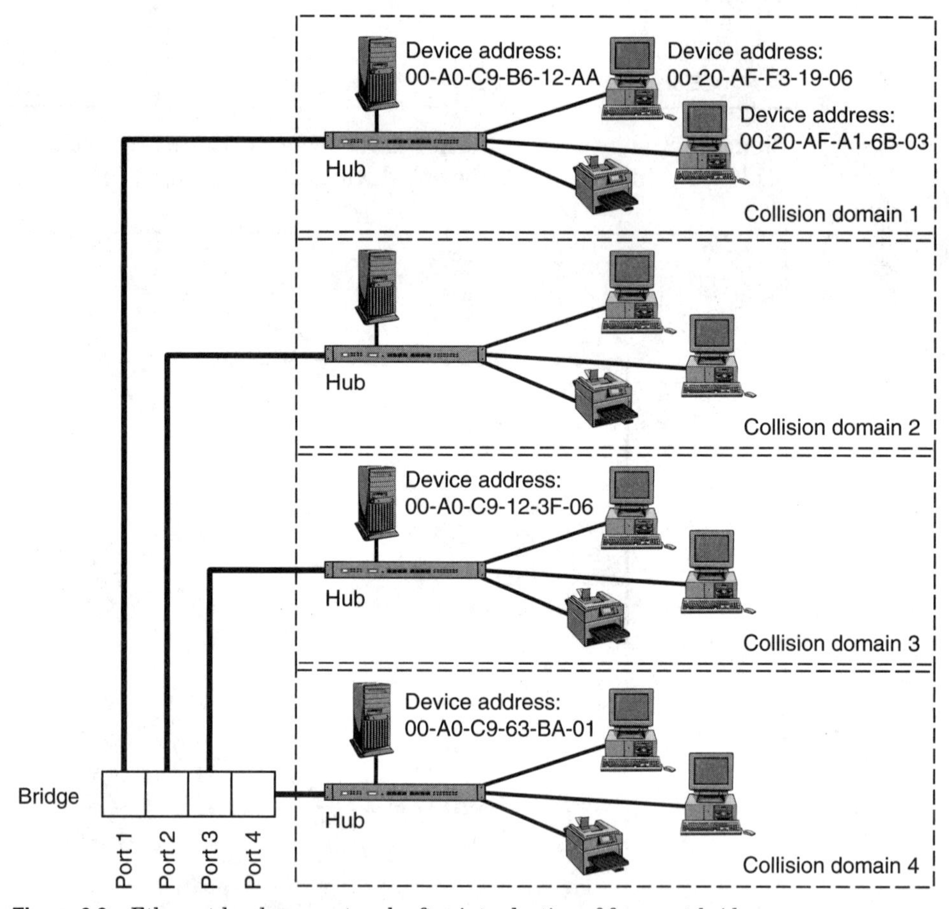

Figure 6.9 Ethernet local area network after introduction of four-port bridge.

The bridge learns device addresses from frames as they pass through the bridge.

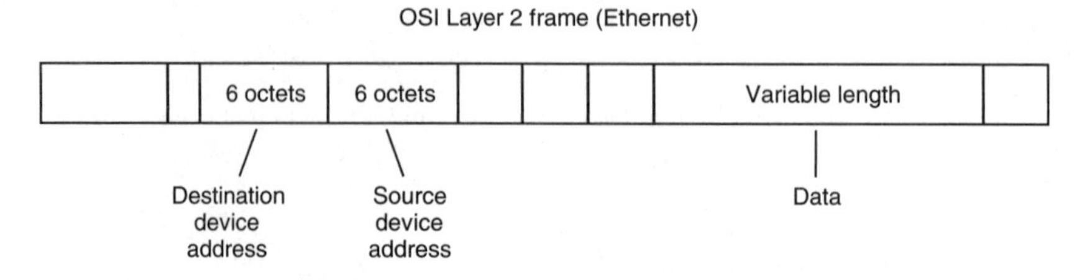

Figure 6.10 Ethernet OSI Layer 2 frame.

While learning the device addresses, the bridge creates its SAT. For the purpose of this example, the following SAT is used by the bridge.

Device address	Port number	Aging time (1,000,000 to 1 seconds)
00-20-AF-F3-19-06	1	186 seconds
00-A0-C9-B6-12-AA	1	296 seconds
00-A0-C9-63-BA-01	4	17 seconds

Frame processing is performed by the bridge based on entries found in its SAT. The following are some examples of how frames are processed by the bridge.

Incoming frame on port 1	
Source device address:	00-20-AF-F3-19-06
Destination device address:	00-A0-C9-63-BA-01
Action by bridge:	Forward the frame out of port 4.

Incoming frame on port 1	
Source device address:	00-20-AF-F3-19-06
Destination device address:	00-A0-C9-B6-12-AA
Action by bridge:	Filter the frame.

Incoming frame on port 1	
Source device address:	00-20-AF-A1-6B-03
Destination device address:	00-A0-C9-12-3F-06
Two actions by bridge:	1. Make SAT entry for unknown source address. 2. Since destination address is unknown, flood the frame out of ports 2, 3, and 4.

Incoming frame on port 4	
Source device address:	00-A0-C9-63-BA-01
Destination device address:	FF-FF-FF-FF-FF-FF (broadcast address)
Action by bridge:	Flood the frame out of ports 1, 2, and 3.

NOTE: Device addresses are described in detail in Chapter 5: Network Communications.

Bridge types

In most cases, all ports on a bridge use the same OSI Layer 2 LAN technology (e.g., Ethernet, token ring, or FDDI). However, bridges can also be used to link different types of technologies.

Transparent bridge. Transparent bridges are intended for use in a common LAN environment, in most cases, Ethernet. It is possible for the bridge ports to be configured for different media, in which case the bridge performs the necessary signal conversions (e.g., between a copper-based collision domain and an optical fiber-based collision domain).

Source Routing Bridge (SRB). SRBs are intended for use on token ring LANs. They can also be used on FDDI networks, since the two technologies are similar in architecture. When SRBs are used, network devices that originate frames are responsible for including path information in each frame. One or more SRBs direct a given frame through the bridged network on the basis of information provided by the source of the frame, eliminating the need for SATs.

NOTE: It is possible for the ports on a transparent bridge or SRB to operate at different speeds. For example, a two-port transparent bridge can connect a 10 Mb/s Ethernet collision domain to a 100 Mb/s Fast Ethernet collision domain. Similarly, an SRB can connect a 4 Mb/s Token Ring Network (TRN) ring to a 16 Mb/s TRN ring. In such cases, the bridges are equipped with additional buffer memory to store the frames waiting to be forwarded from the faster port to the slower one.

Source Routing Transparent (SRT) bridge. An SRT bridge is a hybrid device, capable of operating as an SRB or a transparent bridge, depending on the contents of the frames it processes. If a given frame contains path information, it is processed by the SRB component of the bridge. If there is no path information in the frame, the bridge processes the frame according to the contents of its SAT. SRT bridges are intended for use on token ring or FDDI networks.

Translational bridge. Translational bridges are used to link different LAN technologies (e.g., a TRN ring to an FDDI ring). In such cases, the bridge must perform additional processes, such as altering frame sizes or formats, prior to forwarding frames from one port to another.

Encapsulating bridge. Encapsulating bridges are used when the common LAN environments to be linked (e.g., two Ethernet collision domains) are separated by a different LAN technology (e.g., an FDDI backbone ring). In such cases:

1. The encapsulating bridge connected to one of the Ethernet collision domains places each frame to be forwarded to the other Ethernet collision domain into an FDDI frame.
2. The encapsulated frame is then transmitted out of the port connected to the FDDI network.
3. A second encapsulating bridge—connected to the same FDDI network and to the destination Ethernet collision domain—removes the Ethernet frame from the FDDI frame and forwards it out of the port connected to the Ethernet collision domain.

NOTE: Encapsulating bridges are sometimes described as tunneling bridges, since the forwarded frames are said to tunnel through a different type of network (see Figure 6.11).

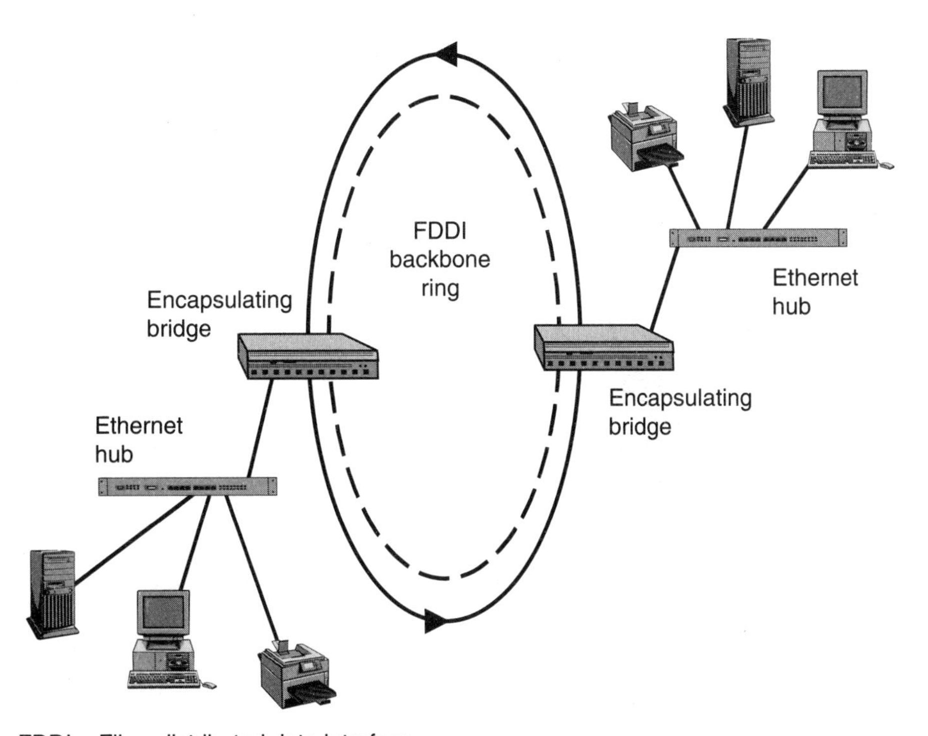

FDDI = Fiber distributed data interface

Figure 6.11 Encapsulating bridges.

Remote bridge. Remote bridges are used when the collision domains to be linked are geographically distant, requiring telecommunications circuits to connect the bridges located at each site (see Figure 6.12). A remote bridge operates in a similar manner to a translational bridge, converting frames from the format used on LANs to the format used on telecommunications circuits. In most cases, the telecommunications circuit operates at a slower speed than the LAN, making it necessary to provide additional buffer memory in the remote bridges.

NOTE: Remote bridges are sometimes described as half-bridges, since they must be used in pairs to connect any two sites.

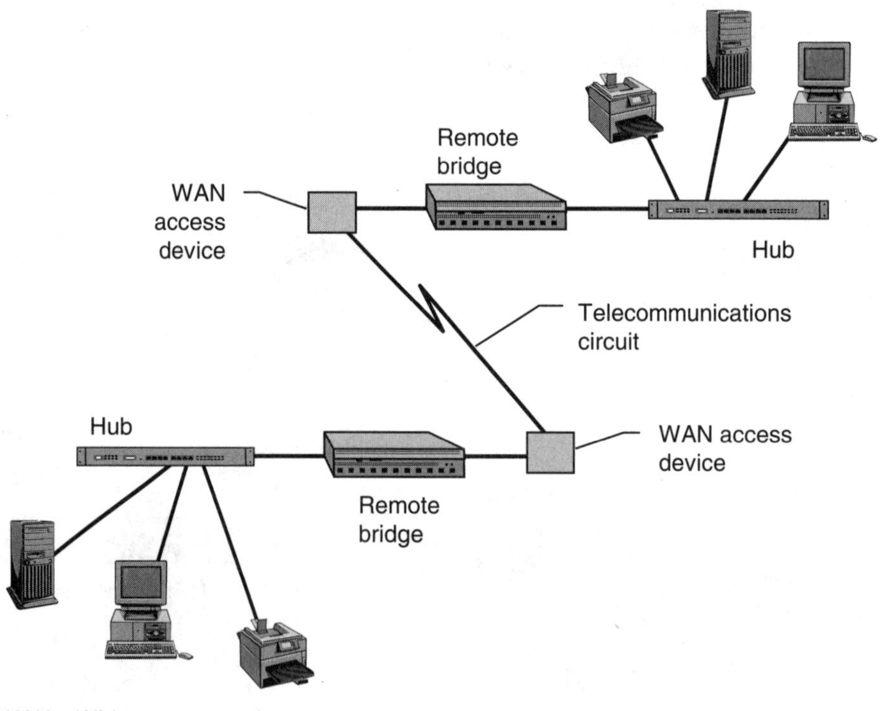

Figure 6.12 Remote bridges.

Spanning Tree Algorithm (STA)

When a bridge is used to link two or more collision domains, the bridge must remain operational at all times. If it fails, the network is split into two or more separate LANs that cannot communicate with each other. A secondary bridge is desirable from a fault-tolerance point of view; however, two transparent bridges cannot be in simultaneous operation between two collision domains. Example 6.2 illustrates the problem, which is referred to as a broadcast storm.

EXAMPLE 6.2 Broadcast storm.

In this example, two bridges link collision domains A and B. Each bridge is a two-port model (see Figure 6.13).

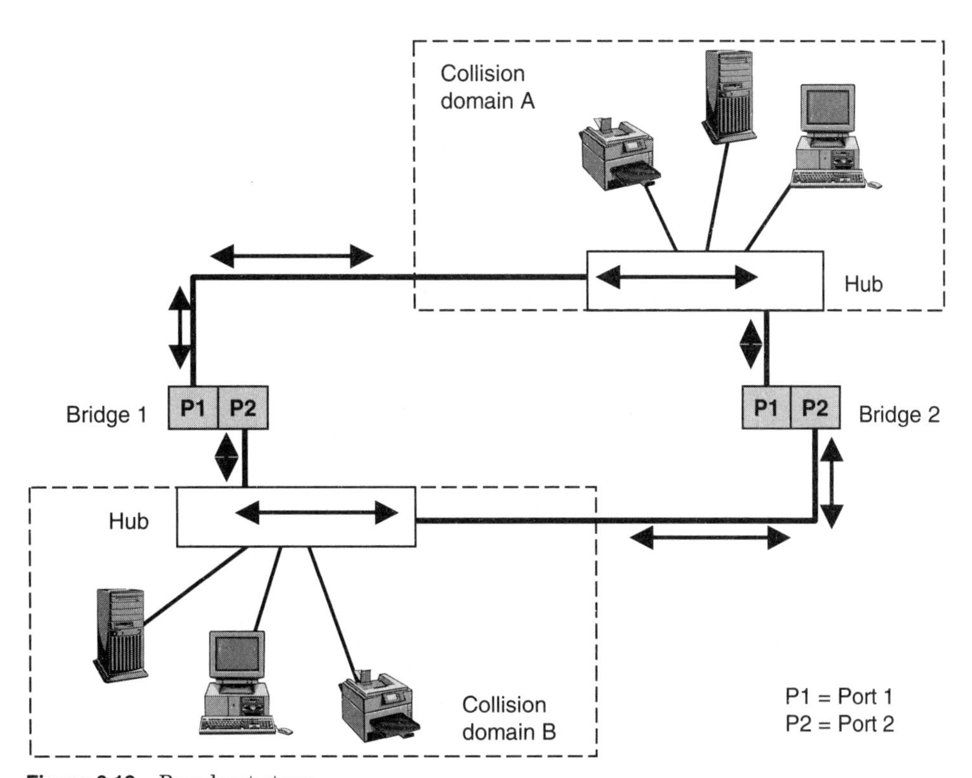

Figure 6.13 Broadcast storm.

In Figure 6.13, if a station in collision domain A issues a broadcast frame, it is received on Port 1 of Bridge 1 and Port 1 of Bridge 2. The result would be as follows:

- Bridge 1 floods the frame out of Port 2, which is received by all devices in collision domain B, including Port 2 of Bridge 2.
- Bridge 2 also floods the frame out of Port 2, which is received by all devices in collision domain B, including Port 2 of Bridge 1.
- Both bridges process the frame as a broadcast frame issued in collision domain B. As a result, each bridge floods the frame out of Port 1 to collision domain A.

This process is repeated until the network stops operating due to the congestion caused by the looping broadcast frames. This condition is called a broadcast storm.

The STA was developed to enable parallel bridging, making available alternate paths between bridged collision domains. STA makes it possible to install two or more bridges between any two collision domains. Using STA, a loop-free set of operational paths is dynamically created between all collision domains. Other paths are designated as backup or standby paths, ready to be used whenever there is failure of an operational path.

STA operates as follows:

1. Each bridge, as well as each port on a bridge, is assigned a unique identifier.
2. Each port is also assigned a value representing the path cost of the port. The path cost is an arbitrary value, and can be assigned or changed to indicate a preference for a given port. For example, a port with a faster operating speed can be assigned a lower path cost to favor its use.
3. One of the bridges is designated as the root bridge.
4. Each of the remaining bridges calculates the least-cost path to the root bridge, which is considered the primary operational path.
5. All other paths are classified as backup or standby paths. Ports associated with such paths will not process frames unless there is failure of a port in the primary path.

Example 6.3 illustrates STA operations on a network made up of five collision domains and five bridges.

EXAMPLE 6.3 STA operations.

The bridged collision domains are physically linked as shown in Figure 6.14:

NOTES: To simplify Figure 6.14, the devices connected to the hubs are not shown.

The bridge identifier can be configured manually.

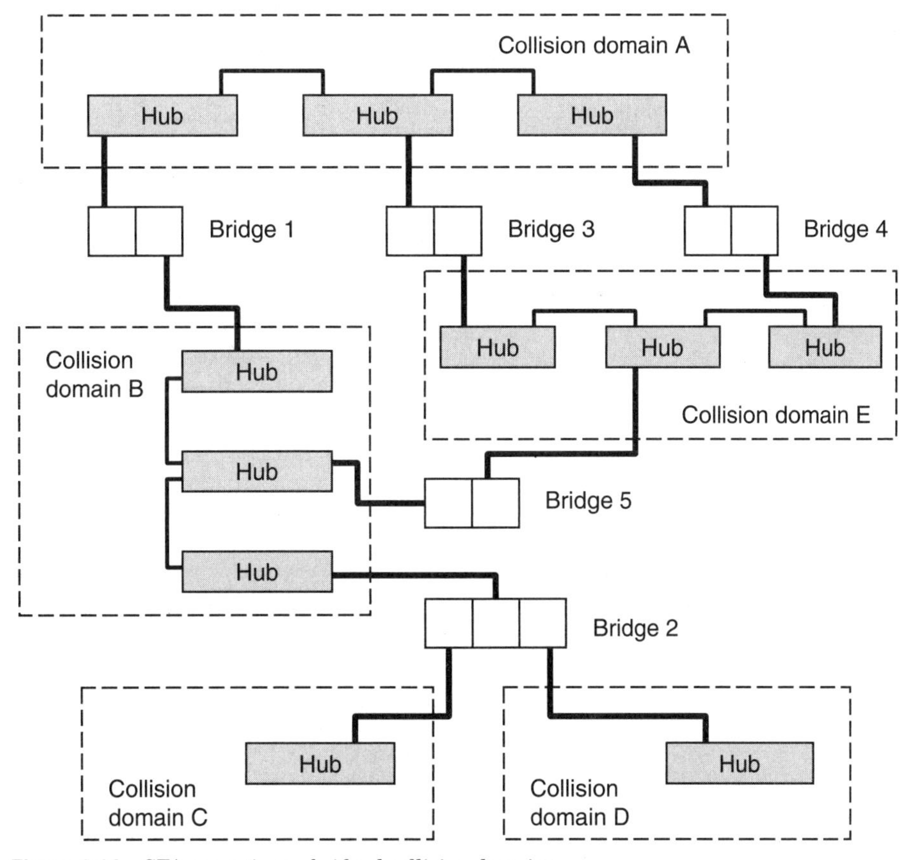

Figure 6.14 STA operations—bridged collision domains.

As part of the network configuration, a path cost is manually or automatically assigned to each port (see Figure 6.15).

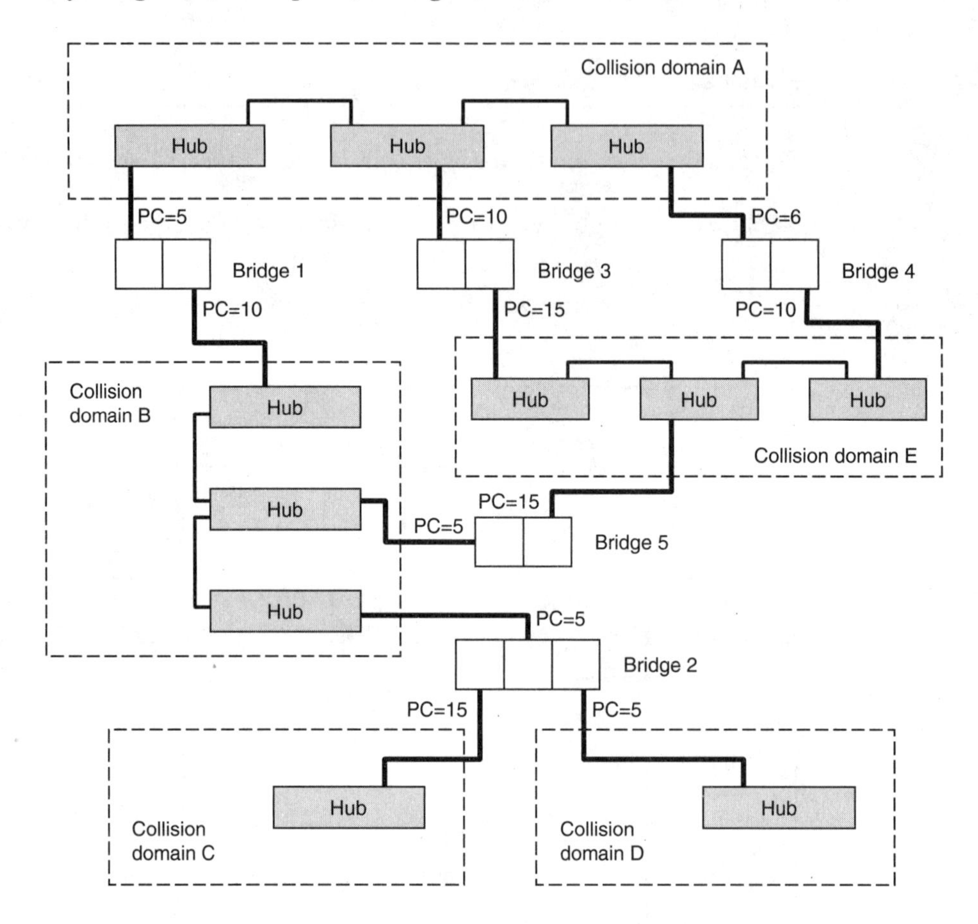

Figure 6.15 STA operations—assigning path costs.

Step 1. STA selects the root bridge by comparing the values of various parameters within each bridge. In this example, assume that Bridge 1 is selected to be the root bridge.

Step 2. The root bridge becomes the designated bridge for the collision domains it links. In this example, it is collision domains A and B.

Step 3. Each of the remaining collision domains must also be assigned a designated bridge. In cases where only a single bridge provides access to a collision domain, this bridge becomes the designated bridge. In this example, Bridge 2 is the designated bridge for collision domains C and D (see Figure 6.16).

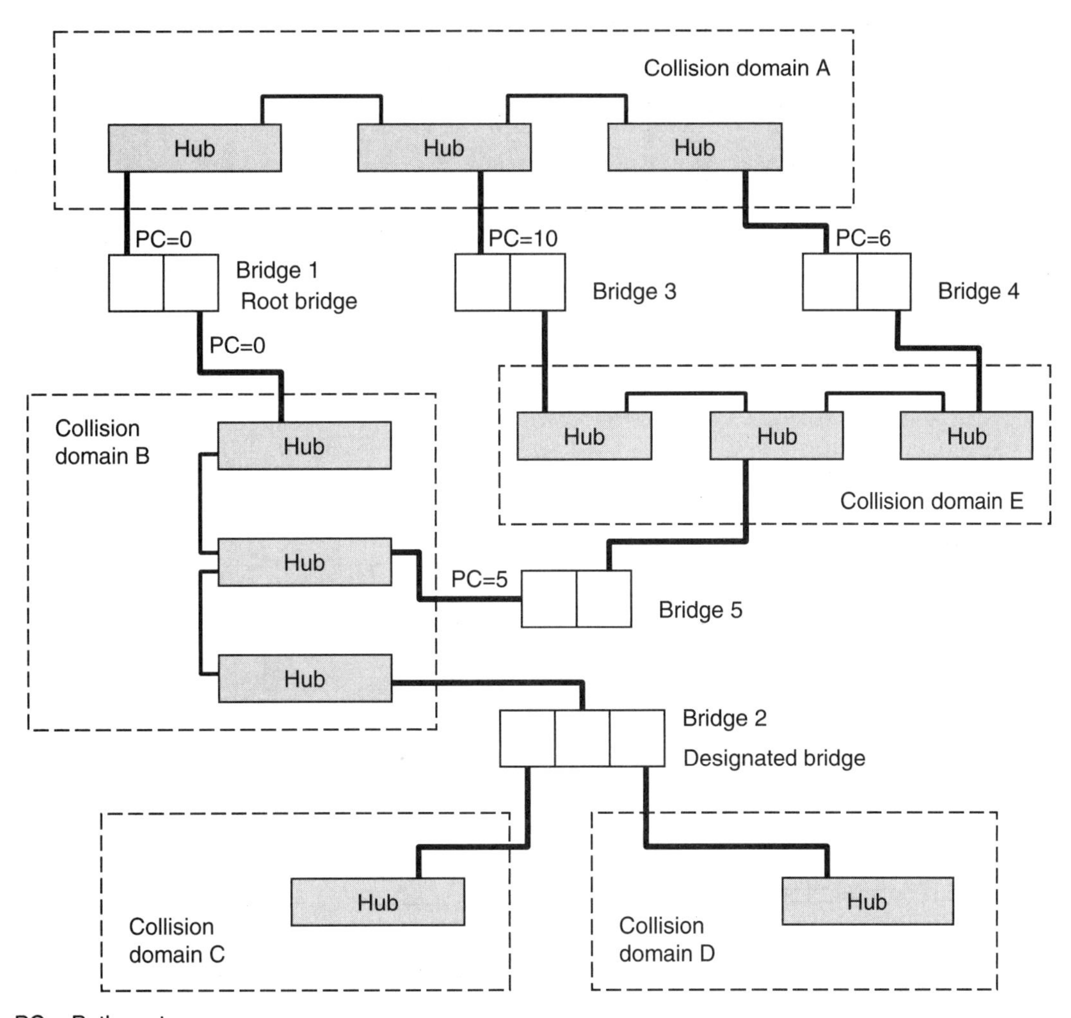

Figure 6.16 STA operations—assigning a designated bridge.

Step 4. In cases where multiple bridges can provide access to a collision domain, a calculation based on the lowest path cost from the root bridge to the collision domain is performed to select the designated bridge. When STA is used, all bridges exchange path cost values for all of their connections, using bridge protocol data units (BPDUs). The root bridge always transmits a path cost of zero for all of its ports. Each bridge always adds a default value of one to the path cost it receives from another bridge prior to transmitting its path cost. In this example, there are three possible paths from the root bridge to collision domain E, as follows:

- The root bridge and Bridge 3 can be used, with a total path cost of 10 + 1 = 11.
- The root bridge and Bridge 4 can be used, with a total path cost of 6 + 1 = 7.
- The root bridge and Bridge 5 can be used, with a total path cost of 5 + 1 = 6.

NOTE: The path cost of the port connected to the destination collision domain is not used in this calculation.

Based on this calculation, the lowest path cost from the root bridge to collision domain E is through Bridge 5.

The tree-shaped operational paths created by STA are shown in Figure 6.17. The dotted lines represent standby paths that can be activated in case Bridge 5 fails.

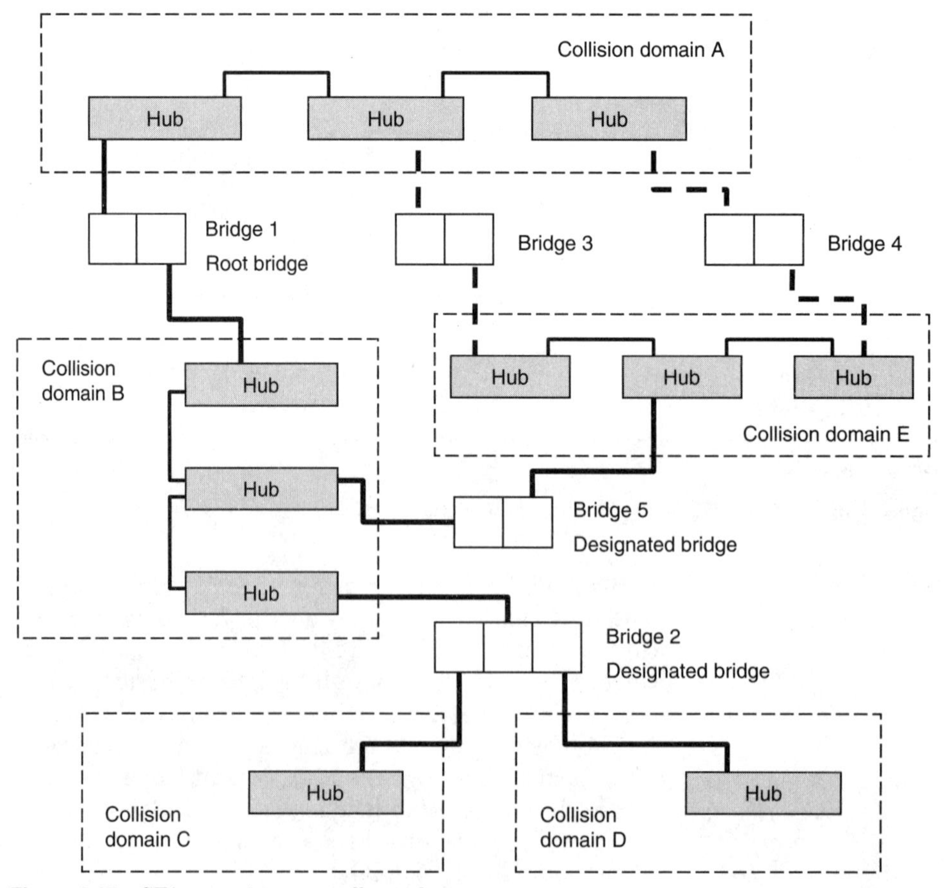

Figure 6.17 STA operations—standby paths.

Layer 2 Switching

General

Switching can be described as the evolution of bridging technologies. A switch is the functional equivalent of a bridge with additional features not available on bridges, such as a large number of ports. Switches replace traditional bridges and in many cases, they replace hubs as well. For example, a switch equipped with 48 ports can be used as the equivalent of a 48-port bridge, with hubs connected to each of its ports. Alternatively, switch ports can replace some of the hubs—providing a dedicated communications channel to each device formerly connected to a hub. From a performance point of view, the ideal LAN configuration uses only switches for LAN access, eliminating all sharing of communications channels.

Two additional benefits of switched connections are full-duplex operations and link aggregation or port trunking. In full-duplex operations, the device connected to a switch port can simultaneously transmit and receive frames. By comparison, a hub connection is half-duplex—the device can either transmit or receive, but not both at the same time. When using full-duplex configuration, microsegmentation is required. Only a single device may be connected to each switch port. Full-duplex operation is particularly suitable for applications such as LAN videoconferencing, where data is simultaneously sent and received at both ends of a videoconference link. When full-duplex operations are enabled, the operating speed of the LAN is considered double the standard speed (e.g., a full-duplex Fast Ethernet port is classified as a 200 Mb/s connection). Full-duplex operations eliminate collision domains, since collisions cannot occur when the devices at both ends of a connection can transmit and receive at the same time.

Link aggregation makes it possible to configure multiple switch ports to operate as a single communications channel, providing a flexible method for improving performance and fault tolerance. For example, two, three, or four full-duplex Fast Ethernet ports can be aggregated to provide a 400, 600, or 800 Mb/s channel, respectively. If there is a failure in any of the ports, the remaining ones continue to provide connections to the attached device. To avoid looping due to parallel paths, STA identifies the aggregated links as a single port.

Switch operations

A switch contains a matrix of connections linking each port to every other port. Each time a frame is received on one port, a temporary or virtual connection is established through the matrix. This links the incoming port to an outgoing port that is connected to the frame's destination. Once the frame is transferred between the ports, the virtual connection is removed.

In most switches, connections and frame transfers between different incoming and outgoing ports can take place simultaneously. A switch capable of simultaneous full-duplex frame transfer from half of its ports to the other half is referred to as a nonblocking switch (see Figure 6.18).

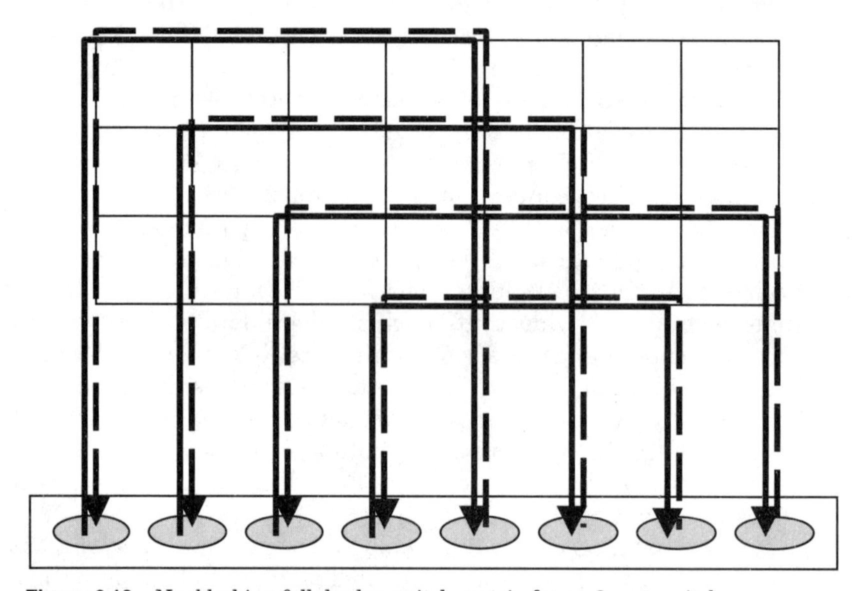

Figure 6.18 Nonblocking full-duplex switch matrix for an 8-port switch.

NOTE: A nonblocking switch must have a switch matrix capacity equal to the sum of the maximum port speeds of the total number of ports, which can extend to the gigabit per second (Gb/s) range (see Table 6.1).

TABLE 6.1 Switch Matrix Capacity

Number of Ports	Maximum Port Speed (Mb/s)	Switch Matrix Capacity Needed (for nonblocking full-duplex operations)
8	10	8 × 10 = 80 Mb/s
16	10	16 × 10 = 160 Mb/s
24	10	24 × 10 = 240 Mb/s
32	10	32 × 10 = 320 Mb/s
48	10	48 × 10 = 480 Mb/s
8	100	8 × 100 = 800 Mb/s
16	100	16 × 100 = 1.6 Gb/s
24	100	24 × 100 = 1.2 Gb/s
32	100	32 × 100 = 3.2 Gb/s
48	100	48 × 100 = 4.8 Gb/s
8	1000	8 × 1000 = 8 Gb/s
16	1000	16 × 1000 = 16 Gb/s
24	1000	24 × 1000 = 24 Gb/s
32	1000	32 × 1000 = 32 Gb/s
48	1000	48 × 1000 = 48 Gb/s

NOTE: The term backplane is also used to describe a switch matrix.

In cases where there is uneven traffic distribution in the switch, large amounts of buffer memory may be required for each of the ports. For example, a 24-port switch may have 23 stations and one server connected to its ports. If all of the stations simultaneously send data to the server, each of the 23 station ports must store data in its buffer memory while waiting for its turn to connect to the single server port and transfer a frame. If the server is connected to a port that operates at a higher speed than the station ports (e.g., the server is connected to a 100 Mb/s port and stations are connected to 10 Mb/s ports), the server will transfer frames faster than any station can receive them, possibly causing the buffer memory to overflow. Buffer overflow makes it necessary to alert the sender that frames have been lost and to request retransmission of the lost frames, which is an inefficient use of network resources.

Flow control is a mechanism used to manage the rate of exchange of frames between full-duplex ports on a switch. The intent is to ensure that the number of frames waiting to be transmitted from a port does not exceed the capacity of the port buffer memory. When flow control is implemented, switch ports—and in some cases, the devices connected to the ports—are capable of generating pause messages to indicate that they are temporarily unable to receive data. A pause frame contains a time delay indicator, which is used as a counter to measure the duration of the pause. At the end of the pause, the receiver of the pause frame continues its transmission, unless it receives another pause frame. During the delay, the port that issued the pause frame clears some or all of its buffer memory by processing the frames it contains. The combination of flow control frames and buffer memory makes it possible to equalize traffic flows within a switch, especially in cases where some devices send or receive most of the frames processed by the switch.

Flow control is implemented differently when a switch port is configured to operate at half-duplex instead of full-duplex. For example, a switch port may be connected to a hub, which cannot operate in full-duplex mode. If the hub generates frames faster than the switch port can forward them, the switch can stop hub transmissions by generating one or more false collision signals. This method—called backpressure—prevents a half-duplex device connected to a port from transmitting data, giving the switch time to forward some or all of the frames stored in its buffer memory.

Switch types

Switch performance ratings are based on the number of frames a switch is able to process in a given unit of time, typically one second. Once a frame arrives at a port, the switch must be capable of reading the destination device address in the frame, performing a search in the SAT, and forwarding the frame out another port before the next frame arrives at the same port. The term latency is used to describe the amount of time it takes a switch to process a frame—the lower the latency, the higher the rated performance of the switch.

Three methods of frame processing can be used, as follows:

- Store-and-forward
- Modified cut-though
- Cut-through

NOTE: The most flexible switches allow each port to be configured to use any of these methods. The least flexible switches use the same method on all ports.

Store-and-forward switching. The traditional method of frame processing is the store-and-forward method, where the entire incoming frame is copied to buffer memory prior to forwarding. In store-and-forward switching, latency values depend on the length of the frames processed—for Ethernet frames, values can range from 64 octets to 1518 octets. If tag headers are included, this range is increased to 1522 octets.

NOTE: Tag headers are explained later in this chapter.

This method is the slowest of the three methods, but also the most reliable. After the frame is copied to buffer memory, it is inspected for errors. If an error is found, the frame is discarded instead of being forwarded to its destination (see Figure 6.19).

Cut-through switching. When cut-through switching is used, part of an incoming frame is copied to buffer memory—up to and including the destination address field. For Ethernet frames, this value is six octets. Once the destination address is processed, the frame is forwarded to its destination without an inspection for errors (see Figure 6.19).

This method is the fastest of the three methods, but also the least reliable. It is best suited for network environments where most of the frames generated by applications are small and arrive at the switch with no errors.

Modified cut-through switching. In modified cut-through processing, the first 64 octets of an incoming frame are copied to buffer memory and inspected for errors. If no errors are found, the frame is forwarded to its destination (see Figure 6.19). This method, like cut-through switching, provides a fixed latency value for the switch and is intended to be a compromise between the store-and-forward (highest latency/highest reliability) and cut-through (lowest latency/lowest reliability) methods.

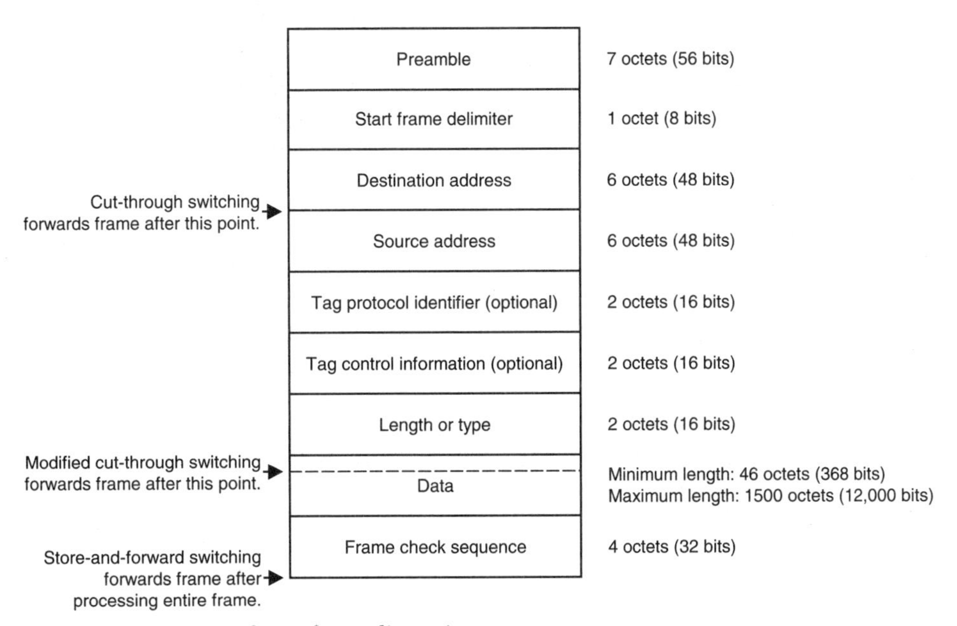

Figure 6.19 Ethernet frame forwarding points.

NOTE: An octet is a grouping of eight bits. It is frequently referred to as a byte.

Some switches have the capability to dynamically change the method they use to process frames. Such devices, sometimes described as adaptive switches, begin operations using the modified cut-through method. If an excessive number of frames is observed with errors in the first 64 octets, the switch begins to use the store-and-forward method. If there are no errors observed over a period of time, the switch can return to the modified cut-through method.

Higher layer switching

The technologies used for switching have also been applied to routers, resulting in devices called Layer 3 switches or routing switches. Historically, routers have processed OSI Network layer datagrams using software-based processes and rules. In Layer 2 switches, similar processes and rules for frame forwarding are packaged in the form of logic embedded in hardware. Layer 3 switches use hardware-based logic for many routing functions, which enables much faster datagram processing than traditional software-based routing. Typically, such devices are less flexible than traditional routers (e.g., they support a limited number of protocols compared to traditional multiprotocol routers).

An incoming frame on a Layer 3 switch is processed on the basis of network address information found in the data field of the frame. In some cases, a greater level of detail can be extracted and used to direct a frame. For example, a Layer 4 switch directs frames on the basis of OSI Transport layer protocol information and a Layer 7 switch processes frames on the basis of the applications used to generate the frames.

Higher layer switching provides great flexibility to network administrators, making it possible for them to direct, modify, and refine network traffic flows on the basis of organizational priorities. This process is sometimes described as traffic engineering or traffic shaping. As an example, a Layer 7 can forward frames containing data generated by a Web browser out of a different switch port than the frames generated by videoconferencing software.

Virtual LANs (VLANs)

General

On a conventional LAN, there is a single broadcast domain, spanning all of the hubs and switches as well as the devices connected to them. This is described as a flat network. Any broadcast frame issued by a device is delivered to all other devices, regardless of the speed or type of connection used. The greater the number of devices on a LAN, the greater the number of broadcasts transmitted. Excessive broadcasts limit the total number of devices on a LAN, even when all connections are switched.

In most cases, there is no need for the broadcast to be sent to every device on the organizational network. For example, a broadcast request for configuration or management data issued by a station in one department is usually intended for a server in the same department. However, if the organizational network is made up entirely of hubs and switches, every station and server in the organization will receive the broadcast.

The traditional way of restricting broadcasts is by using routers to divide a single broadcast domain into multiple smaller ones. Soon after the introduction of switches, vendors began to offer models with similar capabilities, making it possible to form multiple VLAN groups on a single physical LAN. A broadcast frame issued by a device in a VLAN group is only delivered to other devices in the same group by the VLAN switch(es) (see Figure 6.20).

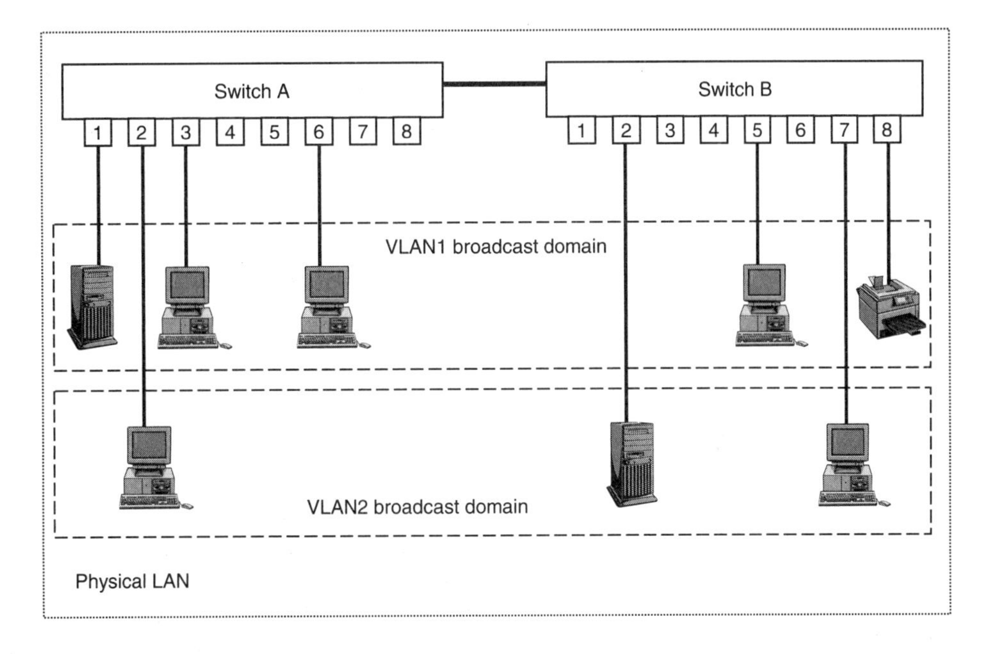

LAN = Local area network
VLAN = Virtual LAN

Figure 6.20 Simple VLAN configuration.

NOTE: In Figure 6.20, VLAN 1 is a broadcast domain consisting of ports 1, 3, and 6 on Switch A and ports 5 and 8 on Switch B. VLAN 2 is another broadcast domain consisting of port 2 on Switch A and ports 2 and 7 on Switch B. Other ports are unassigned. The physical LAN is the combination of both switches and connected devices. If the switches have not been configured for VLAN operation, the physical LAN is also the LAN broadcast domain.

VLANs provide the performance benefits associated with physical LAN segmentation without the need to connect a given station, server, or shared peripheral device to a specific switch port. Assuming all switches on an organizational network provide the same features, any device can be connected to any switch port. The VLAN groups various combinations of ports into a common broadcast domain. Devices in the same VLAN group operate as if they were connected to the same switch and cabling system, even though each device may be physically connected to a different switch using any type of approved cabling. If stations need to be relocated, the switch ports they are transferred to can be assigned to the same VLANs as their previous ports, making the relocation transparent from a network traffic point of view.

On most networks, it is necessary for some devices to have broader access than others. For example, some stations may require access to servers in other VLAN groups. A common way to enable communications between VLANs is to use a router to transfer frames from one VLAN to another. The router can be a separate device as shown in Figure 6.21 or it can be in the form of routing logic integrated into a Layer 2 switch as shown in Figure 6.22.

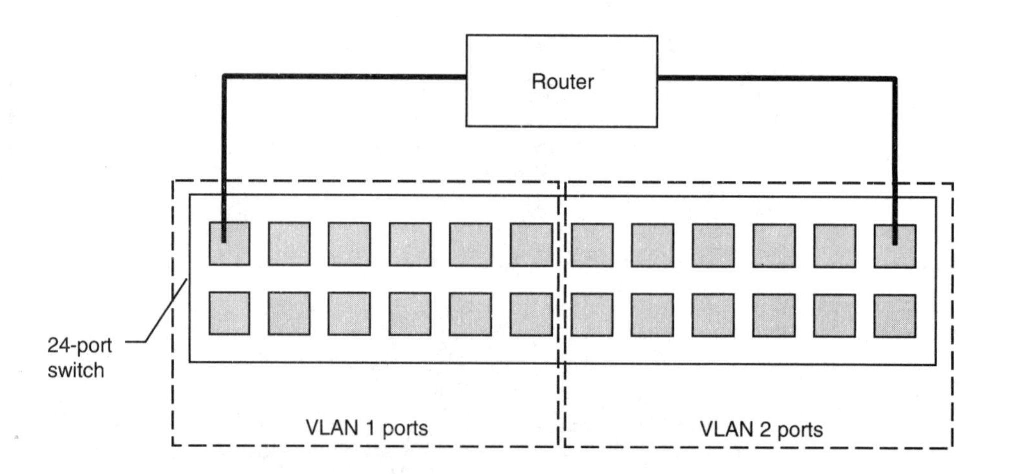

VLAN = Virtual local area network

Figure 6.21 Communications between VLANs using a separate router.

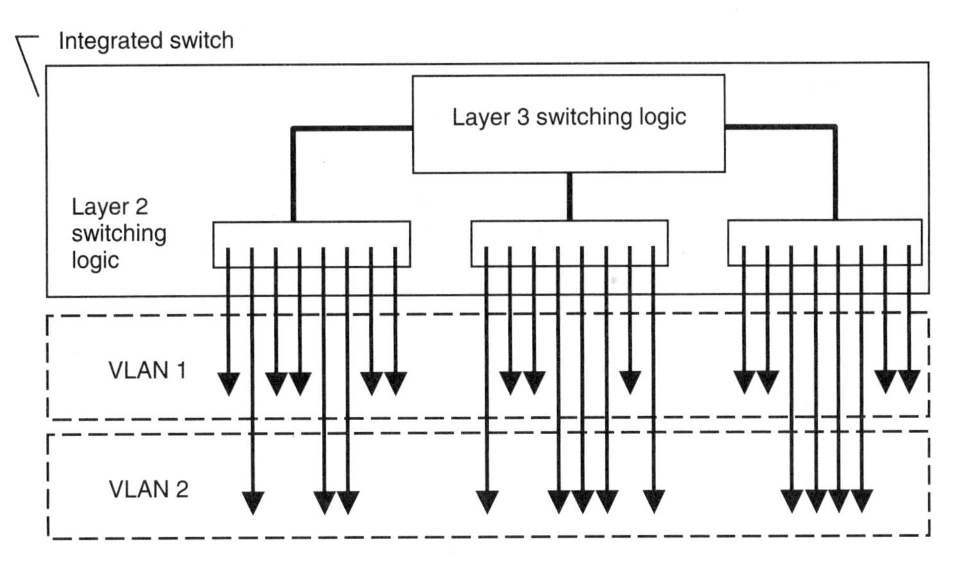

Figure 6.22 Communications between VLANs using an integrated Layer 2/Layer 3 switch.

NOTE: If the organizational network is small and the number of devices requiring connection to multiple VLANs is limited, it is possible to avoid using a router by manually assigning such devices to multiple VLAN groups.

Virtual LAN (VLAN) operations

Prior to the introduction of IEEE standards for VLANs, switch vendors provided various proprietary techniques for creating and operating VLANs. In most cases, switches capable of sharing VLAN data had to be obtained from the same manufacturer. Each vendor identified VLAN associations in a different way.

The introduction of the IEEE 802.1Q standard in 1998 made it possible for VLAN information to be stored in a uniform format in all frames on Ethernet, token ring, and FDDI networks. The process of entering VLAN data into the appropriate fields in a frame is referred to as VLAN tagging.

The fields used to store VLAN information can also carry data used to indicate the priority of a given frame. The IEEE 802.1p initiative was to provide different classes of service in an OSI Layer 2 switched environment, making it possible to give forwarding priority to frames generated by time-sensitive applications.

NOTE: IEEE 802.1p was incorporated into the revised IEEE 802.1D bridging standard issued in 1998 and is not available as a separate document.

Figure 6.23 shows an Ethernet frame, highlighting the fields used for VLAN association and traffic prioritization. At a minimum, switches on the network must be capable of identifying and processing the contents of these fields for VLAN operations to take place.

It is also possible to configure stations or servers with NICs that are capable of tagging frames. This makes it possible for a NIC to issue frames with different priorities, each associated with a different application running on the station or server. If the NIC is not capable of tagging frames, the switch port connected to the station or server must perform the tagging. This can result in the assigning of a single priority to all frames generated by the connected device, regardless of the application used to generate the frame.

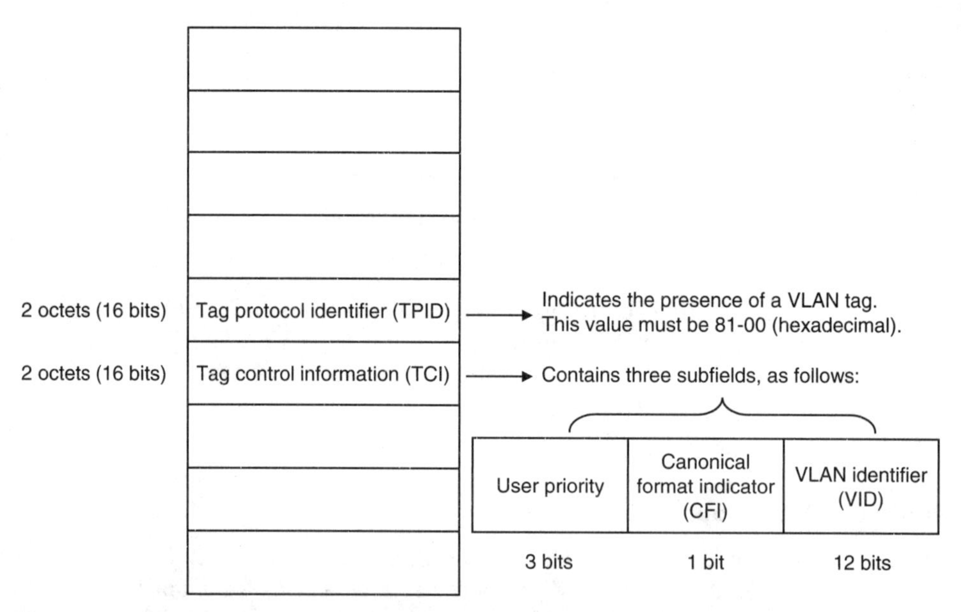

Figure 6.23 VLAN and prioritization fields in an Ethernet frame.

The three bits assigned to the user priority subfield makes it possible to create 2^3 (8) different priority levels for frames, with values ranging from 0 to 7. Switches divide their buffer memory into separate areas called queues or forwarding queues. Each incoming frame is placed into an appropriate forwarding queue based on its priority. Frames in a high-priority port queue are forwarded before those in a low-priority port queue, regardless of the order of arrival into the port.

The single bit assigned to the canonical format indicator (CFI) subfield is used to indicate the arrangement of bits in the frame's address field. For example, Token ring networks transmit address bits in the opposite order of

Ethernet networks. The CFI bit is used when the switching environment contains a mix of LAN technologies.

The 12 bits assigned to the VLAN identifier (VID) subfield makes it possible to create 2^{12} (4096) different VLANs on a single physical network, with values ranging from 0 to 4095. Three VID values have special meaning on the network, leaving 4093 values available for assignment:

- A VID value of zero indicates that the frame does not contain VLAN identification. A switch receiving such a frame will only inspect the user priority subfield for priority level information. The zero VID is also referred to as the null VLAN.
- If the VID value is one, the frame is intended for the default VLAN. If no other VLANs are created on the network, the default VLAN is identical to the physical network.
- The VLAN value of 4095 (FFF hexadecimal) is reserved and cannot be assigned.

A large switched network may require hundreds of VLANs to be established and maintained. To minimize the administrative burden and prevent mistakes, a protocol called the generic attribute registration protocol (GARP) VLAN registration protocol (GVRP) has been developed. GVRP can be described as the switching equivalent of the routing protocols used to identify paths and manage traffic flows between routers on an internetwork.

When GVRP is used, switch ports and NICs routinely exchange VLAN identification data to indicate their VLAN associations. When a switch receives a broadcast frame issued by a device on a particular VLAN, the switch forwards the frame out of all ports on the path to other members of the same VLAN. Without GVRP, VLAN associations have to be configured manually for every port in a switch and changed each time a device is relocated and connected to a different switch port.

Virtual LAN (VLAN) types

To provide additional flexibility to network administrators, most VLAN-capable switches make it possible to set up VLANs using a variety of criteria. Some of these are as follows:

Port-based Virtual LANs (VLANs). The least complex method of assigning VLAN associations is based on port connections. When port-based VLANs are used, any device connected to a given port on a switch becomes a member of one or more VLANs assigned to that port. The connected device can be any of the following:

- Station
- Server

- Shared peripheral device
- Router interface
- Hub
- Another switch

Device address-based Virtual LANs (VLANs). A device address-based VLAN associates the NIC in each switch-attached device with one or more VLANs. A device does not lose its VLAN membership if it is relocated and connected to another port on the switch. When the device starts sending frames from its new port, the switch assigns the new port to the VLAN(s) associated with the NIC.

Port and device address-based Virtual LANs (VLANs). When both port and device addresses are used as the basis for VLAN membership, a central database is used to associate each NIC with a specific port on a switch. Each port is also identified as a member of one or more VLANs. This combination ensures that only approved devices can send and receive frames intended for a particular VLAN. If a port receives a frame from a NIC with a device address not associated with that port, the switch ignores the frame.

Network address-based Virtual LANs (VLANs). A network address-based VLAN is similar to a device address-based VLAN. The difference is the part of the frame that is inspected by the switch for VLAN association. On a network address-based VLAN, the data field of an incoming frame is searched for OSI Network layer address information (e.g., an Internet protocol [IP] destination address). The Network layer address is then used to associate the frame with one or more specific VLANs. Once this is done, the switch assigns the port that received the frame to the VLAN(s) associated with that network address.

Protocol-based Virtual LANs (VLANs). A protocol-based VLAN environment is intended for networks that use multiple higher layer protocols. Protocol-based VLANs operate in the same manner as network address-based VLANs. The only difference is the information used to associate frames with VLANs. While a network address-based VLAN looks for addressing information, a protocol-based VLAN looks for the bits in the frame that indicate the higher layer protocol contained in the frame. Based on the protocol identified, the port that received the frame is made a member of the VLAN(s) associated with that protocol.

Chapter

7

Network Services and Applications

Chapter 7 describes the services commonly installed on networks to assist users and administrators in locating, accessing, and working with available network resources. Emerging voice and video applications are also described, since they are expected to be widely deployed on organizational networks in the future.

Overview

General

From a user's perspective, a network is often seen as extended shared storage of all connected stations. This enables the storage and sharing of applications and files not found on a given station's disk drive(s).

From a software perspective, three types of software are used on most networks:

- Operating system software, used to control hardware such as stations, servers, and network access devices.
- Network service software, used to implement the network communications infrastructure, such as addressing, naming, and cataloging available resources. In many cases, this type of software is integrated into one of the other types.
- Applications software, used to provide users and administrators with the computing tools needed to accomplish various tasks, such as messaging or information analysis.

This chapter describes the services commonly installed on networks to assist users and administrators in locating, accessing, and working with available network resources. Historically, vendors have used proprietary software to provide many or all of these services. However, the popularization of the Internet and its associated standards has provided organizations with the ability to implement a common set of services on all types of networks, without the need to buy all software from a single vendor. The following network services are described in this chapter:

- Addressing services, used to automatically assign addresses to devices at connection time.
- Naming services, used to assign unique names to devices and other network resources, as an alternative to using numeric addresses.
- Directory services, used to build lists of available network resources to simplify the locating of a particular resource.
- Internet services, used to make it possible for organizational staff and external parties to obtain network resources over the Internet.

The introduction of voice and video applications, where devices such as telephones and video cameras are connected to networks traditionally used to link computers and related peripherals, has also promoted the adoption of Internet-based technologies on organizational networks. Originally used to enable real-time conversation between Internet users, voice over Internet protocol (VoIP) technology is described as a potential successor of the existing telephone network infrastructure. Similarly, technologies used to deliver audio and motion video over the Internet—called streaming audio or streaming video—are described as potential successors of radio and television broadcasting as well as practical tools to improve organizational communications. Voice and video applications are described in the last two sections of this chapter.

Addressing Services

General

As described in Chapter 5: Network Communications, multiple levels of addressing are used in networking to identify individual devices, the networks to which they are connected, and the specific protocols or applications they are running. In an environment where multiple local area networks (LANs) are connected to an internetwork, messages must include both a network address and a device address in order to be delivered to the correct destination.

In most cases, the device address of a component such as a station or a server is hardware-based, consisting of the unique identifier assigned to the station's or server's network interface card (NIC) by its manufacturer. Since each NIC has a globally unique address, administrators can use these values without modification, with the knowledge that duplication cannot occur. However, this is not the case with network addresses of devices using the Internet protocol (IP), the most popular of the internetworking protocols available. An IP address is software-based, and identifies both the device and the network or subnet to which it is connected. If a device such as a laptop computer is moved from one location to another, it may require a different IP address to operate as intended. In large organizations with hundreds or thousands of stations—many of them mobile—manual IP address assignment and administration can be time-consuming and the cause of network disruptions whenever incorrect or duplicate addresses are issued. The dynamic host configuration protocol (DHCP) can be used to provide addressing services in place of manual address assignment, which helps minimize both the administrative burden and the possibility of addressing errors.

NOTE: IP addressing and subnets are described in detail in Appendix B: Internet Protocol (IP) Addressing Fundamentals.

Dynamic Host Configuration Protocol (DHCP)

The Internet Engineering Task Force (IETF) issued the DHCP standard in October 1993 as request for comment number 1531 (RFC 1531) and updated it in March 1997 as RFC 2131. The DHCP service uses the terms client and server to describe the devices requesting and supplying addressing and other configuration information, respectively. In most cases, it is not necessary to dedicate one or more servers to DHCP activities. A general-purpose server can be configured to provide DHCP services in addition to its other functions.

On a network equipped with DHCP services, devices do not have to be individually configured with IP addresses. When such devices are powered on, they issue broadcasts requesting address assignments. Selections are made from a pool of available addresses by a DHCP server and sent to the devices.

Three types of address assignments are possible.

- Permanent address allocation, in which an IP address is selected from the pool and assigned without an expiration date to the requesting device.
- Leased address allocation, in which an IP address is selected from the pool and assigned for a limited period of time, which can be hours or days, to the requesting device. This type of address assignment is common in network environments where the total number of stations exceeds the available number of addresses, but only a limited number of stations are connected to the network at any given time.
- Manual address allocation, in which a specific IP address previously selected by an administrator for a given device is assigned to that device when it requests an address.

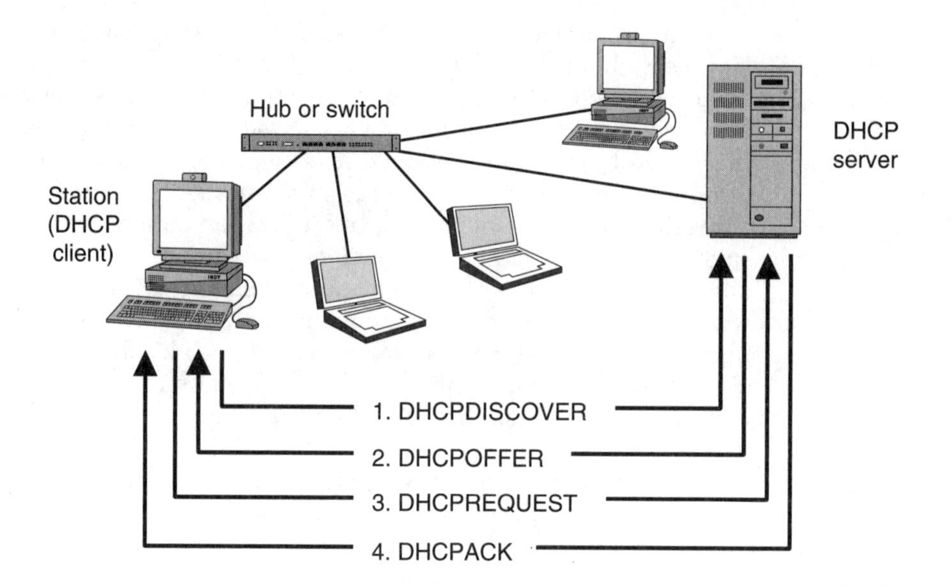

DHCP = Dynamic host configuration protocol

Figure 7.1 DHCP address allocation process.

Figure 7.1 illustrates the initial communications sequence between a DHCP client and a DHCP server.

The process used by DHCP services to manage address assignments when using leased address allocation operates as follows:

1. When the client is powered on, it broadcasts a message called a DHCPDISCOVER message intended for all DHCP servers.
2. Any available DHCP server responds to the DHCPDISCOVER message with a DHCPOFFER message, which contains an IP address.
3. If the client receives more than one DHCPOFFER message, it selects one and broadcasts a DHCPREQUEST message requesting the address offered by that server.
4. The server replies to the DHCPREQUEST message with a DHCPACK acknowledgement containing the lease period of the address.
5. As the lease expiration date and time approaches, the client attempts to renew its lease by communicating with the server that issued the lease. If the lease is not renewed in time, the client is disconnected from the IP network and must go through the process again, beginning with the issuance of a DHCPDISCOVER message.

The DHCP addressing service automates much of the process of obtaining a valid IP address and is particularly useful in environments where many sub-networks exist and stations are moved frequently. Installing DHCP services on multiple servers and assigning a valid pool of addresses to each server can assure administrators that clients will always receive an address consistent with the subnet to which they are connected, with no possibility of address duplication (see Figure 7.2).

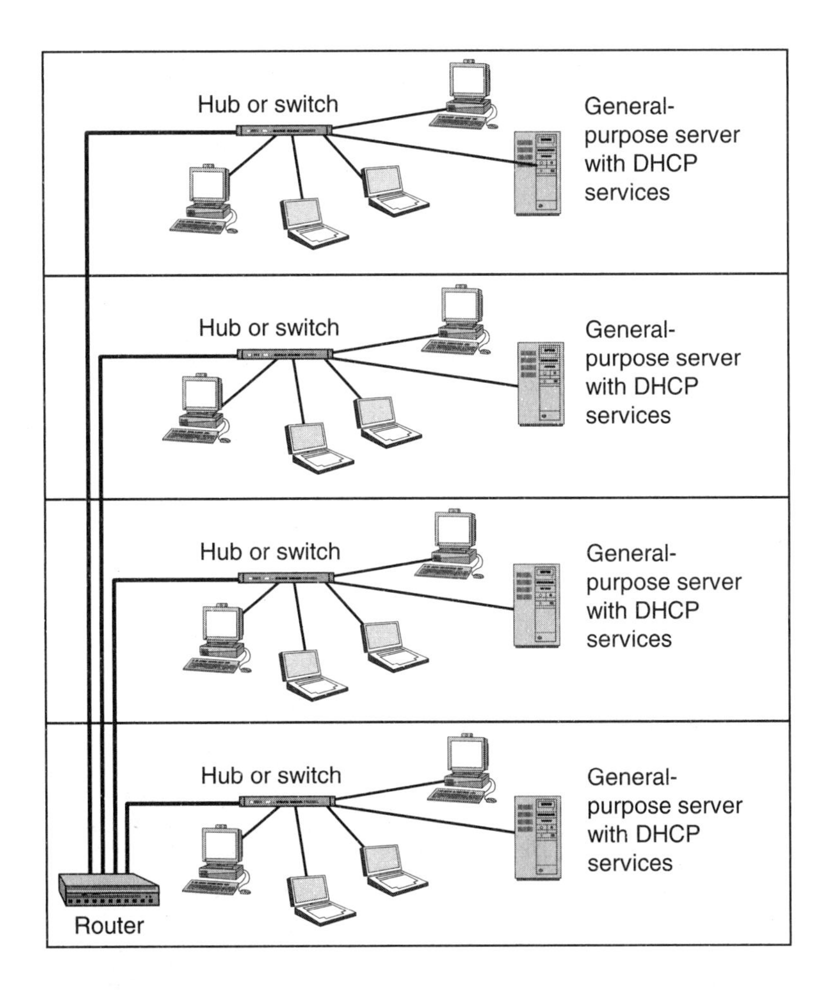

Figure 7.2 Multiple DHCP servers on a network.

Naming Services

General

Naming services are used to make it easier for users and administrators to identify network resources, including devices, applications, data files, and other users. For example, identifying and displaying a network printer using a name such as AccountingColorPrinter instead of a series of numbers simplifies printing operations and management, especially on large networks with thousands of resources.

The largest implementation of naming services is on the Internet, where names such as www.bicsi.org are used to access Internet resources instead of numeric IP addresses such as 206.96.226.20. Using a distributed hierarchical naming architecture, the Internet naming service—called the domain name system (DNS)—makes it possible to access resources such as Web sites using words instead of numbers. Although it is mandatory to use DNS on the Internet for resource naming, the DNS service can also be implemented on internal organizational networks. It enables users to access resources stored on internal servers in the same manner as resources found on the Internet. For example, the names accounting.bicsi.org, marketing.bicsi.org, and conferences.bicsi.org can be used to access three different sets of internal resources, residing on the same or different servers. If any of the servers are moved, replaced, or assigned new IP addresses, users can continue to access them without requiring changes to their station configurations, which simplifies network administration.

Domain Name System (DNS)

DNS was first outlined in RFC 882, dated November 1983. The original set of specifications was issued as RFC 1035 in November 1987. Since that time, a series of updates and extensions have been published, many of them dealing with security-related issues. The ability of DNS to keep pace with the growth of the Internet illustrates the scalability and reliability of its architecture, making it suitable for organizational networks of any size.

As with DHCP, the DNS services can be implemented on one or more dedicated or general-purpose servers. These are identified as DNS servers for naming service purposes. At its core, DNS is a database that can be divided and distributed among multiple servers. All the servers can participate in locating the numeric IP address of a requested resource identified by name. In a small organization located in a single facility, a single general-purpose server can provide all DNS services. Large organizations with multiple sites can implement a DNS server at each site, with each server providing name-to-address translation for local resources. If a user at one site provides the name of a resource located at another site, the local DNS server initiates communications with its counterpart to obtain the necessary information. This type of cooperation among DNS servers eliminates the need to keep all names and addresses in a single server, which improves both response times and fault tolerance.

In order to manage resource naming on a large scale, DNS imposes a naming hierarchy to differentiate between various levels of names, similar to the first, middle, and last names given to individuals. Each entry at a given level is called a domain, and all levels stem from a single root. For example, the Internet has several top-level domains—equivalent to the last names of individuals—such as .COM, .ORG, and .EDU.

When an organization or an individual requests a name for a Web site, they submit their choice of second level name together with the top level name for approval (e.g., bicsi.org or bicsi.com). Once the combination is verified to be unique (not previously assigned) by a registering authority, it becomes available to the organization or individual for use. The DNS database is then updated to associate the new name with the IP address corresponding to the Web server housing the site.

The choice of names assigned to the third and any subsequent levels are at the discretion of the owner of the domain name—www is commonly used for Web site identification (see Figure 7.3). A domain name can be a maximum of 255 characters in length and can contain both alphabetic and numeric characters, with any level name not to exceed 63 characters. No distinction is made between uppercase and lowercase letters. As with numeric IP addresses, periods are used to separate the levels in the name.

When DNS is implemented on an internal organizational network, at least one server must be assigned as the primary (master) DNS server. Additional servers, called secondary DNS servers, can also be configured to provide naming translation services, using copies of the database stored in the master DNS server. Secondary DNS servers are particularly useful in organizations with networks spanning extended geographic distances. In such cases, implementing

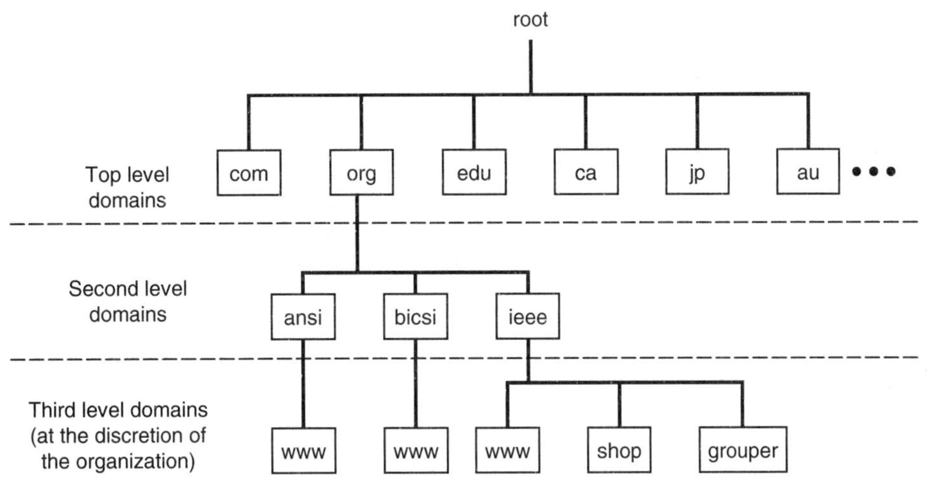

Figure 7.3 Domain name system hierarchy.

a secondary server at each site eliminates the need to access the master server over telecommunications circuits (see Figure 7.4).

The servers used for DNS services can be the same ones used for DHCP address assignments, or they can be different units. In some cases, a service called dynamic DNS (DDNS) can also be implemented. DDNS links the DNS and DHCP services to provide synchronization between the names and the IP addresses of network devices. If a device is assigned a new IP address by DHCP due to lease expiration, the DDNS service automatically updates the DNS database to reflect the new address of the device.

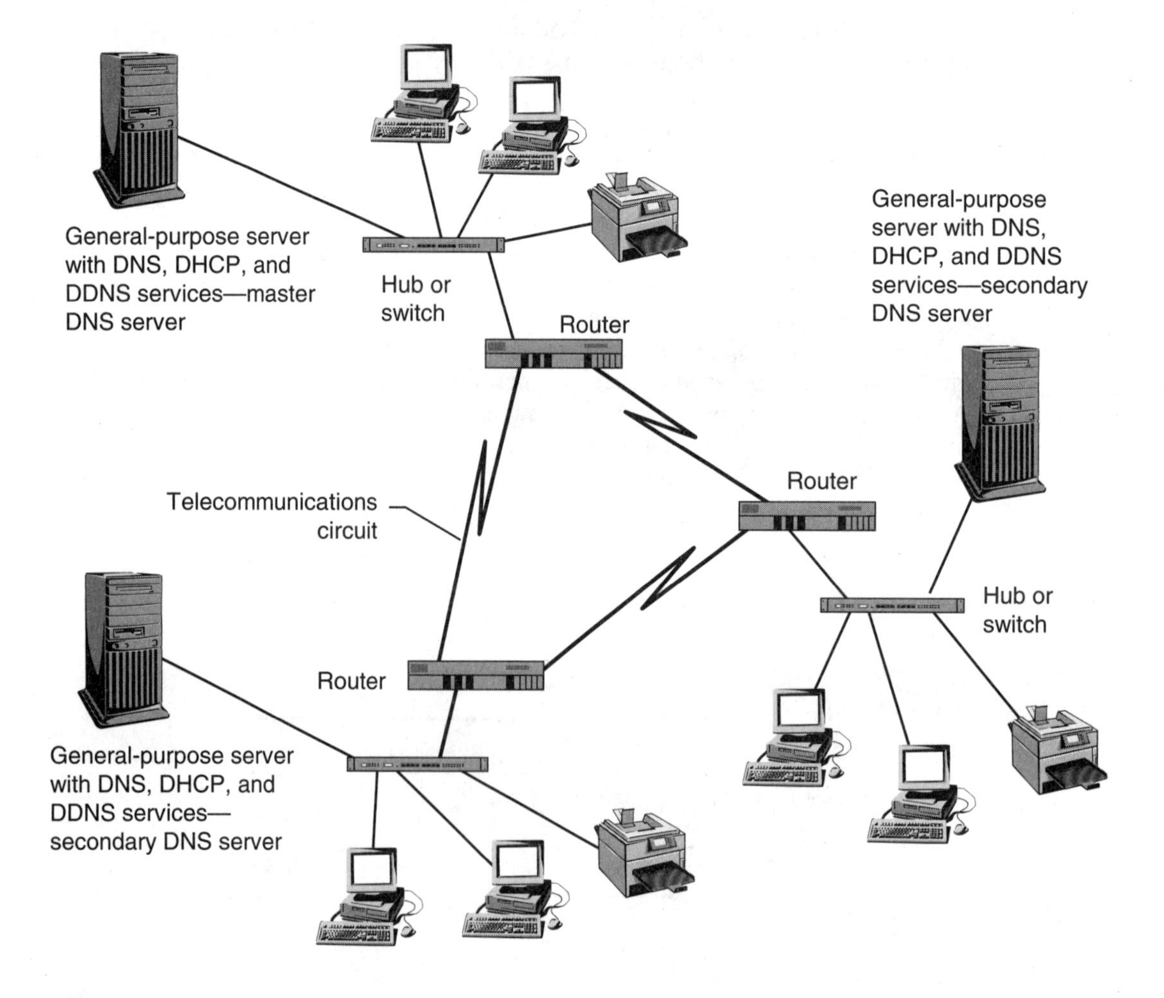

DDNS = Dynamic domain name system
DHCP = Dynamic host configuration protocol
DNS = Domain name system

Figure 7.4 Master and secondary domain name system servers.

Directory Services

General

The function of a network directory service is to present a comprehensive and up-to-date listing of the resources available on a network. Such listings typically include devices, software applications, data files, individual users, and groups of users. A large organizational network can have millions of entries in its directory, making directory service management a complex but critical administrative task. Without a directory, users would be obliged to remember where to find each resource they require on the network (e.g., individual servers, disk drives, and printers). A directory service presents all resources on-screen in a graphical format, using the names assigned to the resources by network administrators. Regardless of where a given resource is physically located, it can be selected and accessed using the directory service.

Directory services are an extension of the numbering and naming services described in previous sections. In all cases, a database is used to store the information used by the service. In the case of numbering services, the database contains the numeric addresses available or assigned to devices. The database used in naming services contains the names assigned to the numeric addresses. The directory services database contains both the addressing and naming information, in addition to many other details, associated with each resource. Additionally, the directory database makes it possible to relate resources to each other, creating resource hierarchies. For example, a software application consists of a collection of files typically stored on a disk drive connected to a server, with access permission granted to some or all network users. The type of relationship between various resources—in this case, a software application, files, a disk drive, a server, and a group of users—can be tracked and viewed on-screen using directory services software.

It is possible to organize a network directory on the basis of different criteria, such as geographic location or organizational structure. Figure 7.5 illustrates a typical network. Figure 7.6 illustrates a directory of resources for the network shown in Figure 7.5, organized on the basis of departmental groupings. The same network is shown in Figure 7.7 organized on the basis of the physical location of resources.

Historically, directory services have been integrated into the network operating system (NOS) installed on each server in the network. Since the NOS directs all resource sharing activities, it is best suited to building and maintaining a resource directory. However, this has led to proprietary implementations of directory services, with each NOS unable to obtain information on resources controlled by another vendor's NOS. To facilitate the exchange of directory-related information, the IETF has published the lightweight directory access protocol (LDAP), which is intended to be used as a universal tool for locating resources on all types of networks, including the Internet.

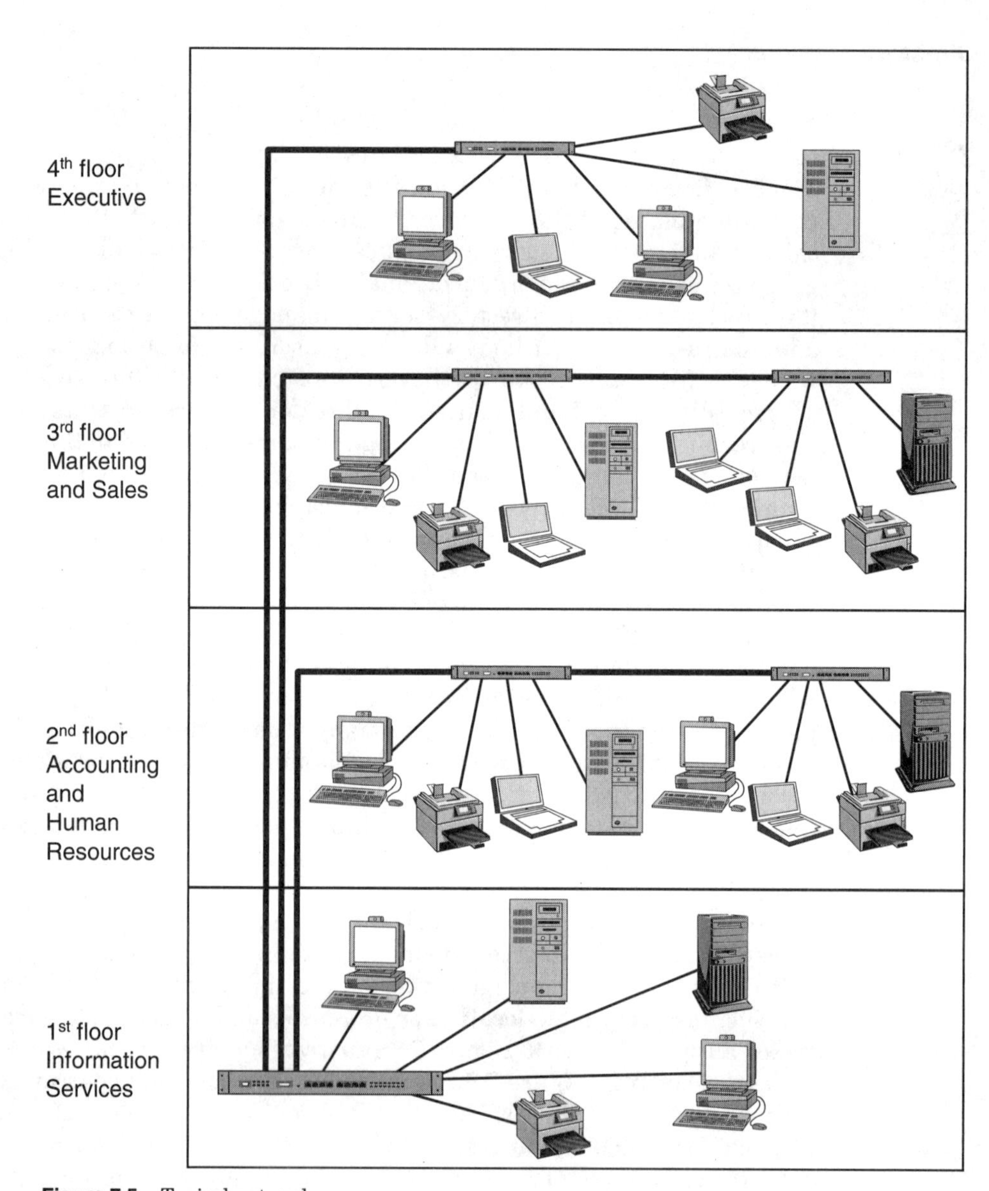

Figure 7.5 Typical network.

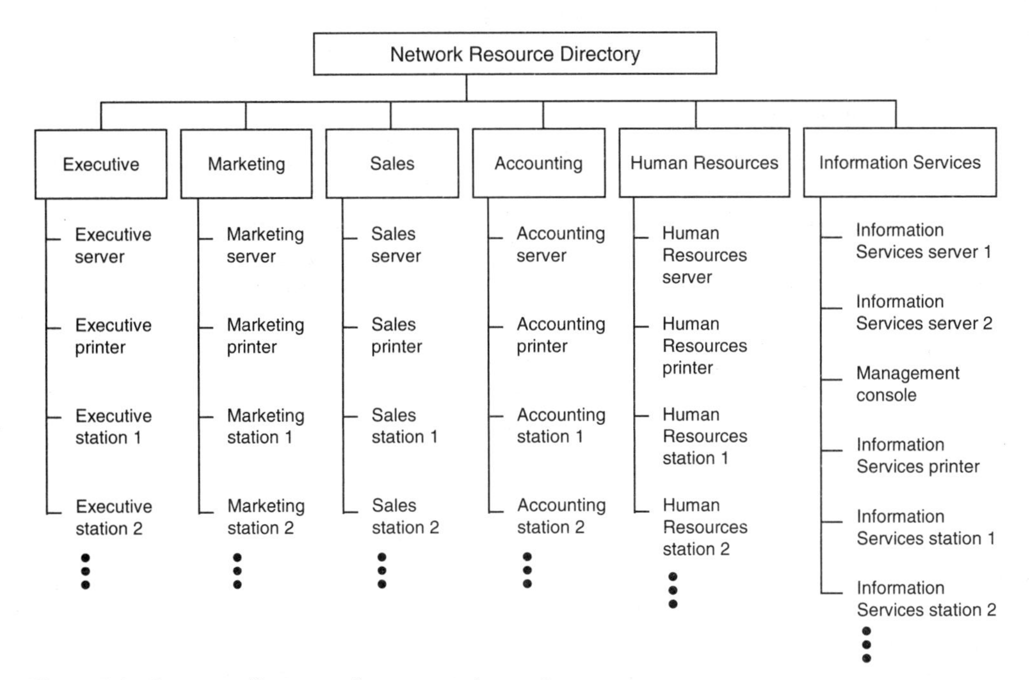

Figure 7.6 Resource directory—departmental grouping.

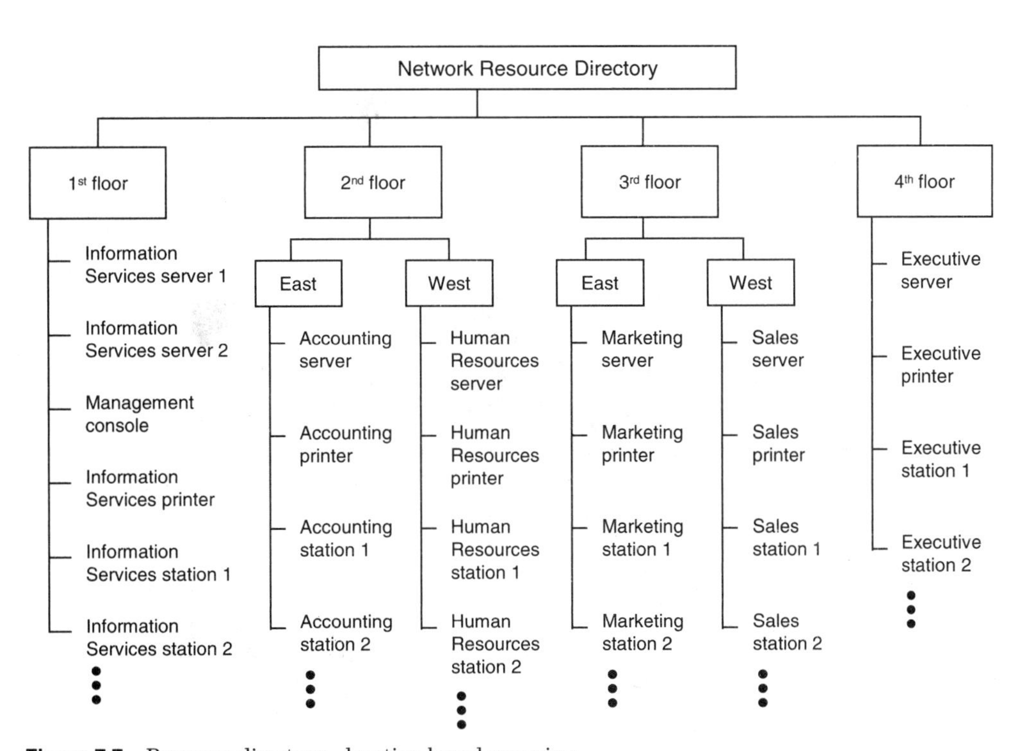

Figure 7.7 Resource directory—location-based grouping.

Lightweight Directory Access Protocol (LDAP)

The initial set of LDAP specifications was published in July 1993 as RFC 1487. An update was issued in March 1995 as RFC 1777 and the current version is LDAP version 3 (LDAPv3), published as RFC 2251 in December 1997. The term lightweight is used to indicate that LDAP is a less complex implementation of the directory access protocol (DAP) used in X.500, the directory standard published by the International Telecommunication Union-Telecommunication (ITU-T).

LDAP is a communications protocol designed to enable all types of applications to access the data contained in a directory database. By making their proprietary directory services LDAP-compliant, vendors make it possible for users to obtain information from their NOS directories using a variety of devices and applications (e.g., any type of Web browser running on a desktop, laptop, or handheld computer). This allows external parties, who may not be using the same applications or NOSs that are used internally, to access an organizational network and obtain listings of resources extracted from the directory for various external groups, such as contact information. Any time the directory information is updated (e.g., when personnel changes are made) the new information appears on the listings presented to all users, including external parties, since all data resides in the same directory database.

All network directory services, including LDAP-compliant ones, must be able to accommodate and track a large number of resources. In many cases, some of the entries in the directory database require frequent updates to reflect the changes in the network environment. When the organizational network contains hundreds or thousands of servers, it is not practical to place all directory information in a single server, from both a response time and a fault tolerance point of view. For this reason, directory services use a distributed architecture, where multiple servers throughout the network are designated to provide directory services. Four types of tools are used to maintain consistency and data integrity in a distributed directory services environment.

- Partition tools are used to extract parts of the complete directory services database for distribution to various servers.
- Replication tools are used to produce copies of the complete database or any of its partitions to improve performance or fault tolerance.
- Directory synchronization tools are used to maintain consistency between the complete database, all partitions, and all replicates. This ensures that the same information is presented to users, regardless of which database is accessed.
- Time synchronization tools are used to ensure that all changes to database entries are processed using a common clock, which is especially important in cases where the servers containing database information are distributed across multiple time zones.

Internet Services

General

Many of the services originally developed for use on the Internet can also be implemented on internal organizational networks. In environments where users require frequent access to the Internet, it is possible to configure the internal network to accommodate the same tools used for Internet activities, such as Web browsers and electronic mail (e-mail) readers. This simplifies network access for the users and network management for the administrators, since a common set of software can be used in place of different applications for internal and Internet access.

Once an organizational network has been equipped with Internet access, users and administrators are typically provided with software tools to enable some or all of the following services:

- Web site access
- E-mail
- Newsgroup access
- File transfer
- Remote terminal

Many organizations use internal Web servers—commonly referred to as an intranet—to increase the efficiency of communications between network users and groups. Internal e-mail and newsgroups are similarly used to exchange messages without geographic or time zone constraints. Large volume file transfers and connecting to network devices using remote terminal services are activities commonly performed by network administrators. In all cases, the applications used to provide the services are the same ones used on the Internet, as follows:

- Hypertext markup language (HTML) and hypertext transfer protocol (HTTP) services enable the creation and operation of Web sites.
- Simple mail transfer protocol (SMTP), post office protocol version 3 (POP3), and Internet message access protocol version 4 (IMAP4) enable the operation of e-mail services.
- Network news transfer protocol (NNTP) services enable the operation of newsgroup postings.
- File transfer protocol (FTP) services enable efficient file transfer between network devices.
- Telnet services enable a network station—described as a remote terminal for the purposes of this service—to use the network in place of a direct cable connection to connect to another network device, such as a router.

Hypertext Markup Language (HTML) and Hypertext Transfer Protocol (HTTP)

HTML can be described as a collection of formatting codes or indicators inserted into files to make them readable by Web browsers. Both the IETF and the World Wide Web Consortium (W3C) publish standards related to HTML, such as RFC 2854, published by the IETF in June 2000 and HTML version 4.01, issued in December 1999 by the W3C. HTML makes it possible to convert the files produced by various applications into a format that can be viewed by any Web browser, eliminating the need to use identical software applications to share files between individuals and organizations.

HTTP is the protocol used to transfer HTML content from one device to another over a network or an internetwork. RFC 2616, published in June 1999, defines the specifications for HTTP version 1.1. The device containing the HTML content to be transferred is referred to as the Web server, and the Web browser issuing an HTTP request for HTML content is called the Web client (see Figure 7.8).

An organization can use both HTML content creation software and Web server software to implement one or more internal Web servers. In most cases, security features in the Web server or NOS software can be used to assign or deny access rights to any part of the internal Web site. The Web server software can be installed on a dedicated device, or it can run as an additional application on an existing general-purpose server with spare processing and storage capacity. It is also common for vendors to implement Web server software in devices such as network printers, to make it easier for administrators to configure and monitor the devices over the network, using familiar Web browser and graphical tools instead of proprietary device command languages.

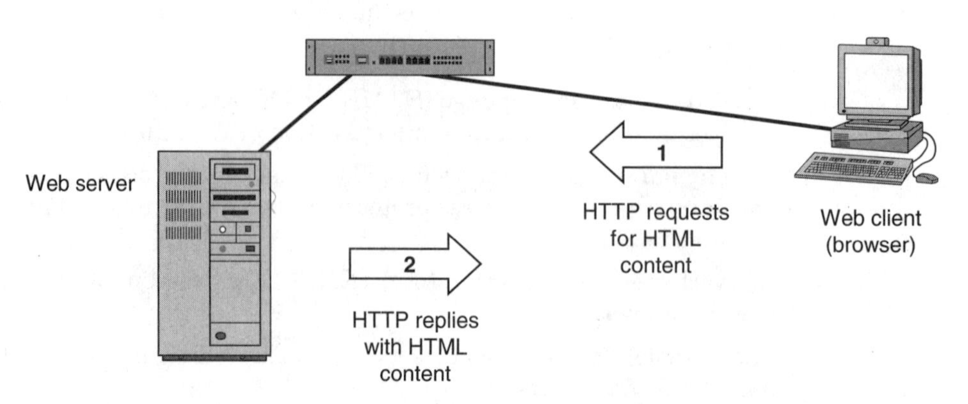

HTML = Hypertext markup language
HTTP = Hypertext transfer protocol

Figure 7.8 Web server and client.

Simple Mail Transfer Protocol (SMTP), Post Office Protocol version 3 (POP3), and Internet Message Access Protocol version 4 (IMAP4)

SMTP is one of the earliest Internet application protocols, published as RFC 788 in November 1981. It has been updated and extended numerous times, most recently as RFC 2821, issued in April 2001. SMTP provides a transport infrastructure for e-mail messages, in which mail servers use the protocol to transfer messages from source to destination servers over a network or internetwork. SMTP provides a common messaging format that can be implemented in any type of messaging software, which makes it possible to exchange messages between different organizational e-mail systems.

The POP3 and IMAP4 protocols are intended for use by e-mail clients such as mail readers to access mail services running on a server. POP3 was issued in November 1994 as RFC 1725 and updated in May 1996 as RFC 1939. IMAP4 was issued in December 1994 as RFC 1730 and revised in December 1996 as RFC 2060. IMAP4 provides more features for manipulating mail messages on servers than POP3, but both are commonly used in mail reader software. When compared to SMTP, POP3 and IMAP4 can be described as protocols used to retrieve e-mail, whereas SMTP can be characterized as a protocol used to send e-mail.

In large organizations, it is possible for various sites or users to be equipped with different types of e-mail software. An internal messaging infrastructure based on SMTP, POP3, and IMAP4 makes it possible for all users to send and receive messages without the need to install a single type of messaging software on all mail servers and user stations (see Figure 7.9).

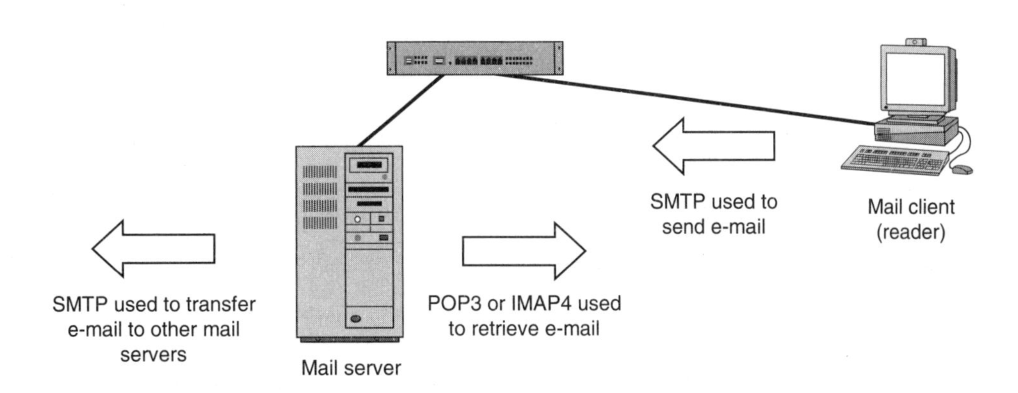

IMAP4 = Internet message access protocol version 4
POP3 = Post office protocol version 3
SMTP = Simple mail transfer protocol

Figure 7.9 Mail server and client.

Network News Transfer Protocol (NNTP)

NNTP was published in February 1986 as RFC 977. It can be described as the equivalent of SMTP for newsgroup postings. A newsgroup is a special type of messaging environment, in which messages are posted to a common environment for public viewing instead of being sent to one or more specific mailboxes, as is the case with e-mail messaging. The term public newsgroup is used to describe any of the thousands of newsgroups accessible on the Internet—the entire public newsgroup environment is also referred to as Usenet.

In an NNTP environment, one or more newsgroups function as the electronic equivalent of bulletin boards or meeting rooms, where posted messages related to various topics are sequenced in chronological order and stored on a news server. The software used on stations to access the newsgroups stored on a news server is called a newsreader. A newsreader can form part of a mail reader or Web browser application or it can be a separate software package. NNTP is used by the newsreader software to post messages to and retrieve messages from the news server(s) over a network or internetwork.

As is the case with Web servers, an organization can implement news server software on one or more of its servers (e.g., the same server(s) containing the organization's internal Web site[s]). This makes it possible for network administrators to set up various public and private discussion areas for some or all network users (see Figure 7.10).

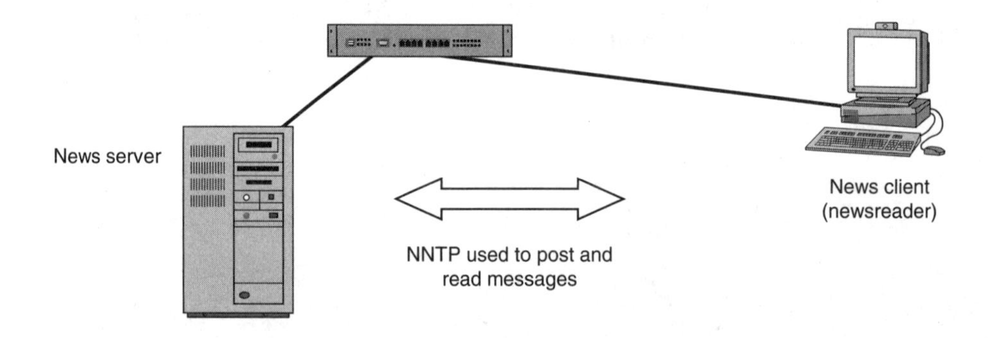

NNTP = Network news transfer protocol

Figure 7.10 News server and client.

File Transfer Protocol (FTP) and Telnet

FTP and Telnet were both initially described in RFCs issued in 1971, making them the earliest Internet applications. Both have been updated many times, notably as RFC 959 in October 1985 for FTP and RFC 854 in May 1983 for Telnet. FTP and Telnet are associated with network management activities. Administrators routinely transfer files from one device to another using FTP and configure various network devices after connecting to them from any station on the network using Telnet.

FTP is a general-purpose file transfer service, as opposed to HTTP, SMTP, and NNTP, which are used to transfer HTML content, e-mail messages, and newsgroup postings, respectively. As is the case with Web, e-mail, and news servers, software to enable FTP server operations can be installed on a dedicated or general-purpose server. The station used to send and receive files from an FTP server must be equipped with FTP client software, which can be found in applications such as Web browsers or installed as a separate utility. Since FTP is not associated with a specific NOS or station type, it can be used to transfer files between devices connected to different types of networks and running different operating systems, which is common on large organizational networks.

Telnet is a general-purpose remote terminal service, designed to make a connection over a network appear to be a direct cable connection between two devices. Historically, specific types of terminals were required to access a given type of centralized computer. Telnet provides a service called a network virtual terminal (NVT), where the two devices to be connected—called the terminal and the host—use a common protocol to exchange messages typically consisting of commands from the terminal to the host. In current implementations, Telnet is commonly used to remotely configure network access devices such as routers, which can be configured to accept Telnet-based connections after a password is provided.

Voice Applications

General

Network voice applications can be described as technologies designed to transport traffic traditionally associated with telephone networks, such as conversations between individuals, voice mail messages, and fax transmissions. Terms such as Internet telephones, IP telephony, and VoIP are used to describe such technologies, since most implementations operate over IP-based networks and internetworks.

When the data traffic being transported consists of digitized fragments of human speech, it is critical for the network to deliver all fragments without loss or corruption and with minimum delay. Failure to do so results in a noticeable degradation of the speech quality heard by the receiver. This type of traffic, characterized as both loss-sensitive and delay-sensitive, is considered a challenge for networks designed to transfer software applications and data files between devices. In order to provide an acceptable level of service—sometimes referred to as toll quality VoIP—networks must be able to identify, prioritize, and dedicate resources to the traffic generated by voice applications, a set of technologies often described as IP quality of service (QoS).

The advantages of implementing voice applications on IP networks include cost savings resulting from a reduction in long distance charges and a unified networking infrastructure for both voice and traditional data applications. Another benefit resulting from the integration is the ability to combine voice and data services within applications. For example, a Web site can make it possible for visitors to speak to customer service staff while viewing a page, using a headset connected to the device running the Web browser application.

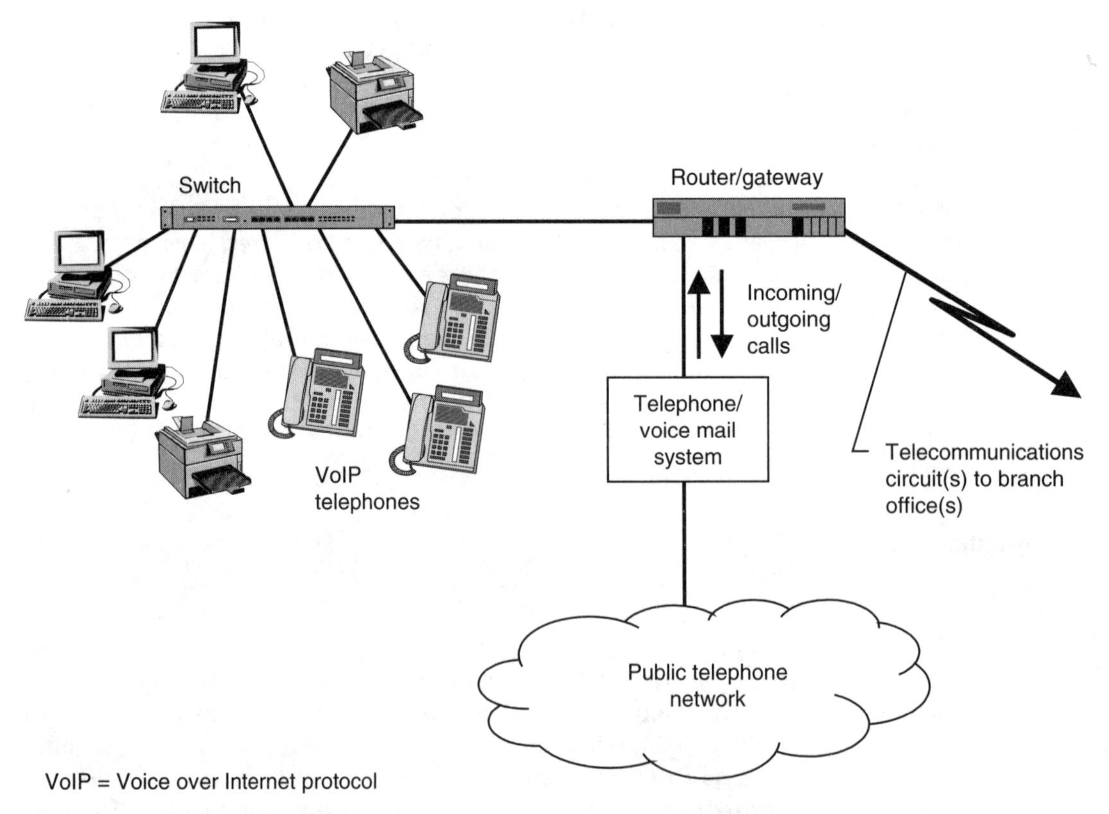

Figure 7.11 Voice over IP infrastructure.

Voice applications can be implemented internally, externally, or both on an organizational IP network (see Figure 7.11), as follows:

- With internal use, all telephone and voice mail communications between staff—within a single facility and between various sites—are removed from the existing voice network and transported over the organizational IP network. Voice mail is stored on servers connected to the IP network.
- With external use, incoming and outgoing telephone calls—those coming from or going to individuals outside the organization—are transferred between the organizational IP network and the public telephone network using hardware and software commonly described as a gateway.

Voice application standards

Standards for the harmonization of telephone networks have traditionally been issued by the ITU-T for implementation by telephone companies in all nations. By comparison, data networking and applications standards are issued by numerous organizations such as the IETF, the Institute of Electrical and Electronics Engineers, Inc.® (IEEE®), national organizations, various industry-based alliances, and individual vendors. Since VoIP applications combine elements found in both voice and data networking, many standards covering a broad range of issues have been proposed, issued, or are under development by a variety of organizations. For example, there are many user productivity features offered by various telephone system vendors that can be programmed into the buttons on a modern office telephone. If VoIP telephones are to replace all existing units, similar features must be standardized and made available on the VoIP telephones and the IP network to make the transition acceptable to users.

Two of the standards associated with network voice applications include:

- G.711, a coding and compression standard for converting speech into a 64 kilobit per second (kb/s) data stream, issued by the ITU-T.
- Real-time transport protocol (RTP) and real-time transport control protocol (RTCP), intended for delay-sensitive traffic on IP networks, issued by the IETF in January 1996 as RFC 1889.

Video Applications

General

Like voice applications, network video applications (also referred to as networked multimedia) are characterized as loss-sensitive and delay-sensitive. An excess level of either condition results in a visible degradation of the video image.

In networked multimedia, digital video applications are used to create and distribute video sequences of various duration, with or without a synchronized audio stream. Such applications can generate sequences in real-time, as in the case of videoconferencing, or can operate in store-and forward mode, where the generated sequence is stored on the network for later viewing. One significant difference between voice and video applications is the much larger amount of storage and data transfer rates required to provide users with video-based services.

Video sequences consist of a series of still images called frames. When displayed in sufficiently rapid sequence—15 to 30 frames per second—a sense of motion is created. A frame rate of 24 to 30 frames per second is described as full-motion video, in which all motion is fluid and smooth, with full synchronization between the audio and video components. Compression is critical in video applications, since a single minute of high-quality uncompressed video can consume 500 megabytes (MB) of storage.

To reduce file sizes and minimum acceptable data transfer rates, digital video sequences are typically displayed in a window that occupies a fraction of the total available display area on a monitor. A commonly used video window measures 352 pixels across by 240 pixels down for a total of 84,480 pixels (352 × 240), where each pixel represents a dot on the screen. If photo-realistic color is required, each pixel must have three bytes of color information to describe its red, green, and blue (RGB) components. Therefore, the 84,480 pixels that make up the video window require 253,440 bytes of storage (84,480 pixels × 3 bytes/pixel) for each frame of video. If the video is intended to be full-motion, 30 frames per second may be required, which translates into a storage requirement of 7.6 MB for each second. A number of alternatives exist to reduce this value.

- The frame rate can be reduced to 15 frames per second, which may result in a less smooth but visually acceptable sequence. This reduces the storage requirement by half, to 3.8 MB for each second.
- The video window can be reduced to 176 pixels by 120 pixels, which is one-fourth the size of the window measuring 352 by 240. Combined with the 15 frames per second frame rate, the storage requirement is reduced to 0.95 MB for each second.
- The video sequence can be compressed prior to storage. For example, if a compression ratio of 6:1 is used, the storage requirement is reduced to 0.16 MB for each second.

Networked multimedia applications can be classified using two criteria.

- The number of user stations simultaneously displaying the same video sequence, which can be one (point-to-point) or more than one (point-to-multipoint).
- The direction of transfer of the video sequence, which can be one-way (unidirectional) or two-way (bidirectional).

Using these two criteria in combination, four types of applications can be described.

- Point-to-point unidirectional applications
- Point-to-multipoint unidirectional applications
- Point-to-point bidirectional applications
- Point-to-multipoint bidirectional applications

Each of these is described in the following sections.

Point-to-point, unidirectional applications

Point-to-point unidirectional applications represent the least complex implementation of networked multimedia. In a point-to-point unidirectional connection, the content transmission is in one direction between two devices—most often, from a server to a station. The content can be produced in real-time or played back from an existing file stored on the server (see Figure 7.12).

Some examples of point-to-point, unidirectional multimedia applications include:

- Live images from remote cameras.
- Multimedia-enabled e-mail.
- Digital documents with embedded multimedia elements.
- Video servers.

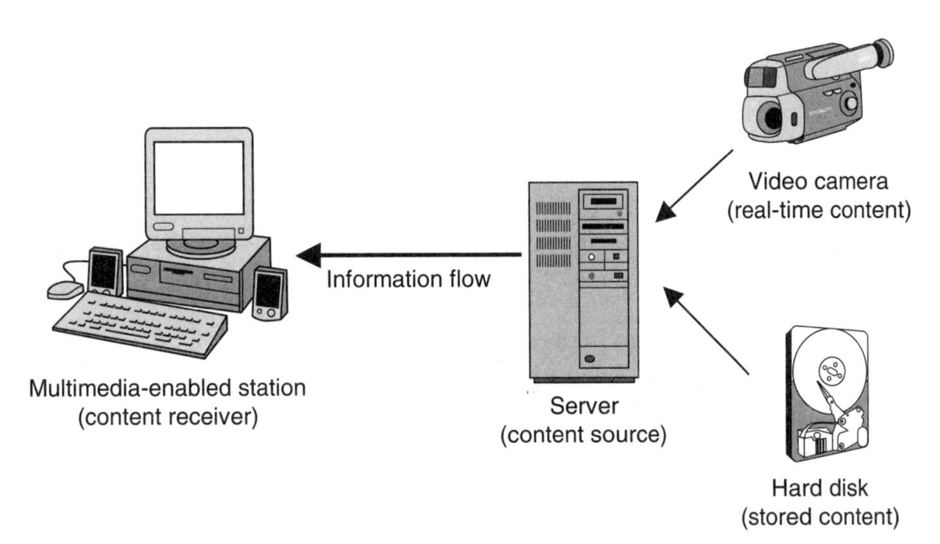

Figure 7.12 Point-to-point, unidirectional transmission.

In a video server application, various video segments are centrally stored in a compressed format. The user initiates the viewing process by choosing the desired video clip from the server. The clip is transmitted to and decompressed at the user's station and displayed on-screen.

Point-to-multipoint, unidirectional applications

In point-to-multipoint unidirectional applications, the same content is transmitted to multiple stations at the same moment in time. As before, the content can be real-time or stored (see Figure 7.13).

Some examples of point-to-multipoint, unidirectional multimedia applications include:

- Video servers.
- Distance learning.
- Digital broadcasts.

Distance learning is a common point-to-multipoint, unidirectional network application. Classes are video taped, digitized, stored, and broadcast at specific times over the network to students at various stations. Digital broadcasting delivers real-time events, such as important announcements, to multiple stations simultaneously over the network.

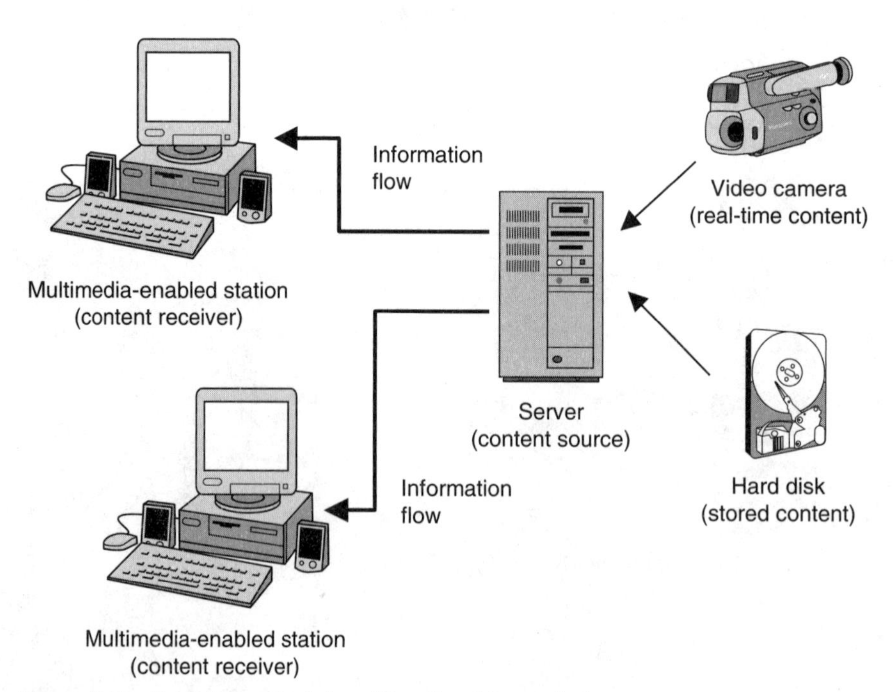

Figure 7.13 Point-to-multipoint, unidirectional transmission.

Point-to-point, bidirectional applications

Point-to-point bidirectional applications simultaneously capture and deliver real-time content at both ends of a connection. Both end devices are capable of generating and receiving multimedia content at the same time (see Figure 7.14).

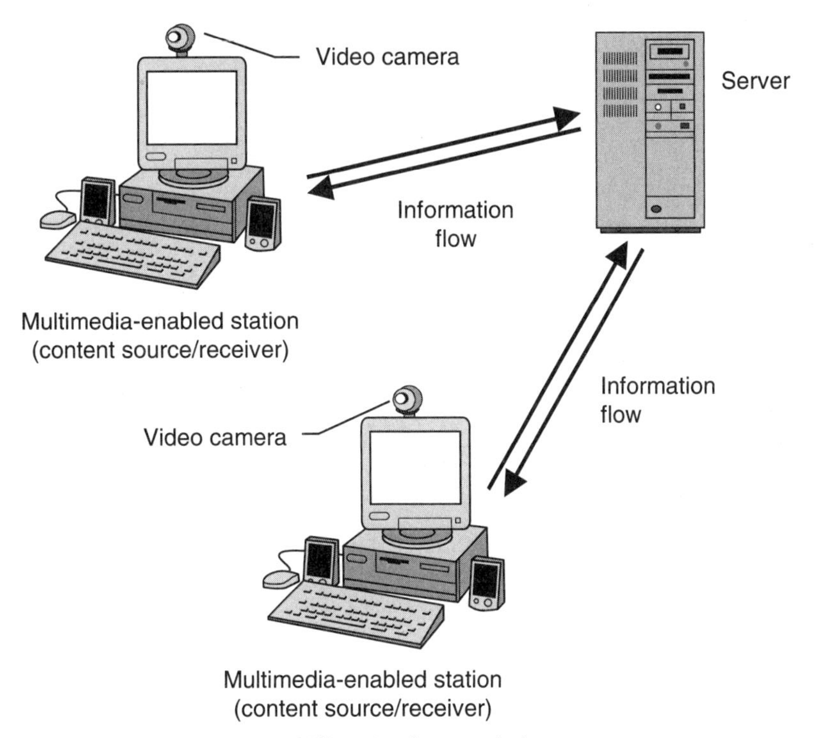

Figure 7.14 Point-to-point, bidirectional transmission.

Some examples of point-to-point, bidirectional multimedia applications include:

- Audio and video conferencing.
- Shared whiteboards.
- Application sharing.

Audio and video conferencing applications provide an interactive environment for communications between the two connected users. Shared whiteboards, also referred to as collaborative workspaces, provide a shared, on-screen window on which either user can enter information. The entry is then immediately displayed on both screens. Application sharing allows the user at one end to launch an application that the user at the other end can also view and manipulate.

Point-to-multipoint, bidirectional applications

In point-to-multipoint, bidirectional multimedia applications, multiple devices are capable of sending and receiving information to and from one another simultaneously in real-time (see Figure 7.15).

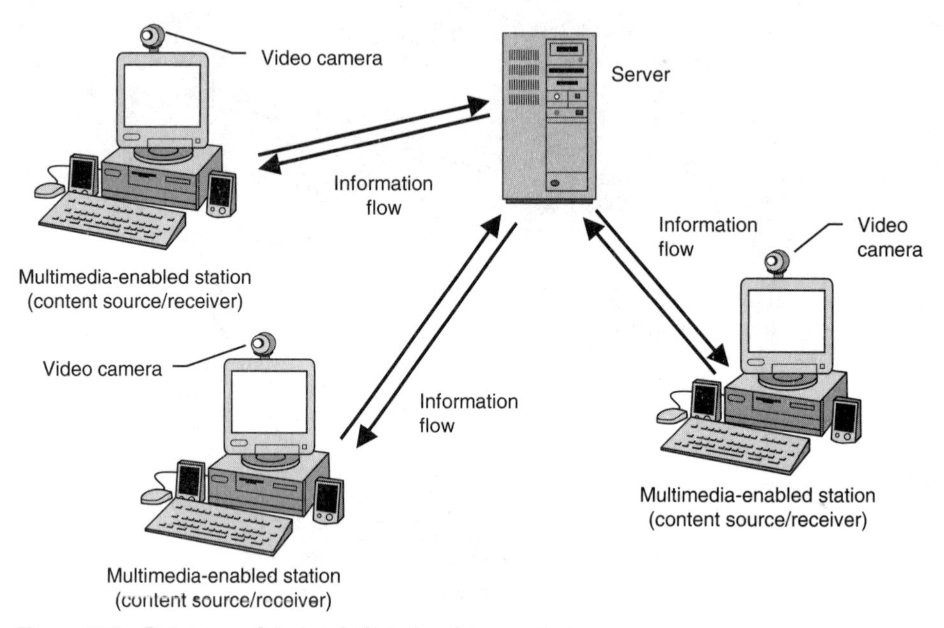

Figure 7.15 Point-to-multipoint, bidirectional transmission.

NOTE: Servers used in the type of point-to-multipoint application shown in Figure 7.15 are sometimes referred to as multipoint conference units (MCUs).

Some examples of point-to-multipoint, bidirectional multimedia applications include:

- Interactive video.
- Multi-party video conferencing.

Interactive video systems, such as video kiosks, are able to deliver video to multiple stations simultaneously. Each viewer can interact with the video session independent of all other viewers (e.g., when responding to an on-screen query during the session). In multi-party video conferencing applications, conference calls between three or more users can be initiated over the network, with each user receiving video and audio transmission from all other participants.

Video application standards

The core set of standards for networked multimedia has been developed by the ITU-T and adopted by most vendors in their applications. The set is made up of four series of specifications.

- The T.120 series of specifications describe application sharing, where multiple users view and manipulate the same on-screen content.
- The H.320 series of specifications describe videoconferencing over a digital telecommunications infrastructure based on integrated services digital network (ISDN) technologies. ISDN is described in detail in Chapter 4: Network Connections.
- The H.323 series of specifications describe video-based services in a LAN environment.
- The H.324 series of specifications describe multimedia applications over an analog telecommunications infrastructure, such as the traditional public switched telephone network (PSTN).

Chapter

8

Network Security

Chapter 8 describes the fundamental elements used to secure organizational network resources, including equipment, software, and data files against unauthorized activities.

Overview

General

Security is a significant consideration in network design, especially in cases where the organizational network is connected to the Internet for inbound or outbound access. The Internet is a shared global internetwork, which is not owned or controlled by any single entity. No liabilities can be assigned if the Internet is used to gain unauthorized access to an organization's internal network.

Regardless of the presence or absence of Internet connections, lack of adequate security makes any organizational network vulnerable to:

- Data theft.
- Unauthorized data modification or deletion.
- Identity theft.
- Disruption of network operations.

The objective of network security is to protect network assets, both tangible and intangible. The goal is to ensure that the equipment, software, and data files on an organizational network are accessible only to authorized individuals. A combination of physical safeguards, technical tools, educational efforts, and organizational policies are used to implement security.

On most networks, security can be classified into two categories:

- Internal security, which prevents unauthorized access to network resources from a local station—a station that is physically connected to the network medium.
- External security, which prevents unauthorized access to network resources from a remote station—a station that is connected over a telecommunications circuit or the Internet.

Messages sent over the Internet are assumed to be accessible to anyone, since they are no longer confined to an internetwork under the control of the organization. In such cases, encryption technologies are used to ensure the following:

- The contents of a message cannot be viewed or modified prior to the arrival of the message at its destination.
- A message can be validated to confirm its source. This prevents forgery and makes it impossible for the sender to deny ownership of the message, referred to as nonrepudiation.

The technologies used to create secure communications channels over the Internet are called virtual private networks (VPNs) and are described later in this chapter. Other security requirements common to most organizational networks include the following:

- The individuals accessing the network must provide proof of their identities before they are granted access to network resources. This process is called authentication.
- After authentication, users are granted access to specific resources—devices, software, Web sites, or data files—based on their identities. This process is called authorization. The database that keeps track of user access privileges is often referred to as an access control list (ACL).
- The network keeps track of all or failed authentication and resource access attempts for review by the network administrator. This process is called auditing or intrusion detection.
- All new and modified files on the network are scanned for viruses. The scanning can be confined to network storage devices or it can cover both network and station storage units. On networks connected to the Internet, both incoming and outgoing files can be scanned, to ensure that neither internal users nor the individuals they communicate with are exposed to harmful programs.
- The term firewall is commonly used to describe the protective equipment and software placed between an organizational network and the Internet.

Encryption

General

Data encryption is a fundamental element of network security. One basic function is to safeguard the passwords and other authentication tools required to gain access to network resources. In cases where a password is stolen and unauthorized access is obtained to the organizational network and data, other forms of encryption can be used to make critical files unreadable to anyone other than authorized users.

Organizational data can be stolen from two places.

- From its stored location
- During its transmission from one device to another

Absolute security can only be achieved when the data is stored on a guarded device, which is completely isolated from any other device. However, this renders the data inaccessible from any other location. Networks were created as a solution to the problem of data isolation. Therefore, the challenge is one of securing data in an environment where resources are shared.

Since the concept of privacy is a contradiction to the concept of common access, cryptographic mechanisms are used to limit the number of individuals who can view the contents of a stored file, even when the file itself can potentially be retrieved by anyone. Privacy is lost by sharing, but it is replaced by confidentiality. This assures the authorized users of the data that only they can view the confidential material.

Cryptographic mechanisms also protect organizational data that is transmitted from one device to another. For all such transfers, the sender needs to be assured that the data:

- Was not modified after being sent and prior to its arrival at the destination.
- Can only be viewed at its destination.

The receiver of the transferred data requires the following assurances:

- All of the received data was transmitted by the sender and not an imposter—the sender cannot claim otherwise.
- The data was not modified after being sent and before its arrival at the destination.
- The data can only be viewed at its destination.

The term data integrity describes the means taken to ensure that received messages are complete and were not altered as they passed through a network. Nonrepudiation refers to proof that the message originated from the claimed source. Cryptographic mechanisms can provide both of these services, in addition to encrypting messages to make them unreadable by unauthorized individuals.

Data integrity is established by encrypting a message after adding a mathematically derived value—called a hash or a message digest—to its contents. When the recipient decrypts the message, a similar calculation is performed on its contents. If the message digest value obtained matches the one inscribed in the message, the recipient is assured that the message was not altered in any way after being sent.

Nonrepudiation requires a mechanism to link a given message to its originator in an indisputable way. With paper documents, an individual's signature establishes proof of ownership. However, a signature alone cannot prove that a document was not altered after being signed. For such assurance, the document can be enclosed in a tamper-proof envelope with a signed seal. A digital signature, combined with the data integrity process, provides this dual assurance.

When messages are exchanged over a network between individuals known to each other, digital signatures and data integrity are sufficient for trusted communications. Most commercial transactions, however, are between buyers and sellers who do not know each other. To avoid fraud, both the buyer and the seller need to verify each other's identity. The cryptographic mechanism that enables this verification is referred to as a digital certificate.

Digital certificates are issued to buyers and sellers by independent, trusted certificate authorities (CAs) after a thorough verification of the identities of the parties requesting certificates. When network communications take place between parties who do not know (or trust) each other, digital certificates are sent with the messages. When such messages arrive, the recipients can confirm the identities of the senders through their certificates. A certificate contains the digital signatures of both the sender and the CA.

A message that includes a digital certificate and data integrity encryption can be trusted to have been sent by the claimed source and not tampered with. The sender cannot deny having sent such a message because of the link between the certificate and the digital signature of the sender.

Encryption can be broadly divided into two categories—private (or symmetrical) key encryption and public (or asymmetrical) key encryption.

In private key encryption, a message is mathematically transformed using a value that only the sender and the receiver possess—the key (see Figure 8.1). It is assumed that the parties involved have obtained the key in such a way as to ensure that security was not compromised (e.g., through a manual exchange at a meeting).

The most popular private key encryption scheme is the data encryption standard (DES), developed by IBM in the 1970s. DES transforms a message into encrypted data called ciphertext. The receiver uses the same key as the sender to decrypt the ciphertext and reconstruct the original message. A more secure version of DES called triple DES (3DES) is also available, but requires more processing to take place in the sending and receiving devices.

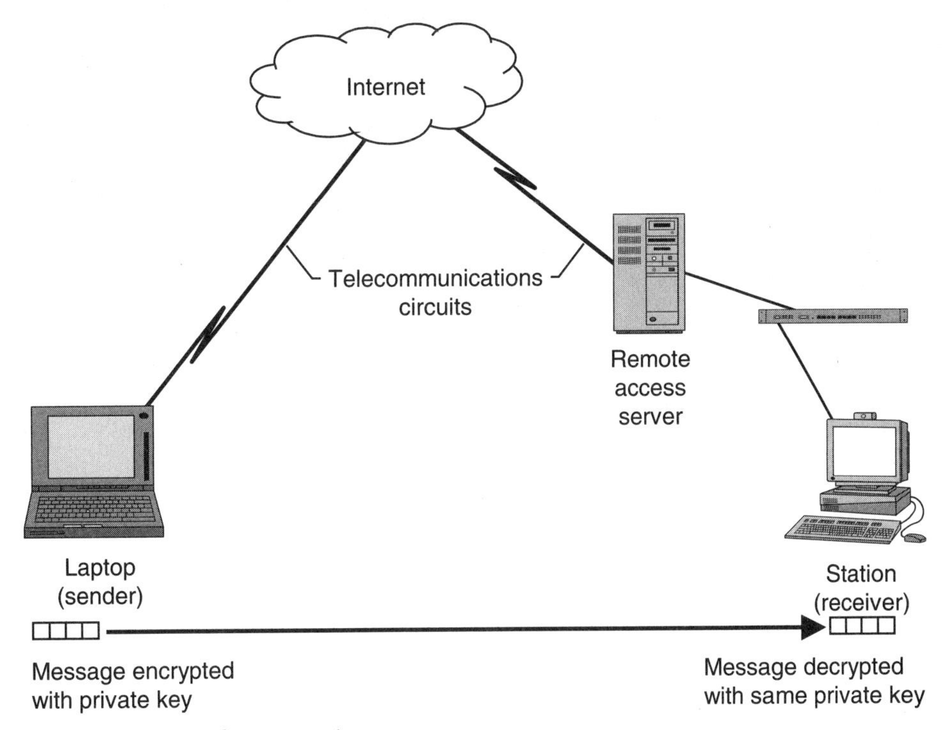

Figure 8.1 Private key encryption.

In public key encryption, each user has two mathematically related keys. One is held privately and the other is placed in a secure public directory. Anyone who wishes to send a message to a given user encrypts the message with the user's public key. When the message is received, the user decrypts it with the private key. In this system, no exchange of keys is ever required, which reduces the possibility of compromising security (see Figures 8.2 and 8.3).

The most popular public key encryption scheme is RSA, named after its developers—Rivest, Shamir, and Adleman. The RSA algorithm is based on the fact that there is no efficient and rapid method to factor very large numbers. Therefore, attempting to discover the value of a private key through manipulation of its corresponding public key would require applying an extraordinary amount of computer processing power over a period measured in years.

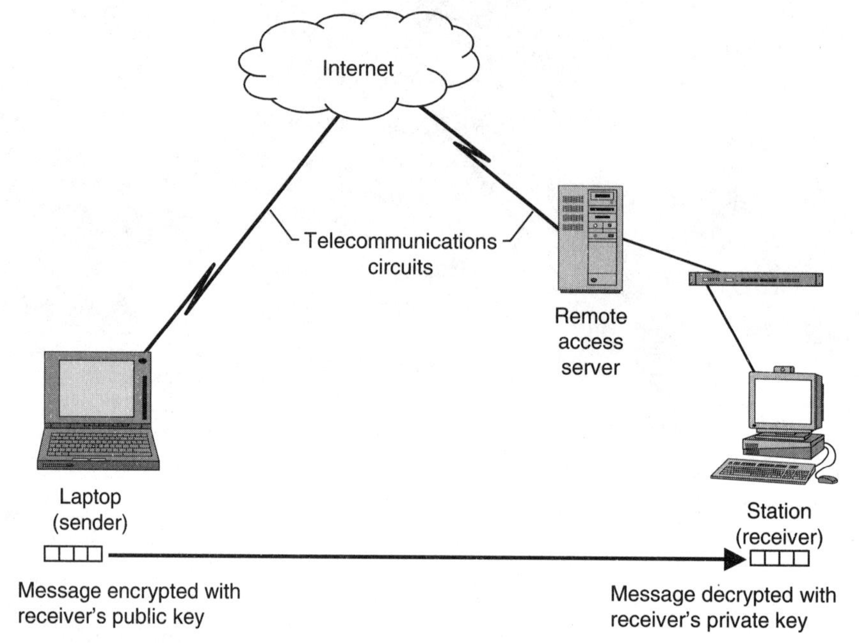

Figure 8.2 Public key encryption.

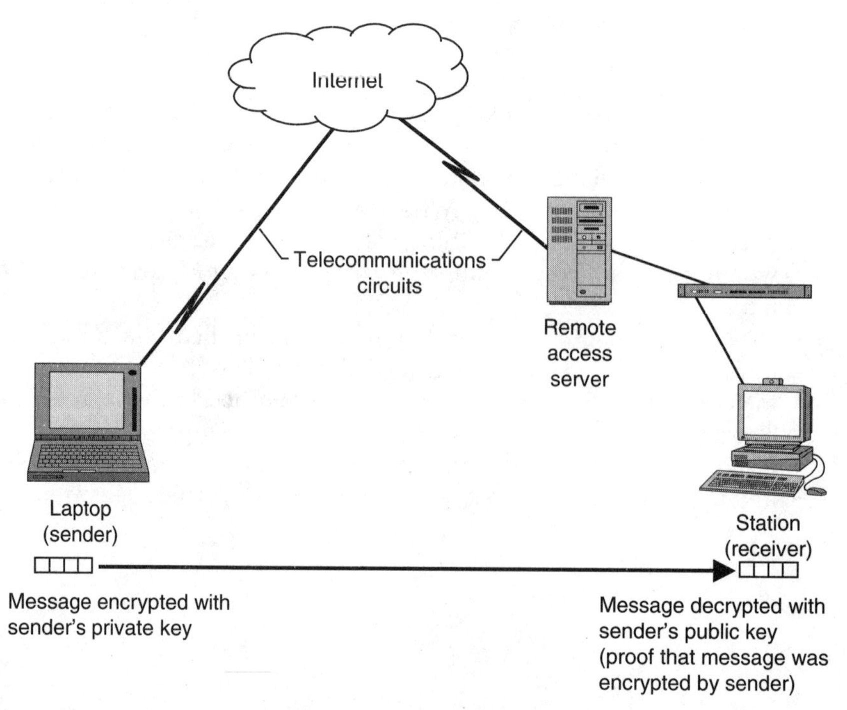

Figure 8.3 Public key encryption with a digital signature.

Authentication

General

Users obtain network services by connecting to a network through stations. Administrators manage networks by observing, tracking, replacing, and modifying network hardware and software. Authentication mechanisms are designed to request, inspect, and verify the identity of a user or an administrator before granting access to:

- A station.
- The network communications channel.
- A shared resource on the network.

Two types of users can request access to the network—known users and unknown users.

Known users, which includes administrators, have preestablished identities on the network. Such identity profiles include user names, e-mail addresses, group memberships, and access privileges. An authentication mechanism for known users asks the individuals who are requesting access to identify themselves and present proof of their identity. This information is then compared to existing network records. If there is a match, the individuals requesting access are admitted to the network.

A known user can be requested to provide identification and proof of identity using one or more of the following elements:

- User name—Also referred to as a login name, this element is assigned to users and devices by the network administrator. Each user name must be unique to ensure the correct delivery of messages to the user or device.
- Password—This element can be created by the user or assigned by the network administrator. Multiple passwords can be assigned to a user, each providing access to a different network resource.
- Physical token—This is an external component that cannot be duplicated. It is presented to and inspected by the network. Examples include smart cards and encrypted diskettes, such as those used for software installation or remote access.
- Biometric data—This is a unique physical characteristic of the individual requesting access that must be presented to and inspected by the network. Examples include fingerprints, facial images, retinal images, and voice recognition.

Unknown users consist of all individuals who do not have identities on the network they are attempting to access. An example is a visitor to an Internet Web site. In these cases, authentication can be any of the following:

- Nonexistent—The unknown users are granted access to all public or guest resources on the network.
- Partial—The unknown users are asked to identify themselves and provide additional information. The data received is used to create a profile for each user, typically including a login name and password, to be used to gain access to the site. This type of authentication is associated with memberships and interest groups, where profiles are used to provide useful information to each member.
- Comprehensive—The unknown users are asked to identify themselves by presenting their digital certificates. These certificates are used as authentication elements and must be obtained in advance from a recognized and trusted authority such as a financial, medical, or educational institute. The network confirms the validity of the certificate and grants access to the appropriate resources on the network. This type of authentication is associated with electronic commerce (e-commerce), where transactions are conducted and funds are exchanged over a network.

Authentication elements can be combined to provide the following two types of assurances to the network:

- Users are who they claim to be through their knowledge (user names and passwords).
- Users are who they claim to be through their possessions (physical tokens, biometric data, and digital certificates).

A variety of authentication technologies have been developed, notably for remote access connections, where it is impossible to verify the identities of individuals requesting network services through their physical presence. The following is a list of technologies widely used for network authentication.

- The point-to-point protocol (PPP), commonly used to connect to the Internet over the public telephone network, provides two authentication protocols—the password authentication protocol (PAP) and the challenge handshake authentication protocol (CHAP).
- Terminal access controller access control system plus (TACACS+).
- Remote authentication dial-in user service (RADIUS).
- Kerberos, an authentication system based on symmetric key cryptography.

In all cases, the role of authentication is to ensure that the devices, software programs, or individuals requesting access to network resources have the necessary credentials to prove their identities. In highly secure environments, the authentication process runs continuously in the background and lasts for the duration of the connection, to ensure that no unauthorized element has substituted itself for the authenticated source after the initial verification.

Authorization

General

Once users have been authenticated, they are granted access to the appropriate network resources. Authorization processes control the activities of each authenticated user on the network. While authentication establishes who the user is, authorization establishes what the user can do.

Any request issued by a user for network services is checked against the authorization profile or ACL for that user. The result is approval or denial of each request. An ACL can exist for users, devices, and software programs—all of these can request resources from the network. The most common ACL is the one built into the network operating system (NOS), which is customized for the organization and maintained by the network administrator. A typical feature of an ACL is the ability to create and manage user and group profiles.

Group profiles permit the administrator to assign a single set of network privileges to many users. Rather than create and maintain a separate user profile for each user on the network, the administrator assigns each user to one or more groups. Through membership, the user immediately gains access to all network resources assigned to the group(s). Since hundreds of resources can exist on a moderate-size network, the creation of groups saves a great deal of time in defining, granting, modifying, and revoking access privileges.

Two of the most popular means of regulating network access are login controls and password restrictions. Both of these are designed to reinforce the first level of access security represented by the authentication process. Each imposes additional conditions for access on known users.

Login controls. Login controls reduce the possibility of unauthorized access to the network by imposing one or more of the following requirements:

- A user can only login during specific hours and days.

  This feature controls the days of the week and the hours of the day that a user can access the network. For example, it is possible to limit a user's access to the network to Monday through Friday from 8:30 a.m. to 6:30 p.m. This ensures that the user cannot access the network on weekends or after working hours.

- A user can only login from a specific station.

  Each station is assigned an identification number by the NOS. This feature defines the station from which a user is allowed to access the network. For example, it is possible to restrict users to their own desktop stations. This means that users can only access the network from their own work areas. Logging in from workstations other than their own could be considered an intrusion.

- A user can only have a limited number of simultaneous logins.

  This feature controls the number of simultaneous active network connections a user can have. For example, limiting users to one login forces each user to only have a single network connection at a given moment. If the connected user wants a login to the network from another station, this individual must first terminate the existing connection to the network. Other login attempts could be deemed as intrusions.

- There is an automatic expiration date on each user account.

  This feature prevents a user from accessing the network after a specific date. Such users can be employees hired for temporary work or students enrolled in a course requiring network access for the duration of the course.

Password restrictions. Password restrictions reduce the possibility of unauthorized access to the network by imposing specific requirements on password owners before they can use network resources. The passwords chosen by users can have one or more of the following restrictions.

- A password must have a minimum length.

  Setting a minimum length for a password makes it more difficult for an intruder to guess a user's password.

- Users must change their passwords at specific intervals.

  This restriction forces every user to choose a new password after a certain period of time (e.g., once a month). This makes it impossible for an imposter who has stolen a password to access the network without detection over an extended period.

- Unique passwords are required (ones not used in the past).

  Each time a new password is chosen by a user, it must be different from a set number of previous passwords. Typically, it is not possible for users to select one of their last ten passwords as their new password.

- There is a limit on the number of grace logins.

  Grace logins allow users to access the network with their old passwords after the password expiration date has passed. When they do, a message is posted that warns users to change their passwords. Limiting the number of grace logins forces users to observe the established password maintenance policies.

- There is a limit on the number of invalid login attempts.

  This helps to prevent an unauthorized individual from accessing the network by attempting various passwords in the hope of guessing a valid one. After a set number of invalid attempts, the user account is disabled.

Multiple levels of authorization can be implemented to limit the range of actions a user can perform on a network resource. For example, a user may be granted or denied the following file-related privileges, or rights to:

- Create a new file in a specific storage area or directory.
- View the contents of a specific file.
- Delete a specific file.
- Modify and save the revised version of a specific file.
- Copy a specific file.
- Rename a specific file.
- Execute a program.
- Share a specific file.

In most cases, an accounting process is linked to the authorization process. Each time a request is made to gain access to a network resource, the attempt is recorded in a log file on the network. The log files can be used at any time to track the network activities of a specific user or the access history of a specific network resource.

The authorization process is also the basis for a form of networking called policy-based networking. In policy-based networking, network resources are granted on the basis of individual or group profiles. For example, one or more users in a department or an organization may be granted higher-speed connections to the Internet, or their traffic may be assigned a higher priority. Policy-based networking uses the tools provided by authorization processes to link organizational priorities to network operations.

Firewalls

General

The purpose of a firewall is to guard against unauthorized access at a specific entry point to network resources. Firewalls are typically used in conjunction with authentication processes to provide controlled access to users whose identities have been validated. The role of the firewall is to prevent any traffic that does not come from an authenticated source from entering the organizational network. In cases where the source cannot be authenticated (e.g., access from the Internet to an organizational Web site) the firewall makes certain the incoming traffic remains confined to the Web site and is not directed at other devices or processes on the network.

A firewall can also be implemented in a reverse configuration, where it enforces the rules governing the traffic exiting the organizational network (e.g., Internet access rules for employees). The term reverse firewall is sometimes used to describe such configurations.

It is possible to run the software for authentication, firewall, and reverse firewall operations on a single device—in some cases, a router. Alternatively, separate devices can be used to validate users and inspect the traffic entering and exiting the organizational network.

Although firewalls are associated with Internet access, they can also be implemented on networks that do not have connections to the Internet. In such cases, a part of the network that generates sensitive data (e.g., research and development, human resources, or legal department) can have its connection to the rest of the organizational network guarded by a firewall. In such cases, connected users may be required to authenticate themselves a second time to gain access to the guarded part of the network.

Since communications take place at many levels on a network, firewalls may also be implemented at many levels. Typically, a user runs an application on a station and communicates with a server to obtain access to various network resources, such as data files, printers, or access to a Web site. At the same time, the NOS may be in the process of sending various service request messages to devices throughout the network in order to synchronize, control, and maintain operations. If an internetwork is in place, the internetwork operating system (IOS) may be initiating similar communications between the routers.

An unauthorized individual with detailed knowledge of operating systems and network protocols can attempt to gain access to an organizational network by impersonating a user, a software application, an NOS or IOS process, or a network device. To be effective, a firewall must be capable of recognizing many different infiltration techniques. However, it must also allow legitimate traffic to pass through to the network on the other side. The greater the variety of applications, operating systems, and protocols, the more difficult it becomes to implement an effective firewall.

Monitoring and intruder detection software is used to ensure the effectiveness of firewalls. This type of software can also be used to test a firewall or the entire network for security weaknesses, a process called vulnerability assessment. The software attempts to bypass the firewall using various techniques and produces a report highlighting potential access points for an intruder. In detection mode, the software uses sophisticated processes to identify suspicious network activities. In addition to passively tracking and logging activities judged suspicious, the software can be configured to initiate an active response to such activities, including:

- Alerting a network administrator by sending a message.
- Severing the suspected connection.
- Refusing subsequent connections from the same source.

Monitoring and detection mechanisms complement authentication, authorization, and firewalls. The latter are intended to prevent unauthorized access to network resources, while the former are used to identify and neutralize individuals who have defeated access security and have gained access to the network.

Firewalls can be grouped into three categories.

- A packet filtering firewall examines each network packet, or datagram, and uses a list of criteria to determine if the inbound or outbound datagram should pass through the firewall or be discarded. It is considered a first level of defense.
- A circuit gateway firewall permits an inbound or outbound connection to take place on the basis of authentication and authorization criteria. This type of firewall does not inspect each datagram—once the device is granted access to the communications circuit, it can send and receive any type of data.
- A proxy firewall takes the place of another network device and initiates all communications on its behalf. All messages sent and received are inspected before being forwarded to the destination network or device. The devices at both ends of the connection never communicate directly—they communicate only through the proxy firewall, which represents itself as one of the devices when communicating with the other.

In addition to inspecting datagrams and connections, a firewall is capable of concealing the network addresses on an internal network from the Internet. This feature, called network address translation (NAT), can also be used to deal with Internet protocol (IP) address shortages by providing greater addressing flexibility to network administrators. A similar feature called port address translation (PAT) provides a similar translation service for higher layer addresses.

NOTE: IP addressing is described in detail in Appendix B: Internet Protocol (IP) Addressing Fundamentals.

Most organizations require two-way access to the Internet. Outbound access permits internal users to access Internet resources. Inbound access makes it possible for Internet users to send and retrieve information from the organizational network over the Internet. In most cases, the resources to be made available to Internet users—sometimes referred to as public resources—are placed on a separate broadcast domain with firewall protection, as shown in Figure 8.4. Several terms are used to describe the separate broadcast domain used to provide public resources, including the following:

- Perimeter network
- Screened subnet
- Demilitarized zone (DMZ) network
- Lobby network (the equivalent of a public lobby in an office building)

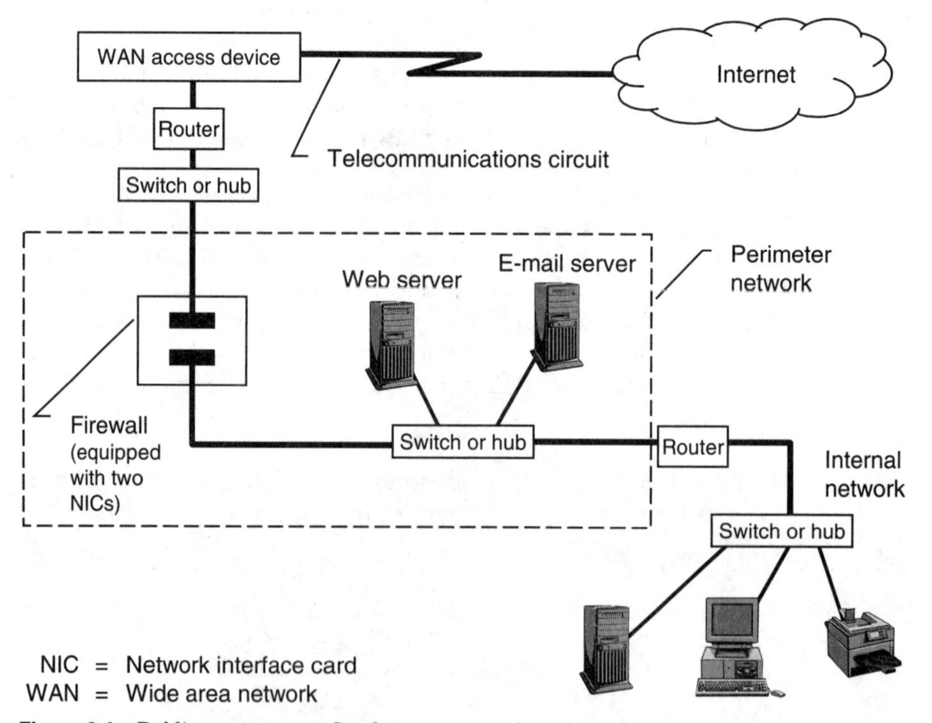

Figure 8.4 Public resources confined to a separate broadcast domain.

In some cases, the volume of traffic or the desire for fault tolerance makes it necessary to provide separate paths for inbound and outbound traffic, as shown in Figure 8.5. Under normal operating conditions, each type of traffic has its own path, which provides the lowest possible delays. In the event of failure of one path, the affected traffic can be temporarily redirected to the other path, which increases overall network availability.

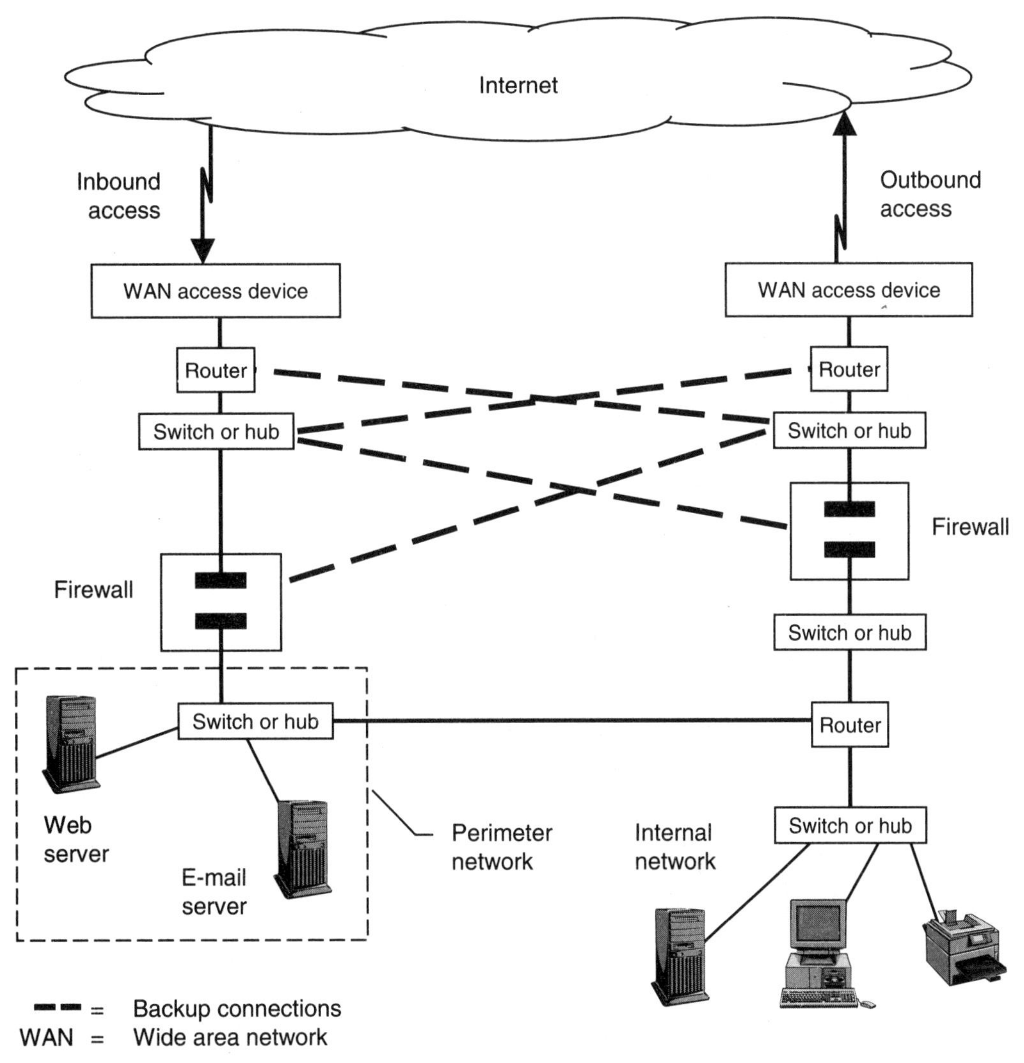

Figure 8.5 Separate inbound and outbound access.

Virtual Private Networks (VPNs)

General

The global availability of the Internet makes it possible for organizations to use Internet connections to link devices over extended distances instead of more expensive point-to-point telecommunications circuits. Most organizations must provide some or all of the following types of connections to their internal networks:

- Mobile access connections, in which the remote station connects to the organizational network from different locations at different times. An example of a typical mobile user is a traveling staff member who periodically connects to the office network through a laptop computer equipped with a modem. If the Internet is not used, this type of connection must use the public telephone network, typically paying long-distance charges for each call or dialing a toll-free number supplied by the organization.
- Remote access connections, in which the remote station connects to the organizational network from a fixed location. An example of a typical remote user is a staff member who routinely telecommutes to the office several days a week through a home computer. If the Internet is not used, this type of connection often requires a high-speed point-to-point digital telecommunications circuit instead of an analog telephone line, to minimize response time and simulate a local connection to the office network.
- Branch access connections, in which the remote station is part of a network located at another organizational site. An example of a typical branch user is a staff member located in a sales office or a warehouse servicing a specific region of the country. If the Internet is not used, the network in the branch office is usually linked to the main office using a router connected to a point-to-point digital telecommunications circuit.

When the Internet is used to transfer data, there is an increased risk of data interception. Since the Internet is a shared public network, an organization cannot obtain exclusive use of an Internet-based communications channel as it can with leased circuits or direct-dial connections. VPN technologies make it possible to securely send and receive data through the Internet. Therefore, a VPN connection is often described as a tunnel.

Data transmitted over a VPN is encrypted, making it unreadable to any unauthorized individuals who may successfully intercept the data as it passes through the Internet. A VPN protects the confidentiality of organizational messages as they pass over the Internet. By comparison, a firewall prevents unauthorized access to the organizational network from the Internet. Figures 8.6, 8.7, and 8.8 illustrate VPN use for mobile, remote, and branch access connections.

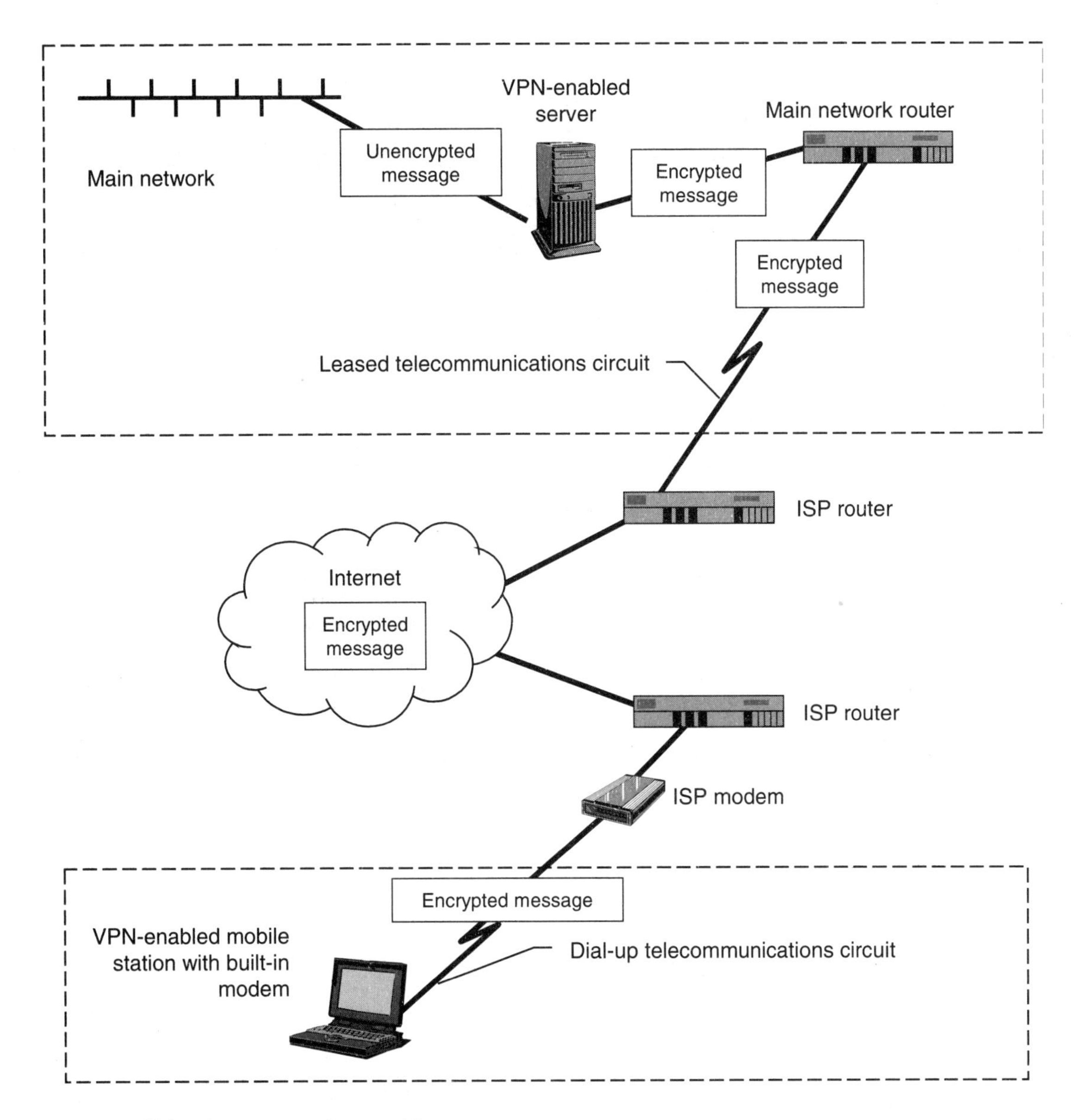

Figure 8.6 Mobile access virtual private network.

NOTE: Figure 8.6 is a simplified representation of a network.

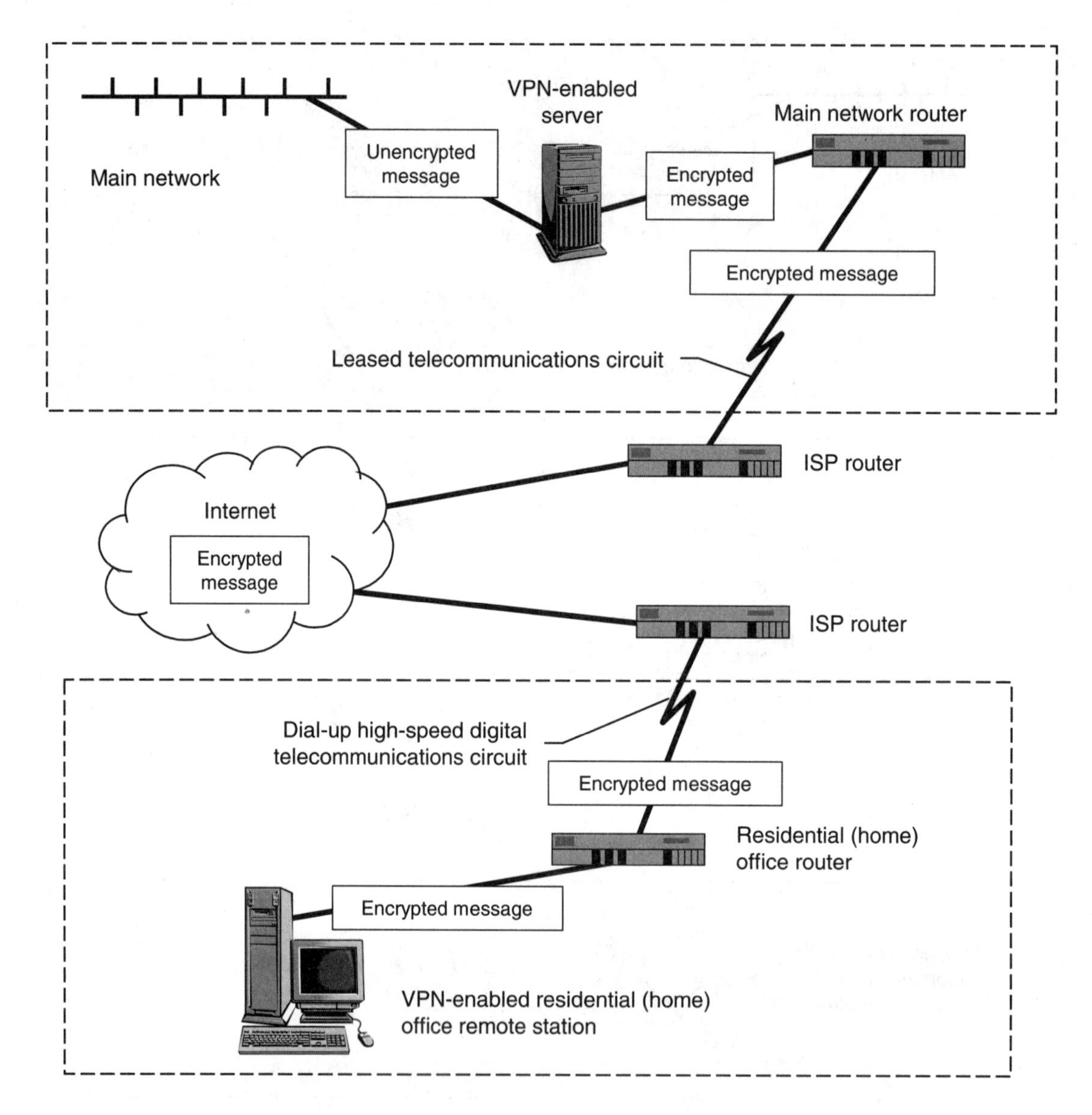

ISP = Internet service provider
VPN = Virtual private network

Figure 8.7 Remote access virtual private network.

NOTE: Figure 8.7 is a simplified representation of a network.

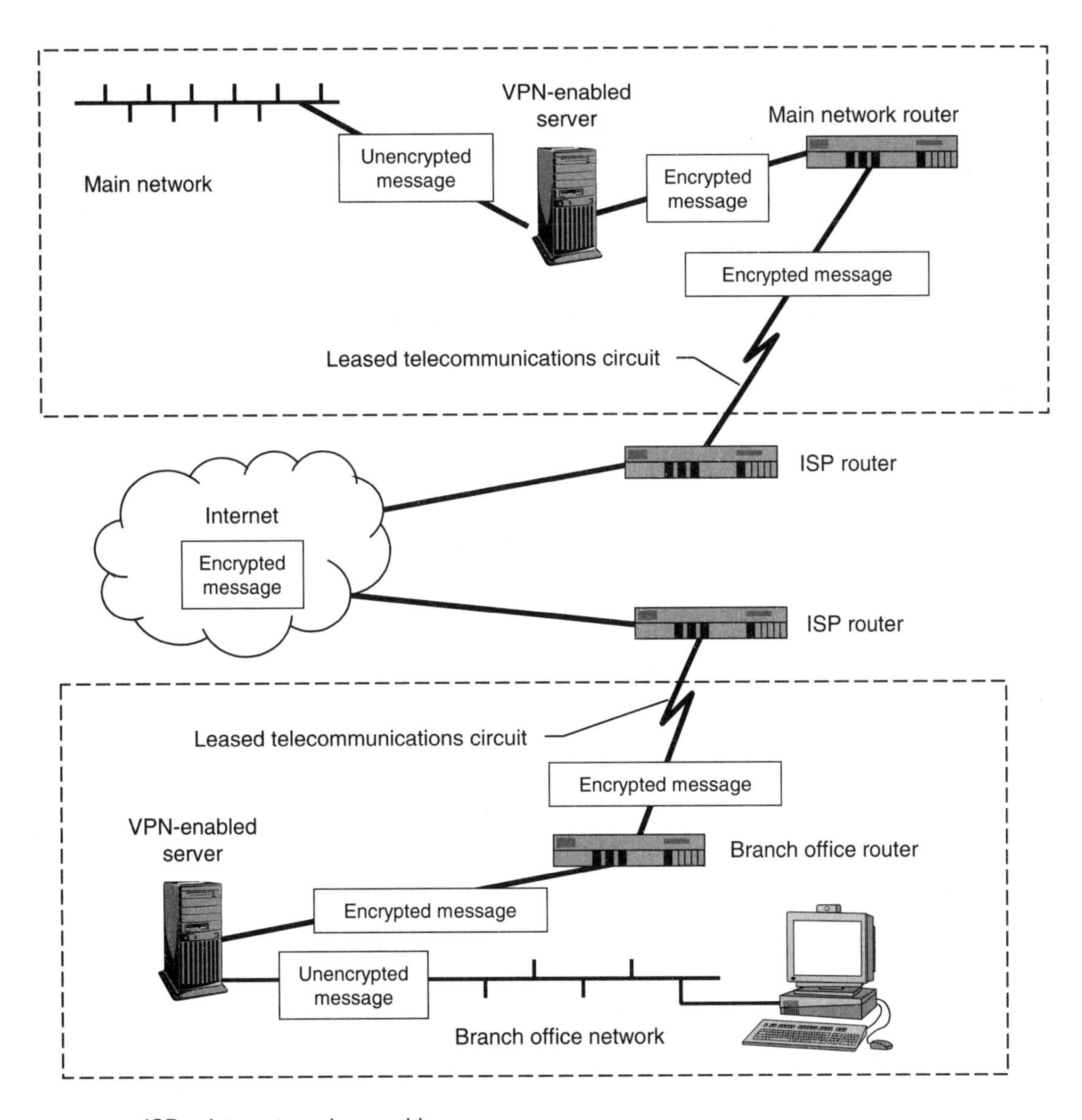

Figure 8.8 Branch access virtual private network.

NOTE: Figure 8.8 is a simplified representation of a network.

A VPN can be implemented using the features available in an organization's router, firewall, or NOS software. It is also possible to obtain VPN services from an Internet service provider (ISP) with nationwide or international operations. In all cases, the VPN must authenticate the user attempting to connect to the organizational network. Once this has been accomplished, one of several available protocols can be used to create the VPN tunnel and encrypt the addressing and data content prior to transmission.

- IP security (IPsec) protocol, which operates at the Network layer or Layer 3 of the Open Systems Interconnection (OSI) model
- Point-to-point tunneling protocol (PPTP), which operates at the Data Link layer or Layer 2 of the OSI model
- Layer 2 tunneling protocol (L2TP), which also operates at the Data Link layer of the OSI model

IPsec is the most comprehensive of the three protocols, developed by the Internet Engineering Task Force (IETF) as the standard security protocol for IP version 6 (IPv6), the update to the existing Network layer protocol used on the Internet. Since IPsec operates at the Network layer, it can be used to secure the datagrams created by all types of applications. Datagrams encrypted with IPsec can be transmitted over a variety of Data Link layer technologies, including local area networks (LANs) and various types of telecommunications circuits.

IPsec fundamentals

The first set of specifications for IPsec was published in August 1995 as request for comment (RFC) 1825 and updated in November 1998 as RFC 2401. IPsec was originally intended for IPv6 datagrams but was later modified to operate with IP version 4 (IPv4), the version of IP in current use. IPsec adds two new components to the standard IP datagram, as follows:

- The authentication header (AH) component of IPsec is responsible for data integrity. It ensures that neither the source address nor the data content of an incoming datagram was modified after it was transmitted by the sending device. A message digest is calculated and inserted into the datagram by the sending device. The receiving device uses the same process and compares its calculated value to the value stored in the datagram. A match indicates that the data was not modified after being placed in the datagram and confirms the identity of the sending device.
- The encapsulating security payload (ESP) component of IPsec is responsible for the data confidentiality. It ensures that the data content of a datagram cannot be viewed by an unauthorized individual who may have obtained access to the datagram. ESP uses private key encryption to secure the contents of datagrams—the sending and receiving devices use a shared key to encrypt and decrypt the data, respectively.

IPsec allows AH and ESP processing to be implemented in two modes, corresponding to two types of networks—private and public.

- IPsec transport mode is intended for use on private organizational networks, where authentication and encryption may be required, but it is not considered necessary to conceal the IP addressing information contained in a datagram. It is assumed that all datagrams are confined to network paths not shared with any other organization (e.g., LANs or leased point-to-point telecommunications circuits).
- IPsec tunnel mode is intended for use in cases where the datagrams generated will pass over a shared network such as the Internet. In tunnel mode, the original datagram, including its addressing fields, is encrypted and placed in a second datagram called a security envelope. The source and destination addresses used on the envelope correspond to the endpoints of the VPN tunnel. If the envelope is intercepted, the addressing information that may have been used to identify and infiltrate an organizational network cannot be viewed.

In summary, IPsec is a standardized technology that makes it possible to transfer data over the Internet without compromising the confidentiality of messages or exposing an organizational network to infiltration. IPsec is optional with IPv4—routers must be configured to recognize datagrams containing IPsec fields—but is a mandatory component of IPv6, the replacement for IPv4 on the Internet.

Security Planning

General

Organizational networks need protection against theft and vandalism in the same way as any other asset. An important step in establishing effective network security is to develop a written plan that defines policies and procedures. Some of the questions that must be asked when preparing a security plan include the following:

- Who is allowed to access the network?
- How will authentication be performed?
- How will intruders be prevented from accessing the network?
- What can an authorized user do on the network?
- How will authorization be granted?
- How can organizational data be made theft-proof and tamper-proof while it is stored and transmitted?
- How can the network be kept virus free?

- How is the network to be monitored for violations to security policy?
- What is the process for dealing with violations to security policy?

While a network security plan does not need to be a lengthy document, it must at least summarize how the network is to be kept secure, subject to the technological environment and the organizational structure. The steps taken should not impair the operations of the organization or prevent users from effectively performing their tasks.

Some of the benefits associated with a written security plan include the following:

- For both users and network administrators, the plan defines what is and is not permitted. It also outlines the consequences of improper conduct. Without clear definitions, it is neither possible to determine when a violation has occurred nor ensure consistency when dealing with individuals who violate security policies.
- A published, official document issued to all staff raises the general level of security consciousness in the organization.
- The plan can be used as a needs analysis document, to be used in evaluating potential technical solutions.

Establishing a security plan

The preparation of a network security plan requires the development of procedures and policies to protect network resources against loss and damage. Generally, the cost of protecting a network against an attack should be less than the cost of recovery from the attack. Without sufficient knowledge of what is being protected or the sources and types of potential attacks, establishing an appropriate level of security is difficult.

Some of the questions that need to be answered in order to establish a security plan include the following:

- What are the resources to be protected?
- How valuable is each resource?
- Who or what does the resource need to be protected from?
- What is the likelihood of encountering an attack of a given type from a given source?
- What measures are to be used to protect the resource against the attack?

When answering these questions, it is important to involve the right people in the process. It is possible that certain user groups within the organization are already involved in establishing network security policies. For example, groups responsible for auditing and control of information systems may have already established a security plan for their own resources. Involving such groups in the development of an organizational security plan can reduce conflicts and result in a stronger security infrastructure.

Implementation of a network security plan calls for all individuals interacting with the network to be made aware of their personal responsibilities for safeguarding organizational resources. In most cases, there are many levels of responsibility associated with a security plan. For example, users should be responsible for protecting their personal passwords, while network administrators can be assigned the responsibility of maintaining security on their segments of the network.

Once a network security plan has been established and implemented, it should be audited periodically to see if all of its objectives are being met by current practices. The audit can also attempt to infiltrate the network, as a test of its ability to resist intrusion and safeguard organizational data.

Risk analysis

Implementing a network security plan must be cost-effective. It is important to clearly understand which network resources are worth protecting and to what extent. Some resources are more important than others. The sources of potential attacks must be identified, both external and internal.

The purpose of risk analysis is to estimate the value associated with the loss of a particular resource. This value is not necessarily a monetary amount. The value of any resource can be calculated by establishing the value of two related factors.

- The importance of the resource
- The risk, which is defined as the probability of an attack on that resource

Mathematically, the value of a resource is expressed as follows:

$$\text{Value (j)} = \text{Importance (j)} \times \text{Risk (j)}$$

where,

Value (j) = The value of resource (j)

Importance (j) = The importance of resource (j)

Risk (j) = The probability of an attack on resource (j)

Typically, the importance of a resource is expressed as a numeric value. For example, a scale of 0 to 10 can be used, where 0 implies that the resource has minimal impact on the operations of the network and 10 implies a critical impact. The risk to a resource is an estimate of the likelihood that the resource will be attacked. Risk can be expressed as a decimal value between 0.00 and 1.00, where 0.00 implies that there is no threat to the resource and 1.00 implies a 100 percent probability of attack.

The concept of risk analysis is illustrated in Example 8.1.

EXAMPLE 8.1 Risk analysis.

On this network (see Figure 8.9), the print and file servers are to be accessed by all network users, while the accounting server is to be accessed only by users in the accounting department. A switch is used to segment the network and prevent the traffic generated by other users from affecting the response times of the accounting department stations. The router is used to provide access to the Internet to all users.

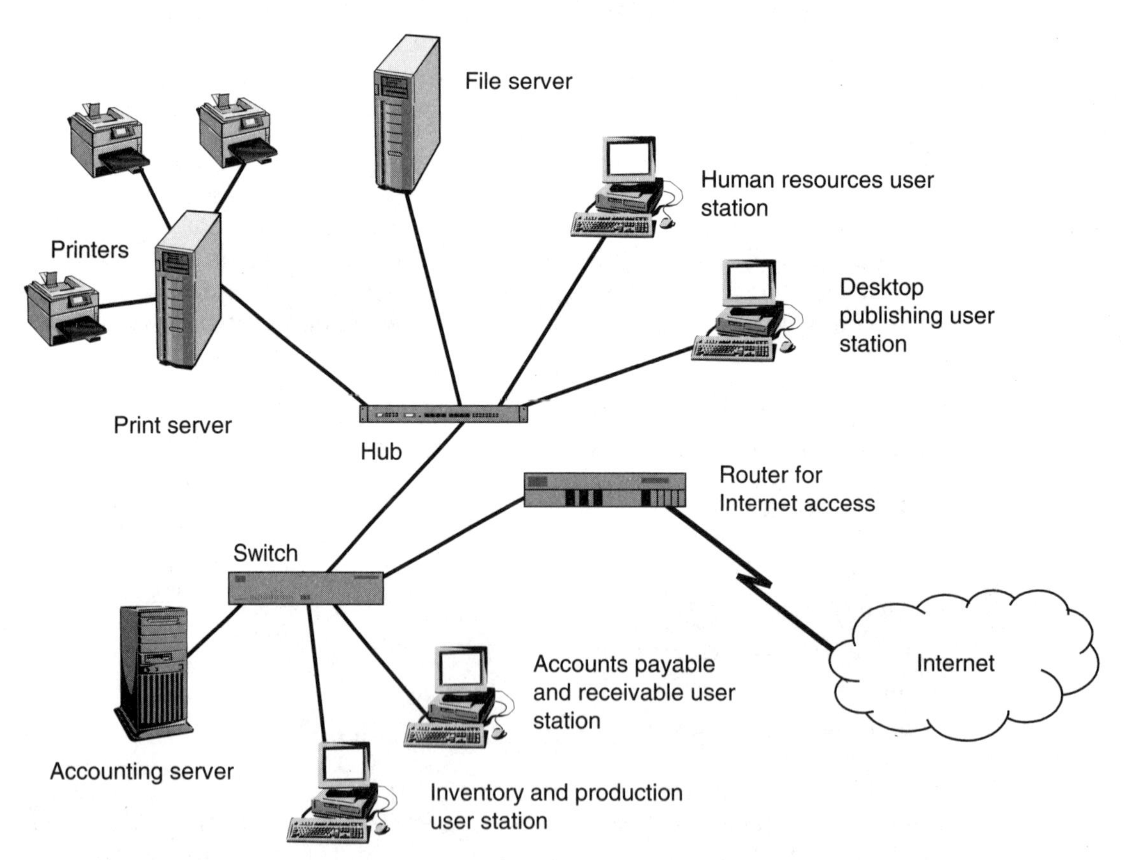

Figure 8.9 Risk analysis.

After careful consideration, the network administrator has assigned the following estimates of importance and risk to each network component (see Table 8.1).

TABLE 8.1 Risk Analysis—Estimating Component Importance and Risk

Component	Importance (1–10)	Risk (0–1)	Comment
Human resources user station	6	0.8	Potential target for internal attack.
Desktop publishing user station	4	0.2	
Accounts payable and receivable user station	7	0.7	Protected by the bridge (internal) and the router (external).
Inventory and production user station	6	0.6	Protected by the bridge (internal) and the router (external).
Print server	4	0.3	
File server	8	0.7	Contains most of the organizational data.
Accounting server	9	0.5	Contains all of the organization's accounting data but is isolated from the rest of the network by the switch.
Hub	6	0.5	Protected from external attack by the router.
Switch	6	0.5	Protected from external attack by the router.
Router	6	1.0	High risk due to potential for attack from external sources. Low importance, since it is used to provide outbound access to various Internet Web sites for research and reference purposes.
Printers	3	0.1	

Based on the estimates in Table 8.1, the resource values shown in Table 8.2 are calculated using the equation of Value (j) = Importance (j) × Risk (j).

TABLE 8.2 Risk Analysis—Calculating Value of Resources

Component	Importance	×	Risk	=	Value
Human resources user station	6	×	0.8	=	4.8
Desktop publishing user station	4	×	0.2	=	0.8
Accounts payable and receivable user station	7	×	0.7	=	4.9
Inventory and production user station	6	×	0.6	=	3.6
Print server	4	×	0.3	=	1.2
File server	8	×	0.7	=	5.6
Accounting server	9	×	0.5	=	4.5
Hub	6	×	0.5	=	3.0
Switch	6	×	0.5	=	3.0
Router	6	×	1.0	=	6.0
Printers	3	×	0.1	=	0.3

Ranking the values in Table 8.2, Table 8.3 shows the results from lowest to highest.

TABLE 8.3 Risk Analysis—Ranking Resources According to Value

Component	Importance	×	Risk	=	Value
Printers	3	×	0.1	=	0.3
Desktop publishing user station	4	×	0.2	=	0.8
Print server	4	×	0.3	=	1.2
Hub	6	×	0.5	=	3.0
Switch	6	×	0.5	=	3.0
Inventory and production user station	6	×	0.6	=	3.6
Accounting server	9	×	0.5	=	4.5
Human resources user station	6	×	0.8	=	4.8
Accounts payable and receivable user station	7	×	0.7	=	4.9
File server	8	×	0.7	=	5.6
Router	6	×	1.0	=	6.0

Based on the values calculated in Table 8.3, the router is the most valuable resource. Its importance is considered to be low since it is only used to allow internal users to access Web sites on the Internet. However, its risk is high since it represents an entry point for unauthorized access from the Internet to the internal network. Therefore, it is critical for this organization to secure its router.

The least valuable resources are the printers. It is unlikely that any special security precautions would be taken to protect these devices.

While this method is subjective with respect to estimating levels of importance and risk, it provides network administrators with an opportunity to consider how each device can adversely impact the operations of the network if it is compromised.

This method of risk analysis can be extended to make estimates of the value of a network as a whole. It can also be used to estimate the value of the software and data on the network. For example, if the information contained in a file is considered critical to the operations of the organization, extra levels of network security can be used to protect the file. Alternatively, the file can be removed from the network and kept on an isolated machine in a secure area.

Chapter

9

Network Management

Chapter 9 describes and defines network management as the combination of planning, procedures, software, equipment, and personnel needed to maintain network operations at maximum efficiency at all times. This is a broad topic that covers a wide range of activities, from network architecture planning to the configuration of an individual device.

Overview

General

Network management is a broad topic, covering the range of activities from network architecture planning to software configuration for an individual device. In many organizations, networks must operate on a continuous basis, making operational management a critical activity. In cases where the network is being expanded in terms of number of connections, new applications, or both—strategic planning is required to successfully manage the changes without disrupting existing operations. Network management can be described as the combination of planning, procedures, software, equipment, and personnel needed to maintain network operations at maximum efficiency at all times.

Several different models can be used to identify and create the various areas of management for a typical organizational network, as follows:

- Equipment model
- Software model
- Operational model
- Organizational model

Equipment model of network management. The equipment model of network management creates areas of management based on equipment types. Software associated with a given type of equipment is placed in the same category as the equipment. For example, a network operating system (NOS) would be included with servers. Using this model, a typical network would be divided into the following equipment groups:

- User devices (e.g., desktop and laptop computers)
- Shared peripherals
- Servers
- Cabling infrastructure
- Telecommunications circuit infrastructure
- Local area network (LAN) access devices (e.g., hubs, switches)
- Internetwork access devices (e.g., routers)

Software model of network management. The software model of network management creates areas of management based on the software and network protocols used in the organization. Hardware associated with a given type of software or protocol is placed in the same category as the software or protocol. For example, routers would be included with internetwork operating systems (IOSs). Using this model, a typical network would be divided into the following software groups:

- Applications software
- Operating systems for user devices
- NOS
- IOS
- LAN protocols (e.g., Ethernet)
- Wide area network (WAN) protocols (e.g., Frame Relay)
- Network layer protocols (e.g., Internet protocol [IP])

Operational model of network management. The operational model of network management divides an organizational network into an operational hierarchy and creates areas of management based on the layers in the hierarchy. All hardware, software, and network protocols associated with a given layer are

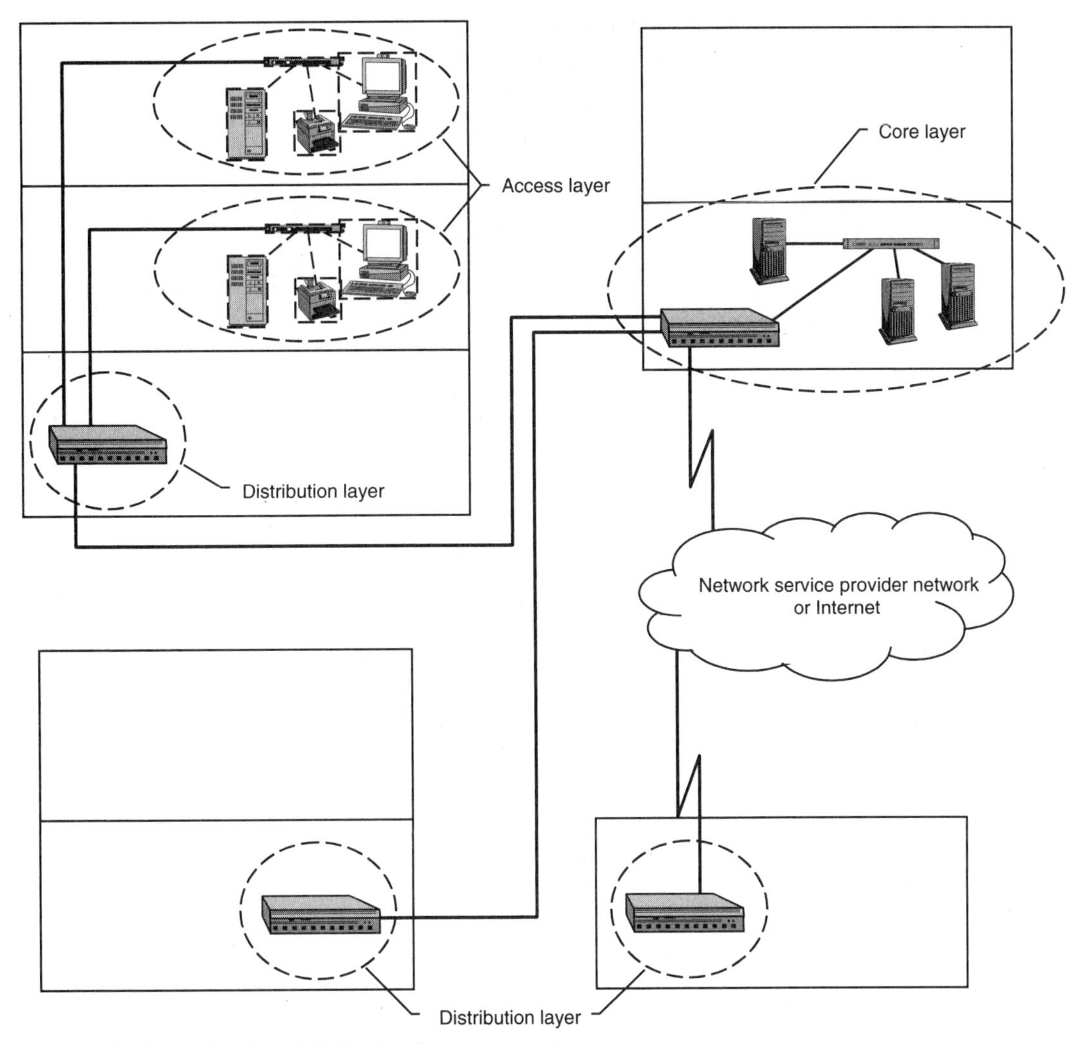

Figure 9.1 Operational model of network management.

placed in the same category as the layer. For example, all equipment, software, and protocols used in a centralized data center would be included in the core layer of the hierarchy. Using this model (see Figure 9.1), a typical network would be divided into the following operational groups:

- Core layer group, made up of the centralized devices accessed by all users (e.g., internal Web or database servers)
- Distribution layer group, made up of devices used to connect individual departments or sites to the core facility (e.g., routers)
- Access layer group, made up of the LANs in each department or site

Organizational model of network management. The organizational model of network management creates areas of management based on organizational units, which may be divided according to one or more of the following criteria:

- Geographic (e.g., regional sites)
- Functional (e.g., sales department)
- Project (e.g., new product development)

In this model, the management personnel assigned to each organizational unit are responsible for all hardware, software, and network protocols used in the unit.

Management goals

Regardless of the model used to create the areas of management, the goal of network management is to maximize:

- Reliability, which is defined as predictable behavior during operation.
- Availability, which is a measure of response time or operating failure in a given period of time. For example, a network may be very reliable when it is functioning, but may function slowly or not at all due to low availability.
- Performance of individual components, which involves optimizing both hardware and software.
- Compatibility, which is the ability of the network to integrate various types of equipment or software.
- Scalability, which is the ability of the network to accommodate growth in traffic.
- Survivability, which affects both reliability and availability. Fault tolerance and disaster recovery are components of survivability, which is defined as the ability of the network to function after minor or major disruption. The cause of the disruption may be software-based or physical (e.g., a fire, equipment theft, or a damaged cable).

Management tools

A variety of software and hardware tools is available for network management, for both strategic planning and operations. Such tools are generally classified into one or more of the following management categories:

- Desktop management, responsible for administering user stations and peripherals
- Systems management, responsible for administering various types of servers

- Network management, responsible for administering overall network operations
- Traffic management, responsible for administering the communications flow generated by network protocols and applications

The simple network management protocol (SNMP) is the most commonly used protocol for managing networks. It allows a network administrator to remotely monitor and manage a network device from a central location, which reduces support costs and increases network availability. SNMP and its related standards are considered essential management tools for most organizational networks. They are described in detail later in this chapter.

Network Management Tools

General

Many types of hardware and software tools are available to assist network administrators. Some of these products (e.g., network design or simulation software) are intended to be used prior to network implementation, to assist in developing an optimized design. Others, such as performance or benchmarking tools, are used to obtain an assessment of an existing network's capabilities for growth planning and troubleshooting purposes.

The most common network management tool is a monitoring station, equipped with the necessary software to present the current—and in some cases, historical—status of the item being monitored. Given the large and varied number of devices, software, cabling, and telecommunications circuits used on most organizational networks, it is a challenge to develop a single monitoring tool capable of providing status information on every component in place. In most cases, a number of tools are used, each focused on one or a limited number of components (e.g., a database monitor or a server management system).

International Organization for Standardization (ISO) recommendations

In addition to developing the seven-layer reference model for network communications, ISO has defined five categories of network management. Most vendors of network management tools have incorporated some or all of the ISO recommendations into their products. The five categories of network management defined by ISO are:

- Configuration and name management.
- Security management.
- Accounting management.
- Fault management.
- Performance management.

Configuration and name management. Configuration management includes the installation, modification, and tracking of the hardware and software settings necessary to enable a given device to operate in the intended manner. A managed device must make it possible for a network administrator to send and receive configuration data to and from the unit.

A typical network consists of various types of personal computers (PCs). Each PC runs an operating system and can contain a variety of internal and external components—including, one or more hard disks and printers, display circuitry, and a monitor. It is possible to configure a PC to perform many different actions. With the appropriate components, the same PC can be configured to operate as a server, a station, or a router. One of the tasks of a network administrator is to determine the appropriate configuration for a given network component. Both the original configuration and any subsequent modifications must be tracked. In the event of problems after making a change, the previous configuration can be restored on the device.

Since changes in hardware and software are routine on most networks, network administrators must be ready to perform any of the following tasks at all times:

- Identify and track new network components and their relationship to existing components.
- Shut down and restart a device or a group of devices on the network after configuration changes—referred to as rebooting.
- Determine which configuration tasks can be performed unattended and then automate the process—referred to as scripting.
- Maintain or change the configurations of network components as needed (e.g., when a device needs to be relocated).
- Update configurations in response to vendor initiatives (e.g., installing upgrades, fixes, patches, or service packs to software).

The management tool most closely identified with configuration management is called inventory or asset management software. This type of software is used to scan, register, and track the configuration status of every station on the network, and in some cases, other devices as well. On most networks, stations are the most difficult items to manage. Unlike most or all servers and network access devices, stations are not physically grouped together in centralized locations. In addition, users can move or make configuration changes to their stations at any time, potentially disrupting network operations. The alternative to using inventory software is to physically inspect each station and record its hardware and software contents, which is impractical on large networks with hundreds or thousands of stations.

Security management. Security management involves the protection of network resources, including physical devices, organizational data, and network operations. A secure network ensures the confidentiality and integrity of stored information and contributes to network availability. The security of a network can be managed using both hardware and software tools.

NOTE: Security is described in detail in Chapter 8: Network Security.

Software-based security management tools include password management systems, auditing software, and encryption key management systems. Passwords are used for authentication and authorization, enabling user access only to approved resources. Auditing keeps track of the resources accessed by users. Encryption modifies the transmitted data, making the contents unreadable to unauthorized individuals.

Hardware-based security management can be as simple as a lock on the door to a room containing network equipment. It is also possible to enhance security by removing or disconnecting hardware (e.g., diskette drives and keyboards from PCs that are intended to run unattended).

Some of the administrative tasks associated with security management include:

- Maintenance and distribution of passwords and other authentication elements, such as smart cards.
- Monitoring access attempts and network activity.
- Controlling access to network management and network security information.
- Collecting and examining audit and security data generated by security management software.

In all cases, network users should be reassured that appropriate security policies are in place and the management of security facilities is ongoing and complete. Network resources should be accessible only to authorized users and network security resources should themselves be very secure.

Accounting management. Accounting management is the tracking, billing, and cost allocation for network resource use. The resource can be a software program, a shared database file, or a network device such as a hard disk or a printer. An organizational network serves many individuals and groups. Accounting management software tools make it possible to charge users for network use. Costs can be allocated to individual users, project teams, departments, divisions, or any other group. Typically, the usage data generated by the accounting management software is used for internal budgeting purposes, where the cost of network operations is distributed among various groups on the basis of usage.

While the main purpose of accounting management software is cost allocation, the data gathered by the software can have other uses, as follows:

- A user or group may be making excessive use of network resources such as disk storage space or Internet access, reducing the resources available to other users.
- One or more individuals may be using the network inefficiently, leading to excessive charges. This information can be used to assign individuals to training programs.
- By tracking resource use over time, network administrators can plan and budget for growth in specific areas.

When configuring accounting management software, network administrators must establish the usage rates for the resources being tracked. It may be possible to assign multiple usage rates to a given resource, with higher rates charged for use during peak periods. Administrators are also responsible for producing reports listing the charges incurred by users and groups.

Accounting management also has a security component, since it is possible to use this type of software to track every keyboard or mouse command issued by a user. Network administrators must advise users that:

- Their authorization materials, such as passwords, must be kept secure to prevent identity theft.
- Specific network activities are tracked, recorded, and have costs assigned to them by software.
- Access to accounting information is restricted.

Fault management. Fault management deals with the detection, isolation, and correction of hardware or software conditions that disrupt normal network operations. A fault condition is persistent and requires action from a network administrator. For example, occasional bit errors are not considered to be faults, since they can be managed by the error correction processes built into network protocols. The goal of fault management is to compensate for a device malfunction and restore network operations as soon as possible.

Both hardware and software tools are used to manage fault conditions. In some cases, the faulty device can be diagnosed using built-in or external programs. In other cases, external testers and analyzers can be used to identify the reason for the fault. A typical fault management process is the:

- Fault is detected.
- Faulty component, device, or network segment is isolated to limit the scope of the disruption.
- Network is reconfigured to minimize the impact of the fault.

- Faulty item is repaired or replaced.
- Network is restored to its prefault state.
- Incident and the resolution process are recorded for future reference and analysis.

Many network analyzers are capable of detecting fault conditions and suggesting possible causes and solutions. In cases where the fault is intermittent, such devices can be configured to monitor the network over an extended period of time. These devices capture and store traffic information to a network storage device for later analysis.

Performance management. Performance management includes the optimization of individual components and the overall network to minimize the time it takes to complete a given task. Since network utilization levels vary with time, the performance of the network (as measured by response time) can be expected to fluctuate within a given range. Performance management software tools allow administrators to set this range for individual devices, software processes, and network data transfer rates. When the measured performance falls outside the range, the software issues an alert to the administrator.

The tools used for performance management can also be used to test various components or the network as a whole by generating data to simulate various levels of network activity. Network administrators can use this feature to assess the limitations of a component or the network, using the results to upgrade components, reconfigure the network, or compare products and technologies. The results of the testing can also be used to set the performance range used as a reference by the management software.

Examples of performance indicators that are typically monitored by network administrators include the following:

- Level of capacity utilization, which indicates the amount of a component's capacity being used or the amount available for use.
- The traffic level on a communications channel, total or divided using various criteria (e.g., according to protocol, software application, or user identification).
- Component or network response times for various types of requests (e.g., file transfers).
- The indicators for any component operating below its expected or historical capabilities.

The reports and statistics produced by performance management tools assist in planning, managing, and maintaining the organizational network. They make it possible to recognize potential problems before users are affected. For example, it is common to require additional network storage capacity in response to increases in the quantities or the average sizes of files being generated by users. Performance management tools can provide the following information to support the request for additional storage devices:

- Historical rate of storage space consumption
- Amount of free space remaining on existing storage devices
- Response times of existing storage devices

Simple Network Management Protocol (SNMP)

General

SNMP and its related standards can be described as a distributed, vendor-independent architecture for managing network components. The first set of specifications for SNMP was issued by the Internet Engineering Task Force (IETF) in May 1990 as request for comment (RFC) 1157. Multiple extensions and updates have been published to incorporate new technologies and increased security requirements. Most network access devices, such as hubs, switches, and routers, as well as servers, stations, and shared peripherals can be managed using SNMP.

The SNMP architecture is comprehensive and is made up of the following elements:

- The structure of management information (SMI) used to describe managed objects, including naming and data type conventions.
- The structure of the database that collects and stores management information for a related group of objects, called the management information base (MIB).
- The protocol used to transfer management information over the network, which is SNMP.
- The structure of the databases that collect and store management information on monitoring devices attached to networks, called remote network monitoring (RMON) MIB and switched network monitoring (SMON) MIB.

Tables 9.1, 9.2, 9.3, and 9.4 are a listing, in chronological order, of the principal RFCs that define the elements in the SNMP architecture.

TABLE 9.1 SMI-Related RFCs

RFC	Date of Issue	Description of Contents
1155	May 1990	SMI
2578	April 1999	SMI version 2 (SMIv2)

TABLE 9.2 MIB-Related RFCs

RFC	Date of Issue	Description of Contents
1156	May 1990	MIB
1213	March 1991	MIB version 2 (MIB-II)

TABLE 9.3 SNMP-Related RFCs

RFC	Date of Issue	Description of Contents
1157	May 1990	SNMP
1441	April 1993	SNMP version 2 (SNMPv2)
1901	January 1996	Community-based SNMPv2 (SNMPv2c)
2571	April 1999	SNMP version 3 (SNMPv3)

TABLE 9.4 RMON-Related RFCs

RFC	Date of Issue	Description of Contents
1271	November 1991	RMON MIB for Ethernet networks
1513	September 1993	Token ring extensions to RMON MIB
2021	January 1997	RMON MIB version 2 (RMON2)
2819	May 2000	RMON MIB updated for SMIv2
2613	June 1999	RMON MIB extensions for switched networks (SMON)

Simple Network Management Protocol (SNMP) fundamentals

When classified using the Open Systems Interconnection (OSI) model, SNMP is considered an Application layer protocol that uses SMI for the Presentation layer formatting of its messages. At the time of its introduction, SNMP used Transport layer user datagram protocol (UDP), Network layer Internet protocol (IP), and Data Link layer Ethernet services. These are shown in Figure 9.2. Subsequent extensions and updates have made it possible to operate SNMP using other protocols and technologies at the lower layers. Regardless of the network environment, the SNMP architecture provides a consistent way of generating, storing, requesting, and transferring management information on a network.

Layer 7	Application	SNMP
Layer 6	Presentation	SMI
Layer 5	Session	
Layer 4	Transport	UDP
Layer 3	Network	IP
Layer 2	Data Link	Ethernet
Layer 1	Physical	

IP = Internet protocol
SMI = Structure of management information
SNMP = Simple network management protocol
UDP = User datagram protocol

Figure 9.2 SNMP and the OSI model.

The relationship between the layers can also be shown using a LAN frame containing an SNMP message (see Figure 9.3).

Distributed management architectures such as SNMP allocate various management responsibilities to multiple entities (see Figure 9.4). Entities used in the SNMP architecture are:

- Managed devices, each containing an agent that is responsible for collecting the object information that is defined in the MIB for the device.
- Proxy agents, responsible for collecting management information from devices that cannot operate their own agents.

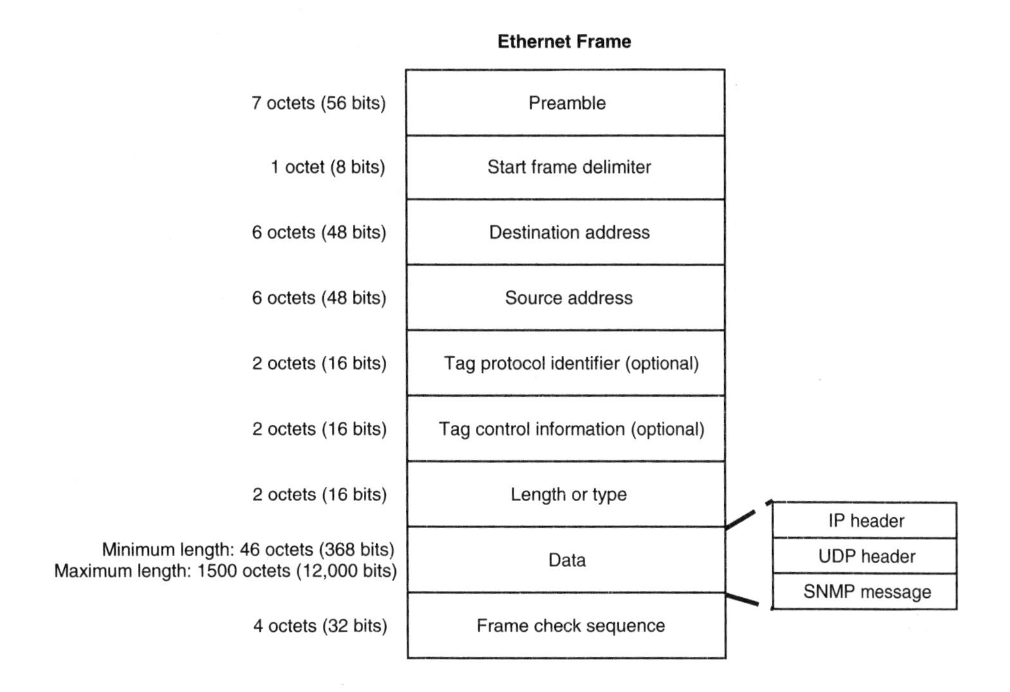

Figure 9.3 SNMP message on an Ethernet local area network.

- One or more network management systems (NMSs), each responsible for requesting MIB contents from various agents and other NMSs it is monitoring. An NMS can also change the value of a MIB object (e.g., resetting the value of a counter in a device).
- Management consoles, the devices that runs one or more NMSs. Network administrators use management consoles to view and manipulate network management information. Typically, such units are high-performance workstations. They must be able to process large amounts of incoming data from various agents and display the results, often using charts, as quickly as possible.

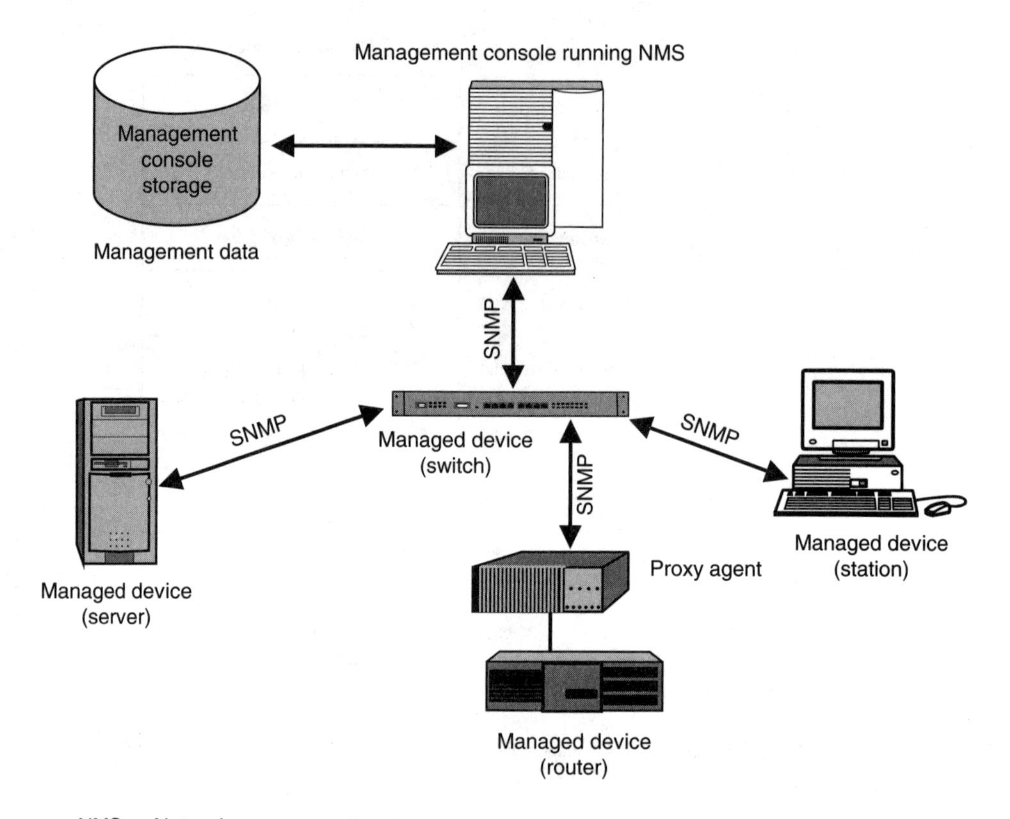

Figure 9.4 SNMP distributed management architecture.

Simple Network Management Protocol (SNMP) messaging

A uniform architecture for managing the quantity and variety of network devices that can exist on an organizational network requires a centralized and highly structured process for classifying MIB objects. Each object in a MIB must have a unique identifier to associate its data with a specific device or event. SNMP uses a hierarchical tree-naming structure approved by ISO for object identification—specifically, the part of the tree administered by the Internet Assigned Numbers Authority (IANA). A partial section of the object identification tree is shown in Figure 9.5.

As an example, using the tree illustrated in Figure 9.5, the numeric identifier for the MIB-2 level of the tree can be expressed as a series of numbers separated by periods. Starting from the root, this results in a value of 1.3.6.1.2.1. Similarly, the ENTERPRISES level has a numeric identifier of 1.3.6.1.4.1. The MIB-2 level contains the Internet standard objects identified in various RFCs

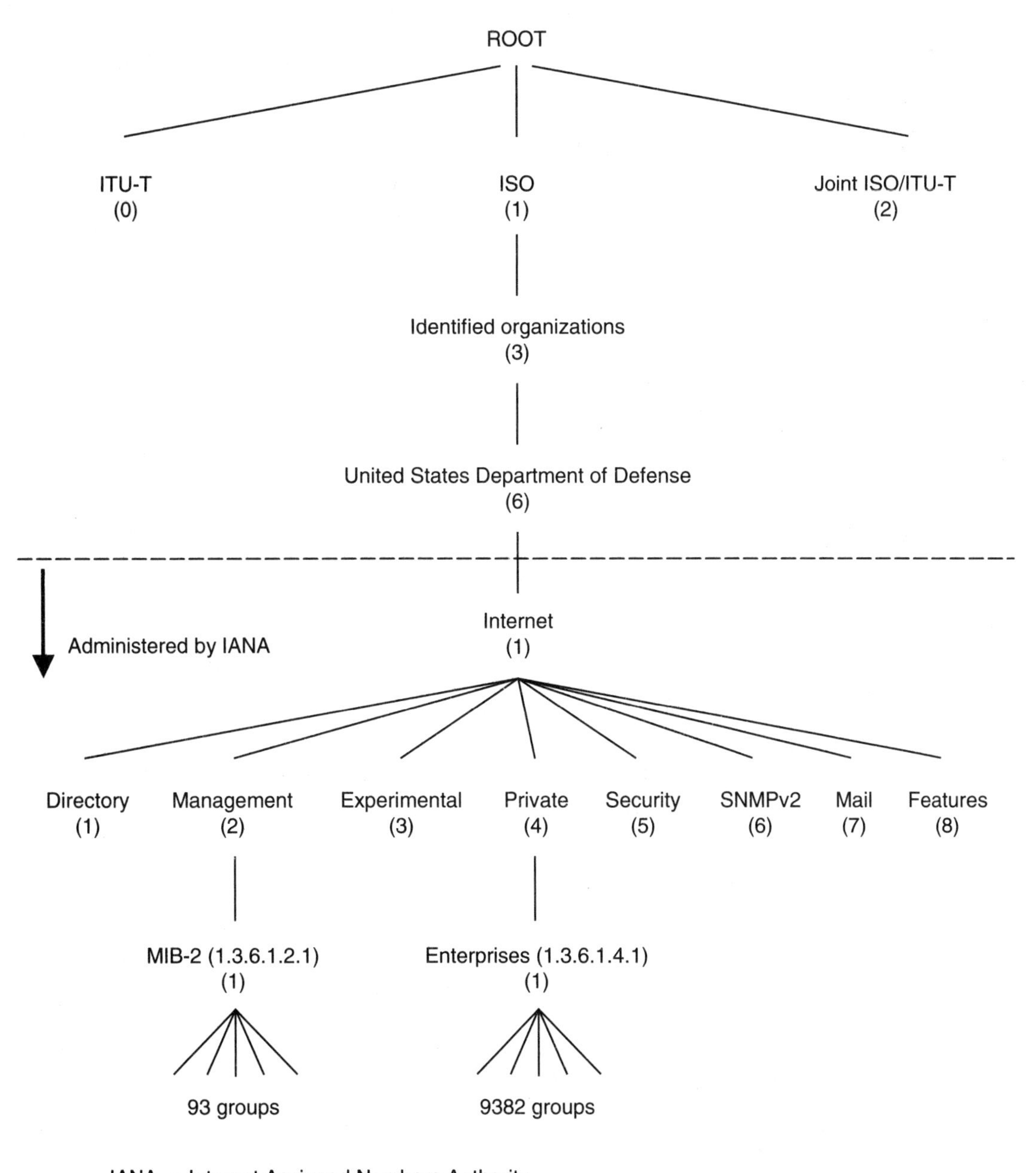

IANA = Internet Assigned Numbers Authority
ISO = International Organization for Standardization
ITU-T = International Telecommunication Union-Telecommunication
MIB = Management information base
SNMP = Simple network management protocol

Figure 9.5 Partial object identification tree.

and the ENTERPRISES level contains the MIBs developed by individual vendors for their network products. Figure 9.6 illustrates some of the groups below the MIB-2 level.

NOTE: As of April 2001, there were 93 groups registered below the MIB-2 level and 9382 groups registered below the ENTERPRISES level. Each ENTERPRISES group represents a different organization.

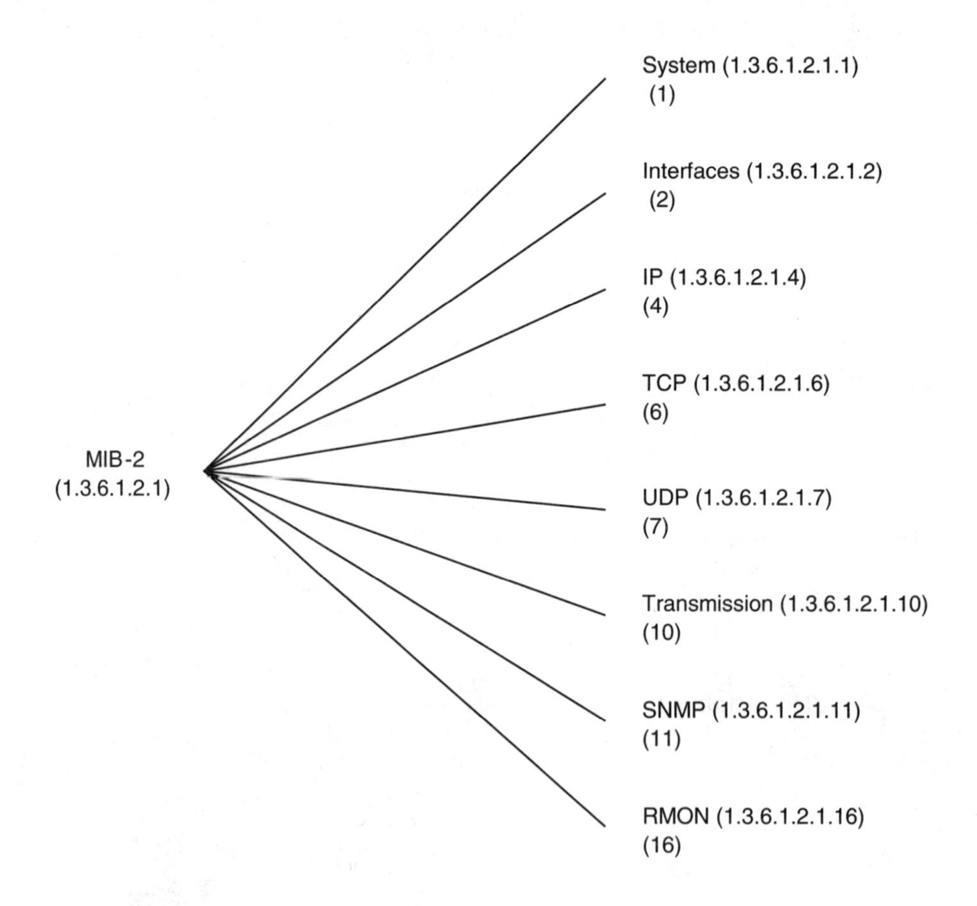

IP = Internet protocol
MIB = Management information base
RMON = Remote network monitoring
SNMP = Simple network management protocol
TCP = Transmission control protocol
UDP = User datagram protocol

Figure 9.6 Partial set of MIB-2 groups.

The following is a brief description of each group illustrated in Figure 9.6:

- The SYSTEM group of objects is used to obtain descriptive information on the device being managed.
- The INTERFACES group of objects is used to obtain information on the interface(s) present on the device being managed.
- The IP group of objects is used to obtain IP datagram-related information from the device being managed.
- The TCP group of objects is used to obtain TCP-related protocol information on the device being managed.
- The UDP group of objects is used to obtain UDP-related protocol information from the device being managed.
- The TRANSMISSION group of objects is used to obtain information on the communications channel used by the device being managed.
- The SNMP group of objects is used to obtain information on the SNMP-related activities performed by the device being managed.
- The RMON group of objects is used to obtain information from a monitoring device placed on a remote network.

For example, an NMS can obtain the name of a device by requesting the value of the object 1.3.6.1.2.1.1.5 from the agent in the device. This numeric identifier corresponds to the object sysName in the SYSTEM group, as follows:

ISO (1)

IDENTIFIED-ORGANIZATIONS (3)

UNITED STATES DEPARTMENT OF DEFENSE (6)

INTERNET (1)

MANAGEMENT (2)

MIB-2 (1)

SYSTEM (1)

sysName (5)

Similarly, to obtain the number of interfaces on a device, the NMS can request the value of the ifNumber object by using the numeric identifier 1.3.6.1.2.1.2.1.

ISO (1)

IDENTIFIED-ORGANIZATIONS (3)

UNITED STATES DEPARTMENT OF DEFENSE (6)

INTERNET (1)

MANAGEMENT (2)

MIB-2 (1)

INTERFACES (2)

ifNumber (1)

The initial version of SNMP specified five types of messages that could be issued by an SNMP application. Three of the message types are intended to be issued by the NMS and two are for use by agents in managed devices. Limiting the number of message types to five makes the protocol easier to implement, one of the reasons for calling it simple. The following is a list of the message types introduced in the first version of SNMP.

- Message types issued by an NMS:
  - GetRequest, typically used to retrieve a single value from a managed object
  - GetNextRequest, typically used to retrieve multiple values stored in a table format
  - SetRequest, used to send a value to be assigned to a managed object
- Message types issued by an agent:
  - Response, used to reply to a GetRequest or GetNextRequest message received from an NMS
  - Trap, used to inform the NMS that a specific event has occurred (e.g., a threshold set by the network administrator for an object has been exceeded). A trap can also be described as an alert, alarm, notification, or exception message.

When SNMPv2 was introduced, it included two significant additions to the message types capable of being issued by an NMS.

- GetBulkRequest, used to retrieve the entire contents of a table with one command.
- InformRequest, used to enable the exchange of messages between different NMSs.

The initial proposals for SNMPv2 included enhancing the functionality and security of network management messaging. Since SNMP can be used to learn about and configure network devices, it is both an essential tool for network administration and a potential threat to network security. Due to the variety of proposals submitted, the security enhancements took a longer time to develop than the functionality enhancements. As a result, several versions of a secure SNMPv2 were considered, including the following:

- SNMPv2c (Community-based SNMP version 2)
- SNMPv2u (User-based SNMP version 2)
- SNMPv2* (SNMP version 2 star)

Of these, SNMPv2c updated the previously issued SNMPv2 with a limited security enhancement. The security models described in SNMPv2u and SNMPv2* were used to develop a complete set of security enhancements for SNMP, documented and published as SNMPv3. The third major version of SNMP provides all of the functionality originally intended for SNMPv2. Specifically, SNMPv3 adds the following elements to the SNMP architecture:

- Authentication, to confirm the origin of a management message.
- Data integrity, to prevent the alteration of the contents of a management message.
- Confidentiality, to encrypt the contents of a management message.
- Key management, to administer the security processes for management messaging.

Remote Network Monitoring (RMON) fundamentals

RMON and its successor, RMON2, have extended the capabilities of SNMP and have made it possible to manage large distributed networks using the SNMP architecture. Like SNMP agents, RMON agents, called RMON probes, are incorporated into managed devices such as servers or routers. It is also possible to connect a dedicated or stand-alone RMON probe device to a network in the same manner as a station, server, or shared peripheral device. In both cases, an NMS is used to configure the probe and specify the RMON MIB objects to be used. The role of the probe is to collect and store data from the network to which it is connected. Once the probe is in operation, it can be accessed using the NMS at any time to provide current or historical data on network operations (see Figure 9.7). The RMON standards enable interoperability between the different types of probes and NMSs offered by vendors.

At the time of its introduction, RMON featured nine groups of objects, intended to be used to manage Ethernet networks. A subsequent update added a tenth group for token ring network management. Figure 9.8 illustrates the ten groups below the RMON level in the object identification tree.

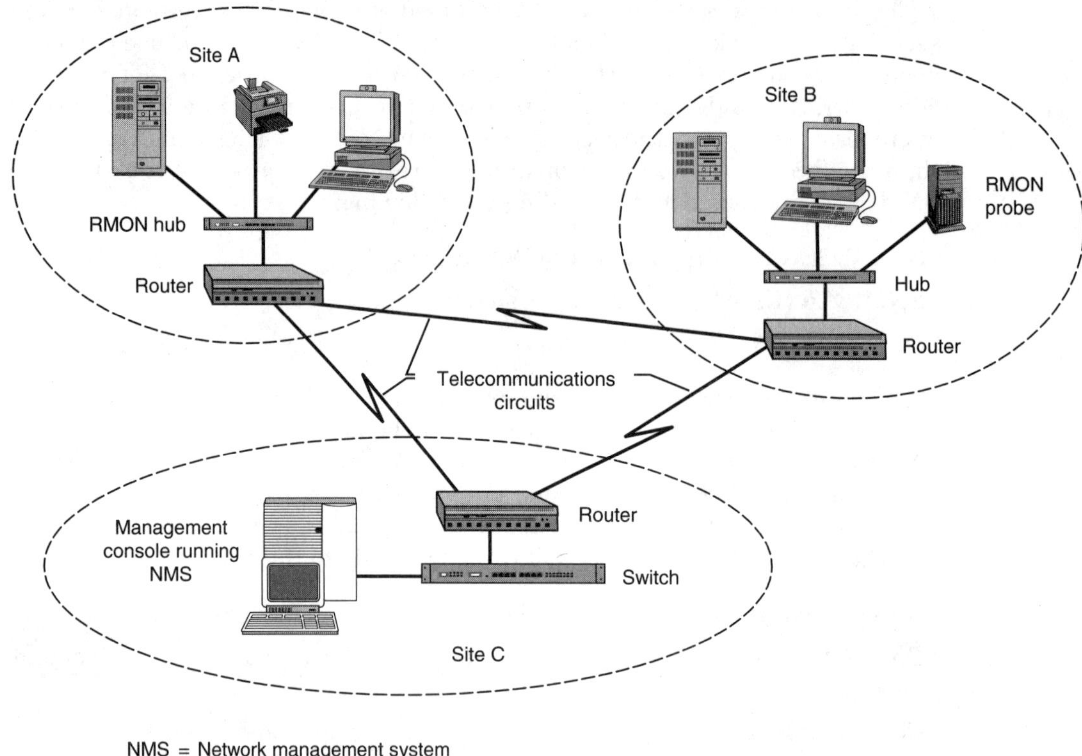

Figure 9.7 RMON-based network management.

The initial set of specifications for RMON made it possible to obtain management data at the OSI Data Link layer, from an Ethernet or token ring LAN. RMON2 was developed to enable the collection of network data at higher layers of the OSI model. For example, network administrators may be interested in tracking the levels of network activity associated with Web access, videoconferencing, or other network applications. In cases where multiple OSI Network layer protocols are used, administrators can use RMON2 to monitor and collect data on each protocol. RMON2 also makes it possible to monitor the exchange of messages between any two devices on the organizational network, which makes it easier to identify the causes of various communications-related errors.

RMON2 adds ten additional object groups below the RMON level in the object identification tree (see Figure 9.9).

When designing an RMON-based management system for an organizational network, a key requirement is to place each RMON probe in a location where it can monitor a significant segment of network traffic. Ideally, the combination of all probes provides a complete view of the network, enabling

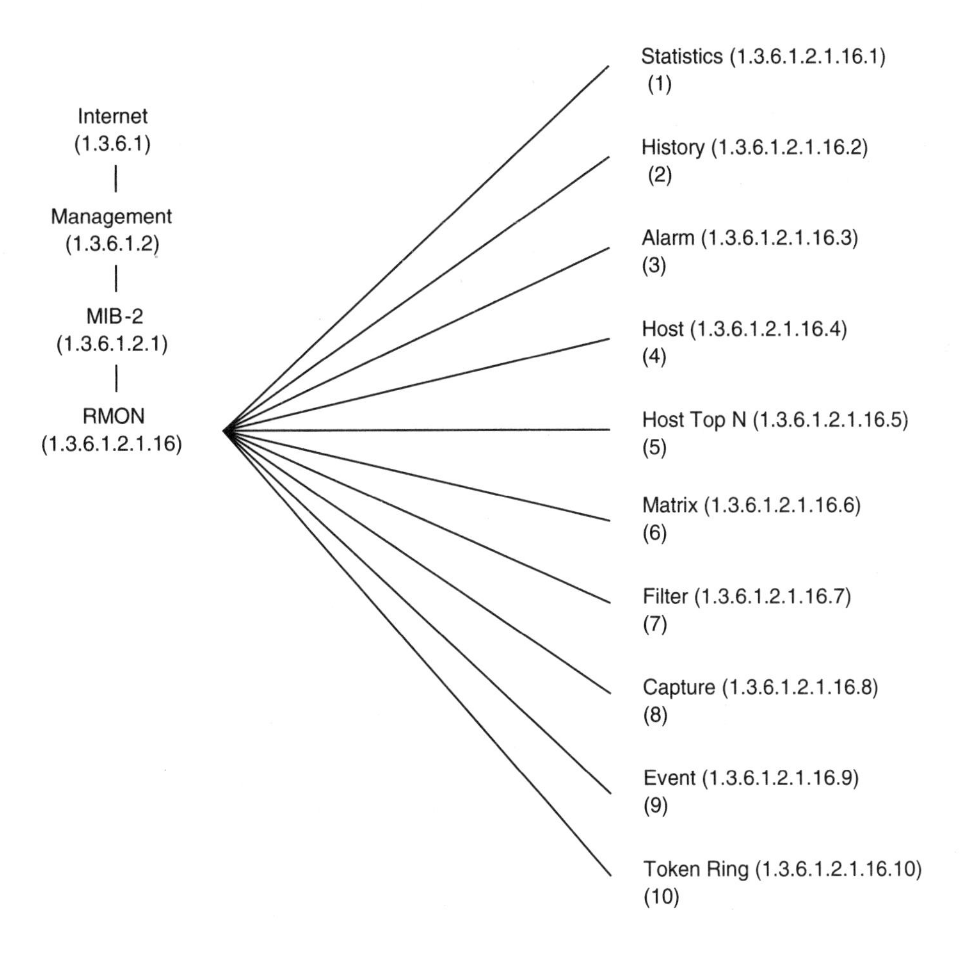

Figure 9.8 Set of RMON groups.

comprehensive monitoring and management. An RMON probe does not typically communicate continuously with the NMS, as is the case with traditional SNMP agents. However, probes continuously collect and analyze information and can issue alerts to the NMS when significant errors are observed. In the event of network failure, data stored in a probe can be used to identify the cause of the failure.

Switching technologies make it more difficult to monitor network communications. A switch creates a temporary point-to-point communications channel between connected devices for the duration of time needed to exchange messages. By comparison, a hub distributes an incoming message from any

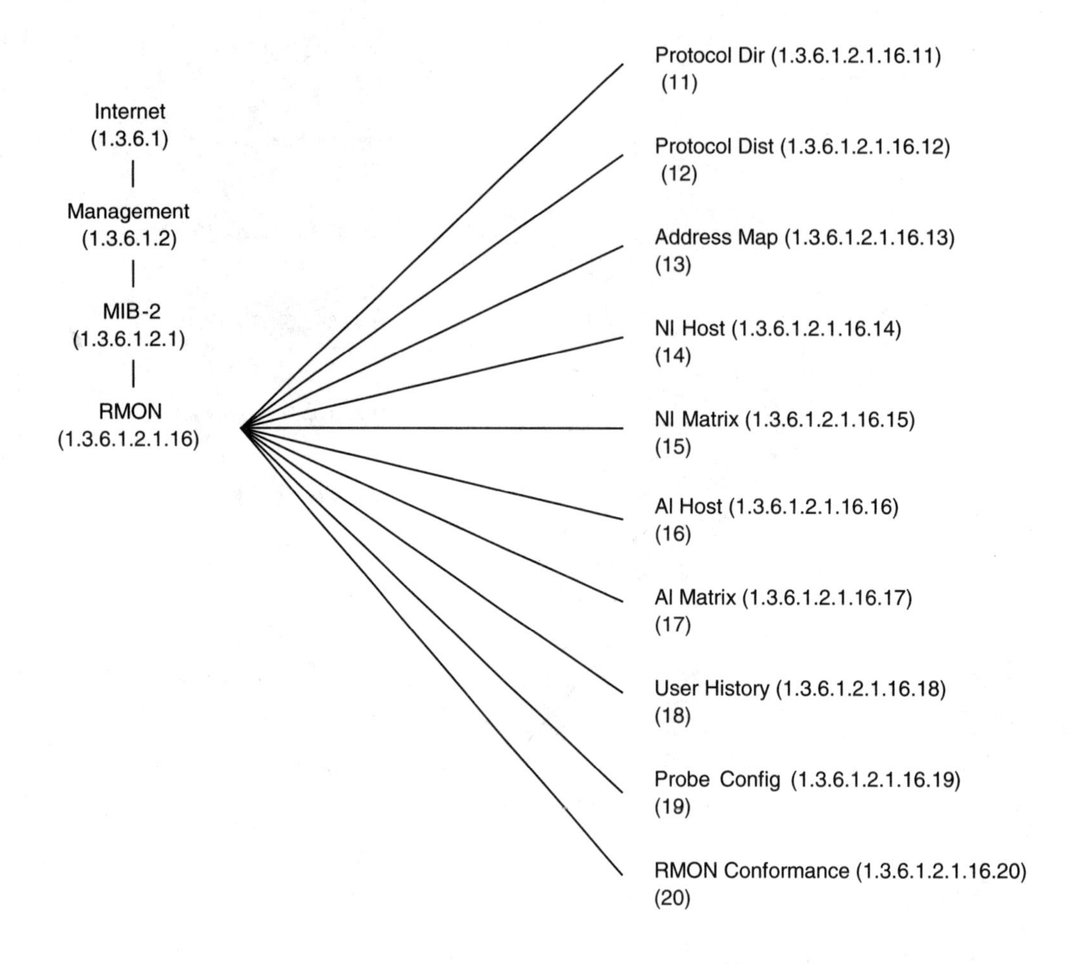

Figure 9.9 Set of RMON2 groups.

connected device to all other connected devices. To monitor a hub-based network, an RMON probe can be connected to any port on a hub, where it will observe all network traffic. In order to monitor a switch-based network, a standard method is required for copying traffic from one or more ports to a management port connected to an internal or external probe, a process referred to as port mirroring. RMON probes intended for use on switched networks use the MIB developed for this purpose, called the SMON MIB.

NOTE: Switching is described in detail in Chapter 6: Switching and Virtual LANs (VLANs).

Appendix

A

Binary and Hexadecimal Numbering Fundamentals

Overview

General

The hexadecimal numbering system (base 16) is commonly used to express values such as network device addresses in a format that is easier to read than the binary numbering system (base 2) format used by devices—although not as familiar as the decimal numbering system (base 10). This appendix shows how to convert between the three systems for the values most commonly converted—the numbers 0 to 255 decimal, corresponding to eight binary digits or bits. A table with all conversions is provided at the end of this appendix.

NOTE: A grouping of eight bits is commonly referred to as a byte or an octet.

Decimal numbering system (Base 10)

In the decimal numbering system, the symbols 0 through 9 are used repeatedly to indicate various values.

- Ten values can be represented using a single symbol (0–9).
- To represent higher values, a second symbol is added to the left of the first, also with a range of 0–9.

- The combination of the two symbols can be used to express one hundred values ($10 \times 10 = 100$), ranging from 00 to 99.
- Adding a third symbol to the left of the second expands the range to one thousand values ($10 \times 10 \times 10 = 1000$), ranging from 000 to 999.

To summarize, each additional symbol increases the number of values that can be represented by a factor of ten, as shown below:

- One digit, a maximum of ten values (0–9)
- Two digits, a maximum of one hundred values (00–99)
- Three digits, a maximum of one thousand values (000–999)
- Four digits, a maximum of ten thousand values (0000–9999)
- Five digits, a maximum of one hundred thousand values (00000–99999)

Hexadecimal numbering system (Base 16)

In the hexadecimal numbering system, each symbol can have sixteen values (zero through fifteen). Since the decimal system provides only ten symbols (0–9), the first six letters of the alphabet (A–F) are used as symbols representing the values 10 through 15. A hexadecimal digit can therefore be any of the following, corresponding to the decimal values 0-15:

- 0, 1, 2, 3, 4, 5, 6, 7, 8, 9, A, B, C, D, E, F

To represent values higher than fifteen, a second symbol is added to the left of the first, also with a range of 0-F. The combination of the two symbols can be used to express 256 values ($16 \times 16 = 256$), ranging from 00 to FF.

Adding a third symbol to the left of the second expands the range to 4096 values ($16 \times 16 \times 16 = 4096$), ranging from 000 to FFF.

To summarize, each additional symbol increases the number of values that can be represented by a factor of sixteen, as shown below:

- One digit, a maximum of 16 values (0–F)
- Two digits, a maximum of 256 values (00–FF)
- Three digits, a maximum of 4096 values (000–FFF)
- Four digits, a maximum of 65,536 values (0000–FFFF)
- Five digits, a maximum of 1,048,576 values (00000–FFFFF)

Binary numbering system (Base 2)

In the binary numbering system, each symbol can have two values (zero or one). To represent values higher than one, a second symbol is added to the left of the first, also with a value of 0 or 1. The combination of the two symbols can be used to express 4 values ($2 \times 2 = 4$), ranging from 00 to 11. Adding a third symbol to the left of the second expands the range to 8 values ($2 \times 2 \times 2 = 8$), ranging from 000 to 111.

To summarize, each additional symbol increases the number of values that can be represented by a factor of two, as shown below:

- One digit, a maximum of two values (0–1)
- Two digits, a maximum of four values (00–11)
- Three digits, a maximum of eight values (000–111)
- Four digits, a maximum of 16 values (0000–1111)
- Five digits, a maximum of 32 values (00000–11111)

Numbering System Conversions

Converting from binary to decimal

To convert a binary value in the range 00000000 to 11111111 to decimal, use Table A.1.

TABLE A.1 Converting from Binary to Decimal

Bit position		8	7	6	5	4	3	2	1
Value of bit position		$2^7=128$	$2^6=64$	$2^5=32$	$2^4=16$	$2^3=8$	$2^2=4$	$2^1=2$	$2^0=1$
Binary value to convert	x								
		↓	↓	↓	↓	↓	↓	↓	↓
Values after multiplication	=		+	+	+	+	+	+	+

Sum of the values after multiplication equals the decimal value.

EXAMPLE A.1 Converting a binary value to decimal

Given a binary value of 10011010, the decimal value is calculated as follows:

Bit position		8		7		6		5		4		3
Value of bit position		128		64		32		16		8		4
Binary value to convert	x	1		0		0		1		1		0
		↓		↓		↓		↓		↓		↓
Values after multiplication	=	128	+	0	+	0	+	16	+	8	+	0

Sum of values = 128 + 0 + 0 + 16 + 8 + 0 + 2 + 0 = 154
Therefore, the binary value 10011010 equals 154 decimal.

Converting from binary to hexadecimal

To convert a binary value in the range 00000000 to 11111111 to hexadecimal, use the following process:

1. Split the eight bits into two groups of four. The first four bits from left to right form the first group. The remaining four bits form the second group.

   NOTE: A grouping of four bits is sometimes referred to as a nibble (or nybble).

2. Convert each group of four bits into decimal, using Table A.2.
3. If the decimal value is between 0 and 9, keep this value (the symbols 0–9 represent the same values in decimal and hexadecimal).
4. If the decimal value is 10, replace 10 with A. For 11, use B. For 12, use C. For 13, use D. For 14, use E. For 15, use F.

TABLE A.2 Converting from Binary to Hexadecimal

Bit position for group of four bits		4	3	2	1
Value of bit position		$2^3=8$	$2^2=4$	$2^1=2$	$2^0=1$
Binary value of bit group to convert	x				
		↓	↓	↓	↓
Values after multiplication	=	+	+	+	

The sum of the values after multiplication equals a decimal value in the range 0 to 15.

- If the resulting value is between 0 and 9, do not modify the value.
- If the resulting value equals 10, replace it with A.
- If the resulting value equals 11, replace it with B.
- If the resulting value equals 12, replace it with C.
- If the resulting value equals 13, replace it with D.
- If the resulting value equals 14, replace it with E.
- If the resulting value equals 15, replace it with F.

EXAMPLE A.2 Converting a binary value to hexadecimal

Given a binary value of 10011010, the hexadecimal value is calculated by first splitting the eight bits into two groups of four bits, from left to right—1001 and 1010.

Each group of bits is converted to a decimal value and then to an equivalent hexadecimal value, as shown below.

First group of 4 bits

Bit position		4		3		2		1
Value of bit position		8		4		2		1
Binary value of bit group to convert	x	1		0		0		1
		↓		↓		↓		↓
Values after multiplication	=	8	+	0	+	0	+	1

Sum of values for first 4 bits = 8 + 0 + 0 + 1 = 9, which equals 9 hexadecimal.

Second group of 4 bits

Bit position		4		3		2		1
Value of bit position		8		4		2		1
Binary value of bit group to convert	x	1		0		1		0
		↓		↓		↓		↓
Values after multiplication	=	8	+	0	+	2	+	0

Sum of values for second 4 bits = 8 + 0 + 2 + 0 = 10, which equals A hexadecimal.

Therefore, the binary value 10011010 equals 9A hexadecimal.

Converting from decimal to binary

To convert a decimal value in the range 000 to 255 to binary, use Table A.3.

NOTE: Calculations in this table are performed on a column-by-column basis starting with the left most column (bit position 8).

TABLE A.3 Converting from Decimal to Binary

Bit position		8	7	6	5	4	3	2	1
Decimal value to convert*									
Subtract value of bit position	-	$2^7=128$	$2^6=64$	$2^5=32$	$2^4=16$	$2^3=8$	$2^2=4$	$2^1=2$	$2^0=1$
		↓	↓	↓	↓	↓	↓	↓	↓
Decimal remainder after subtraction**	=								
Conversion to binary***									

* Enter the entire decimal value into the first column.

** If the "decimal remainder after subtraction" value is greater than zero, it is moved to the "decimal to convert position" in the next right most column (as indicated by the arrow). If it is less than zero, the previous value in the "decimal to convert position" is used.

NOTE: Once the "decimal remainder after subtraction" equals zero, conversion to bits is no longer necessary. Any remaining bit positions are set to zero.

*** Convert to binary by examining the decimal remainder after subtraction. If this value is greater than or equal to zero, enter 1; otherwise, enter 0.

EXAMPLE A.3 Converting a decimal value to binary

Given a decimal value of 154, the binary value is calculated as follows:

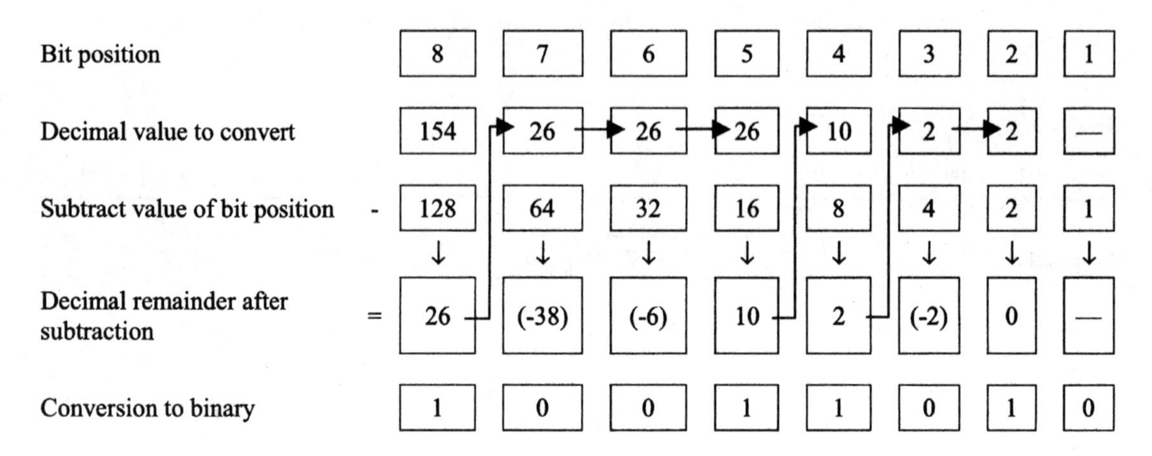

Therefore, the decimal value 154 equals 10011010 binary.

Converting from decimal to hexadecimal

To convert a decimal value in the range 000 to 255 to hexadecimal, use the following process:

1. Convert decimal to binary using Table A.3.
2. Convert binary to hexadecimal using Table A.2.

EXAMPLE A.4 Converting a decimal value to hexadecimal

Convert a decimal value of 230 to hexadecimal.

Step 1: Convert decimal value to binary.

Bit position		8	7	6	5	4	3	2	1
Decimal value to convert		230	102	38	6	6	6	2	—
Subtract value of bit position	-	128	64	32	16	8	4	2	1
		↓	↓	↓	↓	↓	↓	↓	↓
Decimal remainder after subtraction	=	102	38	6	(-10)	(-2)	2	0	—
Conversion to binary		1	1	1	0	0	1	1	0

Therefore, a decimal value of 230 equals a binary value of 11100110

Step 2: Convert first group of 4 bits to hexadecimal.

First group of 4 bits

Bit position		4		3		2		1
Value of bit position		8		4		2		1
Binary value (of bit group) to convert	x	1		1		1		0
		↓		↓		↓		↓
Values after multiplication	=	8	+	4	+	2	+	0

Sum of values for the first 4 bits = 8 + 4 + 2 + 0 = 14, which equals E hexadecimal.

Second group of 4 bits

Bit position		4	3	2	1
Value of bit position		8	4	2	1
Binary value (of bit group) to convert	x	0	1	1	0
		↓	↓	↓	↓
Values after multiplication	=	0 +	4 +	2 +	0

Sum of values for the first 4 bits = 0 + 4 + 2 + 0 = 6, which equals 6 hexadecimal.

Therefore, the decimal value 230 equals E6 hexadecimal.

Converting from hexadecimal to decimal

To convert a hexadecimal value in the range 00 to FF to decimal, use Table A.4.

TABLE A.4 Converting from Hexadecimal to Decimal

Hexadecimal position		2	1
Value of hexadecimal position		$16^1 = 16$	$16^0 = 1$
Decimal value of hexadecimal symbol*	x		
		↓	↓
Values after multiplication	=	+	

* Hexadecimal values between 0 and 9 are entered unchanged while A =10, B = 11, C = 12, D =13, E = 14, and F = 15.

The sum of the values after multiplication equals the decimal value.

EXAMPLE A.5 Converting a hexadecimal value to decimal

Given a hexadecimal value of E6, the decimal value is calculated as follows:

Hexadecimal position		2		1
Value of hexadecimal position		16		1
Decimal value of hexadecimal symbol	x	E = 14		6
		↓		↓
Values after multiplication	=	224	+	6

Sum of values = 224 + 6 =230.
Therefore, the hexadecimal value E6 equals 230 decimal.

Converting from hexadecimal to binary

To convert a hexadecimal value in the range 00 to FF to binary, use the following process:

1. Convert hexadecimal to decimal using Table A.4.
2. Convert decimal to binary using Table A.3.

EXAMPLE A.6 Converting a hexadecimal value to binary

Given a hexadecimal value of BC, the binary value is calculated as follows:

Step 1: Convert hexadecimal value to decimal.

Hexadecimal position		2		1
Value of hexadecimal position		16		1
Decimal value of hexadecimal symbol	x	B = 11		C = 12
		↓		↓
Values after multiplication	=	176	+	12

Sum of values = 176 + 12 = 188
Therefore, the hexadecimal value BC equals 188 decimal.

Step 2: Convert decimal value to binary.

Bit position		8	7	6	5	4	3	2	1
Decimal value to convert		188	60	60	28	12	4	—	—
Subtract value of bit position	-	128	64	32	16	8	4	2	1
		↓	↓	↓	↓	↓	↓	↓	↓
Decimal remainder after subtraction	=	60	(-4)	28	12	4	0	—	—
Conversion to binary		1	0	1	1	1	1	0	0

Therefore, the hexadecimal value BC equals 10111100 binary.

Conversion Table

TABLE A.5 Binary/Decimal/Hexadecimal Conversion Table

Binary Value		Decimal Value	Hexadecimal Value
0000	0000	0	00
0000	0001	1	01
0000	0010	2	02
0000	0011	3	03
0000	0100	4	04
0000	0101	5	05
0000	0110	6	06
0000	0111	7	07
0000	1000	8	08
0000	1001	9	09
0000	1010	10	0A
0000	1011	11	0B
0000	1100	12	0C
0000	1101	13	0D
0000	1110	14	0E
0000	1111	15	0F

TABLE A.5 Binary/Decimal/Hexadecimal Conversion Table (continued)

Binary Value		Decimal Value	Hexadecimal Value
0001	0000	16	10
0001	0001	17	11
0001	0010	18	12
0001	0011	19	13
0001	0100	20	14
0001	0101	21	15
0001	0110	22	16
0001	0111	23	17
0001	1000	24	18
0001	1001	25	19
0001	1010	26	1A
0001	1011	27	1B
0001	1100	28	1C
0001	1101	29	1D
0001	1110	30	1E
0001	1111	31	1F
0010	0000	32	20
0010	0001	33	21
0010	0010	34	22
0010	0011	35	23
0010	0100	36	24
0010	0101	37	25
0010	0110	38	26
0010	0111	39	27
0010	1000	40	28
0010	1001	41	29
0010	1010	42	2A
0010	1011	43	2B
0010	1100	44	2C
0010	1101	45	2D
0010	1110	46	2E
0010	1111	47	2F

TABLE A.5 Binary/Decimal/Hexadecimal Conversion Table (continued)

Binary Value		Decimal Value	Hexadecimal Value
0011	0000	48	30
0011	0001	49	31
0011	0010	50	32
0011	0011	51	33
0011	0100	52	34
0011	0101	53	35
0011	0110	54	36
0011	0111	55	37
0011	1000	56	38
0011	1001	57	39
0011	1010	58	3A
0011	1011	59	3B
0011	1100	60	3C
0011	1101	61	3D
0011	1110	62	3E
0011	1111	63	3F
0100	0000	64	40
0100	0001	65	41
0100	0010	66	42
0100	0011	67	43
0100	0100	68	44
0100	0101	69	45
0100	0110	70	46
0100	0111	71	47
0100	1000	72	48
0100	1001	73	49
0100	1010	74	4A
0100	1011	75	4B
0100	1100	76	4C
0100	1101	77	4D
0100	1110	78	4E
0100	1111	79	4F

TABLE A.5 Binary/Decimal/Hexadecimal Conversion Table (continued)

Binary Value		Decimal Value	Hexadecimal Value
0101	0000	80	50
0101	0001	81	51
0101	0010	82	52
0101	0011	83	52
0101	0011	83	53
0101	0100	84	54
0101	0101	85	55
0101	0110	86	56
0101	0111	87	57
0101	1000	88	58
0101	1001	89	59
0101	1010	90	5A
0101	1011	91	5B
0101	1100	92	5C
0101	1101	93	5D
0101	1110	94	5E
0101	1111	95	5F
0110	0000	96	60
0110	0001	97	61
0110	0010	98	62
0110	0011	99	63
0110	0100	100	64
0110	0101	101	65
0110	0110	102	66
0110	0111	103	67
0110	1000	104	68
0110	1001	105	69
0110	1010	106	6A
0110	1011	107	6B
0110	1100	108	6C
0110	1101	109	6D
0110	1110	110	6E
0110	1111	111	6F

TABLE A.5 Binary/Decimal/Hexadecimal Conversion Table (continued)

Binary Value		Decimal Value	Hexadecimal Value
0111	0000	112	70
0111	0001	113	71
0111	0010	114	72
0111	0011	115	73
0111	0100	116	74
0111	0101	117	75
0111	0110	118	76
0111	0111	119	77
0111	1000	120	78
0111	1001	121	79
0111	1010	122	7A
0111	1011	123	7B
0111	1100	124	7C
0111	1101	125	7D
0111	1110	126	7E
0111	1111	127	7F
1000	0000	128	80
1000	0001	129	81
1000	0010	130	82
1000	0011	131	83
1000	0100	132	84
1000	0101	133	85
1000	0110	134	86
1000	0111	135	87
1000	1000	136	88
1000	1001	137	89
1000	1010	138	8A
1000	1011	139	8B
1000	1100	140	8C
1000	1101	141	8D
1000	1110	142	8E
1000	1111	143	8F

TABLE A.5 Binary/Decimal/Hexadecimal Conversion Table (continued)

Binary Value		Decimal Value	Hexadecimal Value
1001	0000	144	90
1001	0001	145	91
1001	0010	146	92
1001	0011	147	93
1001	0100	148	94
1001	0101	149	95
1001	0110	150	96
1001	0111	151	97
1001	1000	152	98
1001	1001	153	99
1001	1010	154	9A
1001	1011	155	9B
1001	1100	156	9C
1001	1101	157	9D
1001	1110	158	9E
1001	1111	159	9F
1010	0000	160	A0
1010	0001	161	A1
1010	0010	162	A2
1010	0011	163	A3
1010	0100	164	A4
1010	0101	165	A5
1010	0110	166	A6
1010	0111	167	A7
1010	1000	168	A8
1010	1001	169	A9
1010	1010	170	AA
1010	1011	171	AB
1010	1100	172	AC
1010	1101	173	AD
1010	1110	174	AE
1010	1111	175	AF

TABLE A.5 Binary/Decimal/Hexadecimal Conversion Table (continued)

Binary Value		Decimal Value	Hexadecimal Value
1011	0000	176	B0
1011	0001	177	B1
1011	0010	178	B2
1011	0011	179	B3
1011	0100	180	B4
1011	0101	181	B5
1011	0110	182	B6
1011	0111	183	B7
1011	1000	184	B8
1011	1001	185	B9
1011	1010	186	BA
1011	1011	187	BB
1011	1100	188	BC
1011	1101	189	BD
1011	1110	190	BE
1011	1111	191	BF
1100	0000	192	C0
1100	0001	193	C1
1100	0010	194	C2
1100	0011	195	C3
1100	0100	196	C4
1100	0101	197	C5
1100	0110	198	C6
1100	0111	199	C7
1100	1000	200	C8
1100	1001	201	C9
1100	1010	202	CA
1100	1011	203	CB
1100	1100	204	CC
1100	1101	205	CD
1100	1110	206	CE
1100	1111	207	CF

TABLE A.5 Binary/Decimal/Hexadecimal Conversion Table (continued)

Binary Value		Decimal Value	Hexadecimal Value
1101	0000	208	D0
1101	0001	209	D1
1101	0010	210	D2
1101	0011	211	D3
1101	0100	212	D4
1101	0101	213	D5
1101	0110	214	D6
1101	0111	215	D7
1101	1000	216	D8
1101	1001	217	D9
1101	1010	218	DA
1101	1011	219	DB
1101	1100	220	DC
1101	1101	221	DD
1101	1110	222	DE
1101	1111	223	DF
1110	0000	224	E0
1110	0001	225	E1
1110	0010	226	E2
1110	0011	227	E3
1110	0100	228	E4
1110	0101	229	E5
1110	0110	230	E6
1110	0111	231	E7
1110	1000	232	E8
1110	1001	233	E9
1110	1010	234	EA
1110	1011	235	EB
1110	1100	236	EC
1110	1101	237	ED
1110	1110	238	EE
1110	1111	239	EF

TABLE A.5 Binary/Decimal/Hexadecimal Conversion Table (continued)

Binary Value		Decimal Value	Hexadecimal Value
1111	0000	240	F0
1111	0001	241	F1
1111	0010	242	F2
1111	0011	243	F3
1111	0100	244	F4
1111	0101	245	F5
1111	0110	246	F6
1111	0111	247	F7
1111	1000	248	F8
1111	1001	249	F9
1111	1010	250	FA
1111	1011	251	FB
1111	1100	252	FC
1111	1101	253	FD
1111	1110	254	FE
1111	1111	255	FF

Appendix

B

Internet Protocol (IP) Addressing Fundamentals

Overview

General

Devices on a local area network (LAN) require both device addresses and network addresses for their interfaces. Devices with multiple network interfaces—such as servers with multiple network interface cards (NICs) or routers—must be assigned separate addresses for each interface.

This appendix describes the network addressing format developed for use with Internet protocol version 4 (IPv4). IP is the Open Systems Interconnection (OSI) Network layer protocol used on the Internet, and must be implemented on any device connected to the Internet. The popularity and growth of the Internet has also convinced many organizations to use IP throughout their internal networks for consistency, making IPv4 a near-universal Network layer protocol. To accommodate future needs, an updated version of IP called IP version 6 (IPv6) is intended to eventually replace IPv4.

NOTE: All references to IP in this appendix should be interpreted as IPv4.

IP address format

An IP address is 32 bits in length—making 2^{32} (or 4,294,967,296) unique combinations possible for use as addresses. For convenience, the 32 bits are divided into four groups of 8 bits. Since 8 bits can represent 2^8 (or 256) unique combinations, a 32-bit IP address can be expressed using four values, each in the range 0 to 255. For clarity, the four values are separated by periods or dots, making this the dotted decimal format of IP addressing (see Table B.1).

TABLE B.1 Binary and Dotted Decimal Formats of an IP Address

32-bit IP address:	11001111001100010110100001010000
32 bits divided into four groups:	11001111.00110001.01101000.01010000
Dotted decimal equivalent:	207.49.104.80

NOTE: Binary to decimal conversion is described in Appendix A: Binary and Hexadecimal Numbering Fundamentals.

Various combinations of the 32 bits are used to assign values to two components of an IP address.

- The network identification (netid) bits in an address uniquely identify the LAN broadcast domain in which the device is located.
- The host identification (hostid) bits in an address uniquely identify the device within a given LAN broadcast domain.

To indicate the number of bits used for netid numbering in an address, the slash symbol (/) followed by the number of netid bits is commonly appended to an address, as shown in Example B.1.

EXAMPLE B.1 Identifying the netid and hostid bits

Using the address shown in Table B.1, assume that the first 24 bits represent the netid and the last 8 bits represent the hostid.

32-bit IP address:	11001111001100010110100001010000
32 bits divided into four groups:	11001111.00110001.01101000.01010000
Dotted decimal equivalent:	207.49.104.80
netid bits:	11001111.00110001.01101000 or 207.49.104
hostid bits:	01010000 or 80

Therefore, the dotted decimal format of this address can be expressed as 207.49.104/24, where /24 identifies the number of bits assigned to netid numbering.

IP Address Classes

Overview

In some organizations there are many interconnected LANs, each with fewer than fifty devices. In other cases, the reverse is true—there are a small number of networks, each with thousands of devices. To accommodate organizational networks of various sizes, three IP address classes were incorporated into the original design of IP. Each address class assigns a different number of bits to netid, as follows:

- In a Class A address, the first 8 bits are assigned to netid (/8), leaving 24 bits for hostid numbering.
- In a Class B address, the first 16 bits are assigned to netid (/16), leaving 16 bits for hostid numbering.
- In a Class C address, the first 24 bits are assigned to netid (/24), leaving 8 bits for hostid numbering.

TABLE B.2 IP Address Classes and netid Numbering

	Class A	Class B	Class C
Number of bits assigned to netid numbering	8	16	24
Subtract number of reserved bits within netid (these cannot be changed)	–1	–2	–3
Number of bits available for netid	7	14	21
Number of unique netid values	$2^7 = 128$	$2^{14} = 16{,}384$	$2^{21} = 2{,}097{,}152$
Subtract number of reserved netid values (0 and 127 are reserved and cannot be assigned)	–2	–0	–0
Number of available netid numbers	126	16,384	2,097,152
Ranges of available IP addresses within each class (xxx represents hostid address)	1.xxx.xxx.xxx/8 through 126.xxx.xxx.xxx/8	128.0.xxx.xxx/16 through 191.255.xxx.xxx/16	192.0.0.xxx/24 through 223.255.255.xxx/24

TABLE B.3 IP Address Classes and hostid Numbering

	Class A	Class B	Class C
Number of bits available for hostid numbering	24	16	8
Number of unique hostid values	2^{24} = 16,777,216	2^{16} = 65,536	2^{8} = 256
Subtract number of reserved hostid values (0 and 255 [binary 00000000 and 11111111] cannot be assigned in the last group)	–2	–2	–2
Number of available hostid numbers	16,777,214	65,534	254
Ranges of available IP addresses within each class (xxx represents netid address)	xxx.0.0.1 through xxx.255.255.254	xxx.xxx.0.1 through xxx.xxx.255.254	xxx.xxx.xxx.1 through xxx.xxx.xxx.254

TABLE B.4 IP Address Ranges by Class in Binary Form

Class A (/8)	00000001. (netid) 00000000.00000000.00000001 (hostid) through 01111110. (netid) 11111111.11111111.11111110 (hostid)
Class B (/16)	10000000.00000000 (netid) 00000000.00000001 (hostid) through 10111111.11111111 (netid) 11111111.11111110 (hostid)
Class C (/24)	11000000.00000000.00000000. (netid) 00000001 (hostid) through 11011111.11111111.11111111 (netid) 11111110 (hostid)

NOTES: The first netid bit in a Class A address is reserved and cannot be changed. Therefore, all Class A addresses begin with binary 0.

The first two netid bits in a Class B address are reserved and cannot be changed. Therefore, all Class B addresses begin with binary 10.

The first three netid bits in a Class C address are reserved and cannot be changed. Therefore, all Class C addresses begin with binary 110.

TABLE B.5 IP Address Ranges by Class in Dotted Decimal Form

Class A	Class B	Class C
1.0.0.1/8 through 126.255.254/8	128.0.0.1/16 through 191.255.255.254/16	192.0.0.1/24 through 223.255.255.254/24

Address classes were developed to assign Internet addresses on the basis of organizational network size.

- The Class A address range was intended to identify 126 large organizational networks. Each Class A network can contain up to 16,777,214 devices.
- The Class B address range was intended to identify 16,384 mid-sized organizational networks. Each Class B network can contain up to 65,534 devices.
- The Class C address range was intended to identify 2,097,152 small organizational networks. Each Class C network can contain up to 254 devices.

NOTE: There is also a Class D and a Class E range of IP addresses. Class D addresses are reserved for multicast applications and Class E addresses are classified as experimental. The addresses in these ranges cannot be assigned to devices as network addresses.

Private Addresses

To avoid the possibility of address duplication on the Internet, a central authority called the Internet Corporation for Assigned Names and Numbers (ICANN) oversees address distribution.

An IP address visible or accessible over the Internet is called a public address. In cases where no connections to the Internet are anticipated, any IP address can be assigned to a device. However, all devices may require renumbering if it becomes necessary to connect to the Internet at some future time.

To provide flexibility in cases where future Internet connections are a possibility, a range of addresses has been reserved in each class for private use. Private addresses cannot be accessed over the Internet. They can only be used on internal networks. The private address ranges are as follows:

- Class A private address range: 10.0.0.1 through 10.255.255.254, representing one broadcast domain (netid 10) containing a maximum of 16,777,214 devices (hostid .0.0.1 through hostid .255.255.254).
- Class B private address range: 172.16.0.1 through 172.31.255.254, representing 16 broadcast domains (netid 172.16 through netid 172.31), each containing a maximum of 65,534 devices (hostid .0.1 through hostid .255.254).
- Class C private address range: 192.168.0.1 through 192.168.255.254, representing 256 broadcast domains (netid 192.168.0 through netid 192.168.255), each containing a maximum of 254 devices (hostid .1 through hostid .254).

Subnetwork Addressing

Overview

The growth of the Internet has made it necessary to allocate addresses efficiently to accommodate all requests from individuals and organizations. The original set of three address classes was conceived in the 1970s before the introduction of LANs and global Internet access, and lacks the flexibility needed in many organizational environments.

Subnetwork addressing, also called subnetting, makes it possible to use any number of bits in an IP address as netid, rather than the 8, 16, or 24 provided by Class A, B, and C ranges, respectively. However, this also reduces the maximum number of devices that can be included in a broadcast domain. For example, an organization assigned a single Class C address can identify one broadcast domain containing a maximum of 254 devices. If this organization wants to create additional broadcast domains, it can use subnetting to classify any of the eight hostid bits in its Class C address as netid bits, as shown in Example B.2.

EXAMPLE B.2 Subnetting a Class C address

The Class C address used as an example is: 207.49.104.0/24

This consists of one netid address, which contains 254 hostid addresses.

The address of the single broadcast domain in dotted decimal and binary formats is as follows:

- 207.49.104.0 (11001111.00110001.01101000.00000000 in binary form)

A total of 254 addresses can be assigned to devices within this single broadcast domain. These addresses (in dotted decimal and binary formats) are as follows:

- 207.49.104.1 through 207.49.104.254
  (11001111.00110001.01101000.00000001 through
  11001111.00110001.01101000.11111110 in binary)

NOTE: The hostid address containing all one bits (e.g., 11111111 [or .255 in dotted decimal format]) is used to indicate all devices in the broadcast domain. It is used when the same datagram needs to be sent to every device.

Therefore, prior to subnetting, the address can be described as follows:

24 netid bits in the address: → 207.49.104 (11001111.00110001.01101000)
8 hostid bits in the address: → .1 through .254 (.00000001 through .11111110)
All-device broadcast address: → 207.49.104.255 (11001111.00110001.01101000.11111111)

If two of the eight hostid bits are reassigned as netid bits, four subnet broadcast domains can be created using the same address, as described below:

The modified Class C address becomes: 207.49.104.0/26

This consists of four netid addresses, each of which contains 62 hostid addresses.

The addresses of the four subnet broadcast domains (netid bits are shown in boldface) are as follows:

- 207.49.104.0 (**11001111.00110001.01101000.00**000000)
- 207.49.104.64 (**11001111.00110001.01101000.01**000000)
- 207.49.104.128 (**11001111.00110001.01101000.10**000000)
- 207.49.104.192 (**11001111.00110001.01101000.11**000000)

NOTE: In early implementations of subnetting, netid values ending (the values after the last dot) with all zeros or all ones were not permitted. Applied in this example, this rule would allow only two of the four broadcast domains shown above—207.49.104.64 and 207.49.104.128. Current versions of routing software permit subnets ending in all zeros or all ones.

The 62 hostid addresses that can be assigned to devices in the 207.49.104.0 subnet broadcast domain (hostid bits are shown in boldface) are as follows:

- 207.49.104.1 through 207.49.104.62 (11001111.00110001.01101000.00**000001** through 11001111.00110001.01101000.00**111110**)
- 207.49.104.63 (11001111.00110001.01101000.00**111111**)—the all-device broadcast address for this subnet.

The 62 hostid addresses that can be assigned to devices in the 207.49.104.64 subnet broadcast domain (hostid bits are shown in boldface) are as follows:

- 207.49.104.65 through 207.49.104.126 (11001111.00110001.01101000.01**000001** through 11001111.00110001.01101000.01**111110**)
- 207.49.104.127 (11001111.00110001.01101000.01**111111**)—the all-device broadcast address for this subnet.

The 62 hostid addresses that can be assigned to devices in the 207.49.104.128 subnet broadcast domain (hostid bits are shown in boldface) are as follows:

- 207.49.104.129 through 207.49.104.190 (11001111.00110001.01101000.10**000001** through 11001111.00110001.01101000.10**111110**)
- 207.49.104.191 (11001111.00110001.01101000.10**111111**)—the all-device broadcast address for this subnet.

The 62 hostid addresses that can be assigned to devices in the 207.49.104.192 subnet broadcast domain (hostid bits are shown in boldface) are as follows:

- 207.49.104.193 through 207.49.104.254 (11001111.00110001.01101000.11**000001** through 11001111.00110001.01101000.11**111110**)
- 207.49.104.255 (11001111.00110001.01101000.11**111111**)—the all-device broadcast address for this subnet.

The following figures illustrate the subnetting example described in Example B.2. In Figure B.1, the Class C network is shown before subnetting, while Figure B.2 illustrates the same network after it has been subnetted.

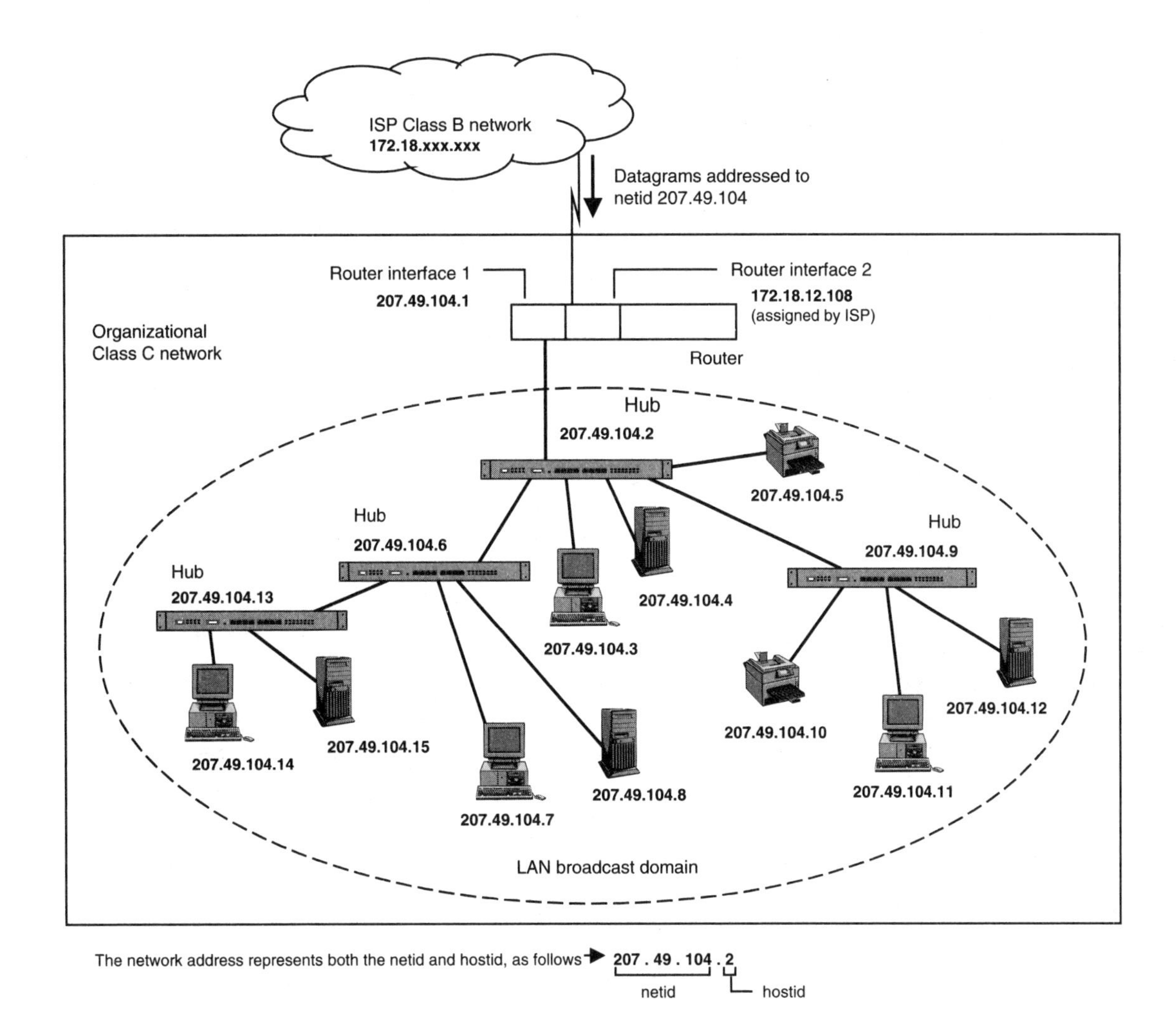

ISP = Internet service provider
LAN = Local area network

Figure B.1 Class C network prior to subnetting.

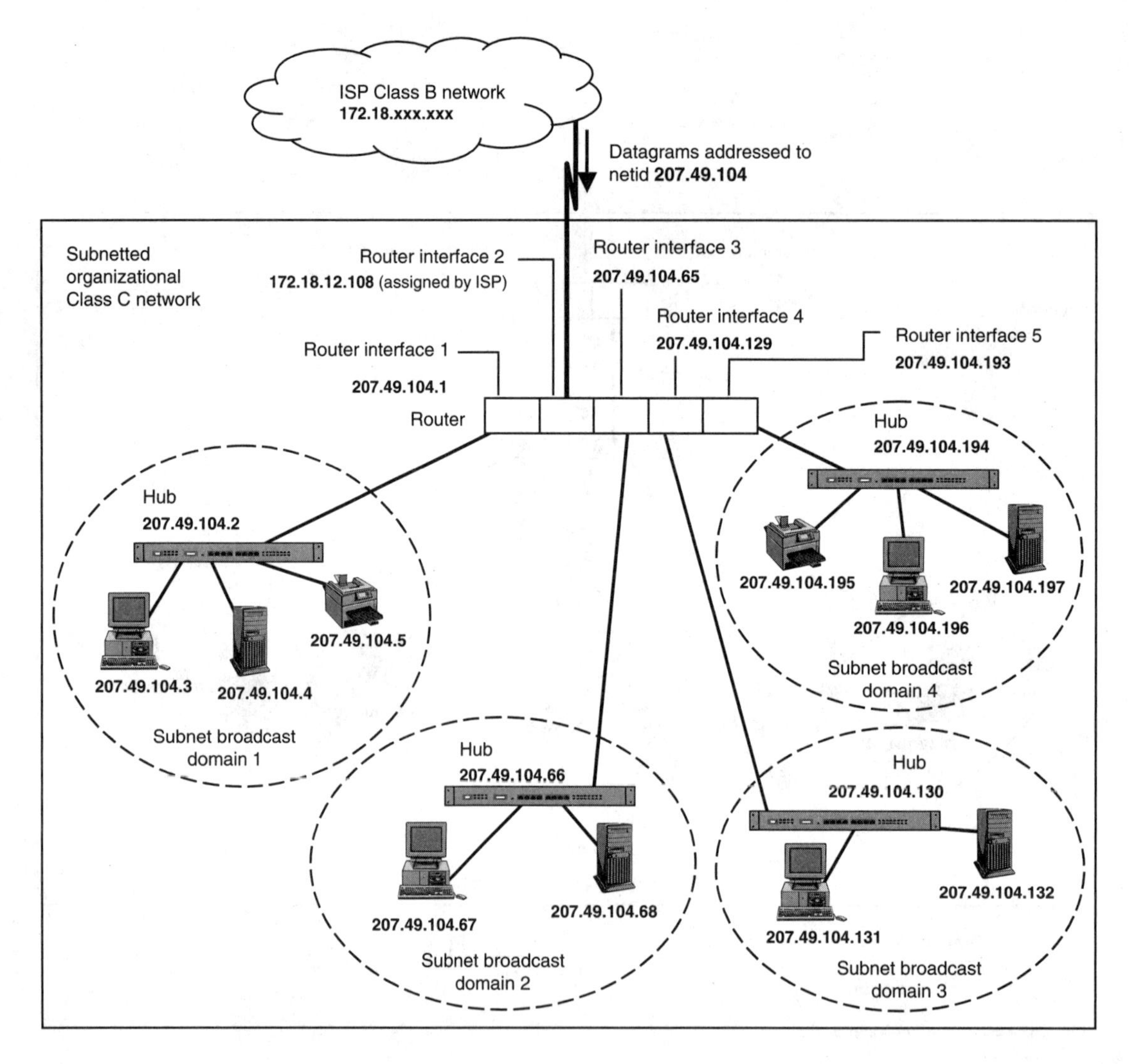

ISP = Internet service provider
LAN = Local area network

Figure B.2 Subnetted Class C network.

Appendix

C

Network Design Examples

Overview

General

Networking can be classified into four levels (see Figure C.1). Each represents a fundamental part of an organizational network, as follows:

- Level one consists of the cabling and telecommunications circuits used on a network.
- Level two consists of the devices connected to the cabling and telecommunications circuits.
- Level three consists of the network architectures used to enable communications between devices.
- Level four consists of the technologies and practices used by administrators to configure and manage the network, as well as the software applications made available to network users.

The following sections describe the initial design and subsequent expansion of a typical organizational network, from a single departmental local area network (LAN) to a wide area network (WAN) linking multiple sites. Each design is divided into four parts to illustrate the four levels in the network.

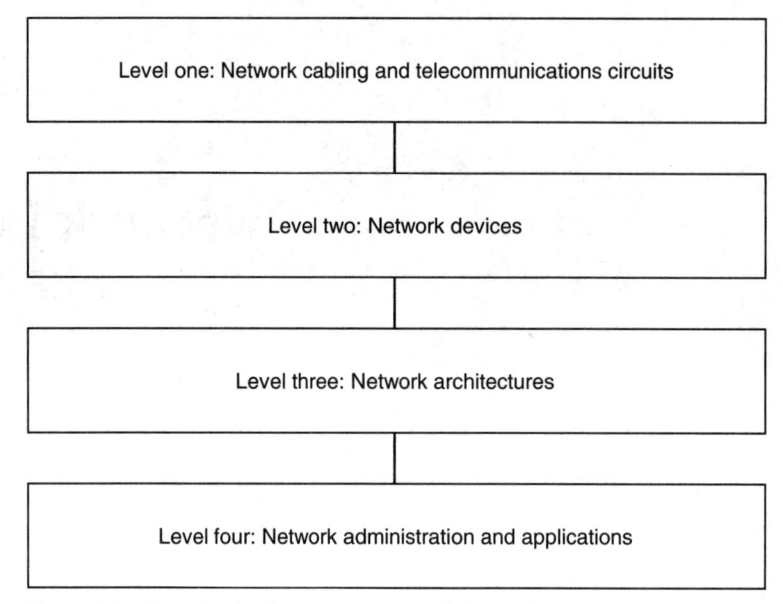

Figure C.1 Four levels of an organizational network.

Local Area Network (LAN)

General

In this section, the first LAN is designed and implemented. This network serves a group of users in a single department, located on one floor of a multi-floor building.

Level 1: communications paths

A structured cabling system is typically used to provide the communications paths for a LAN (see Figure C.2).

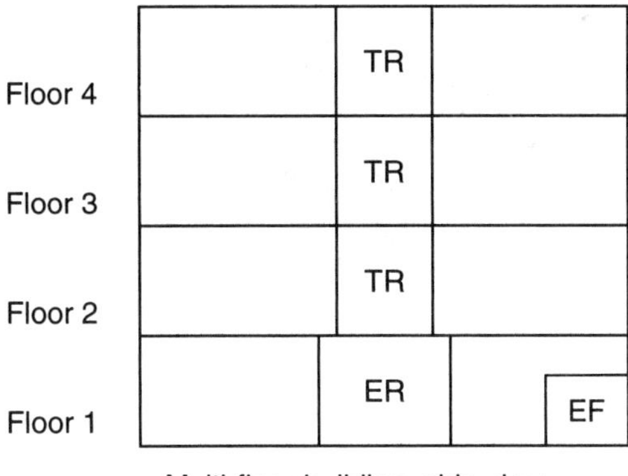

Multi-floor building, side view

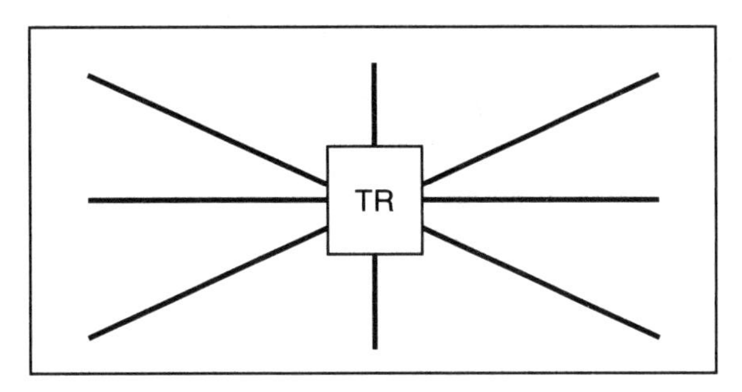

Floor 4, top view

——— = Cabling to work areas
EF = Entrance facility
ER = Equipment room
TR = Telecommunications room

Figure C.2 Local area network communications paths.

Level 2: devices

The basic hardware components of a LAN are stations, servers, shared peripheral devices (e.g. printers), network interface cards (NICs), and network access devices (e.g. hubs, switches [see Figure C.3]).

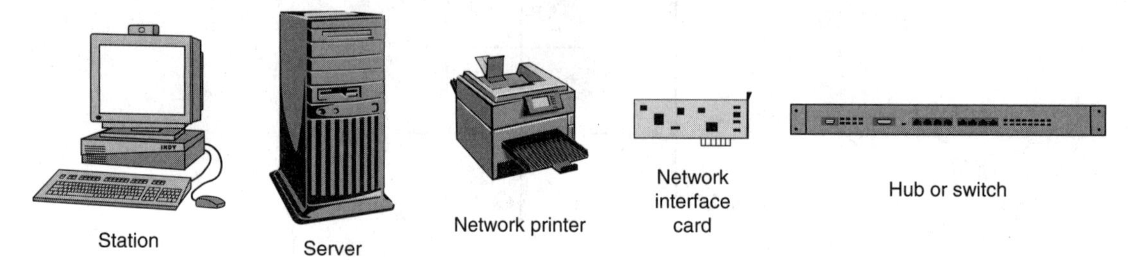

Figure C.3 Local area network devices.

Level 3: architectures

Multiple network architectures are available, each specifying how devices are linked and the methods used to exchange messages over the network communications path(s) (see Figure C.4).

Level 4: administration and applications

From a user's perspective, network applications should be useful and easy to access. From a network administrator's perspective, a network should be secure and easy to manage.

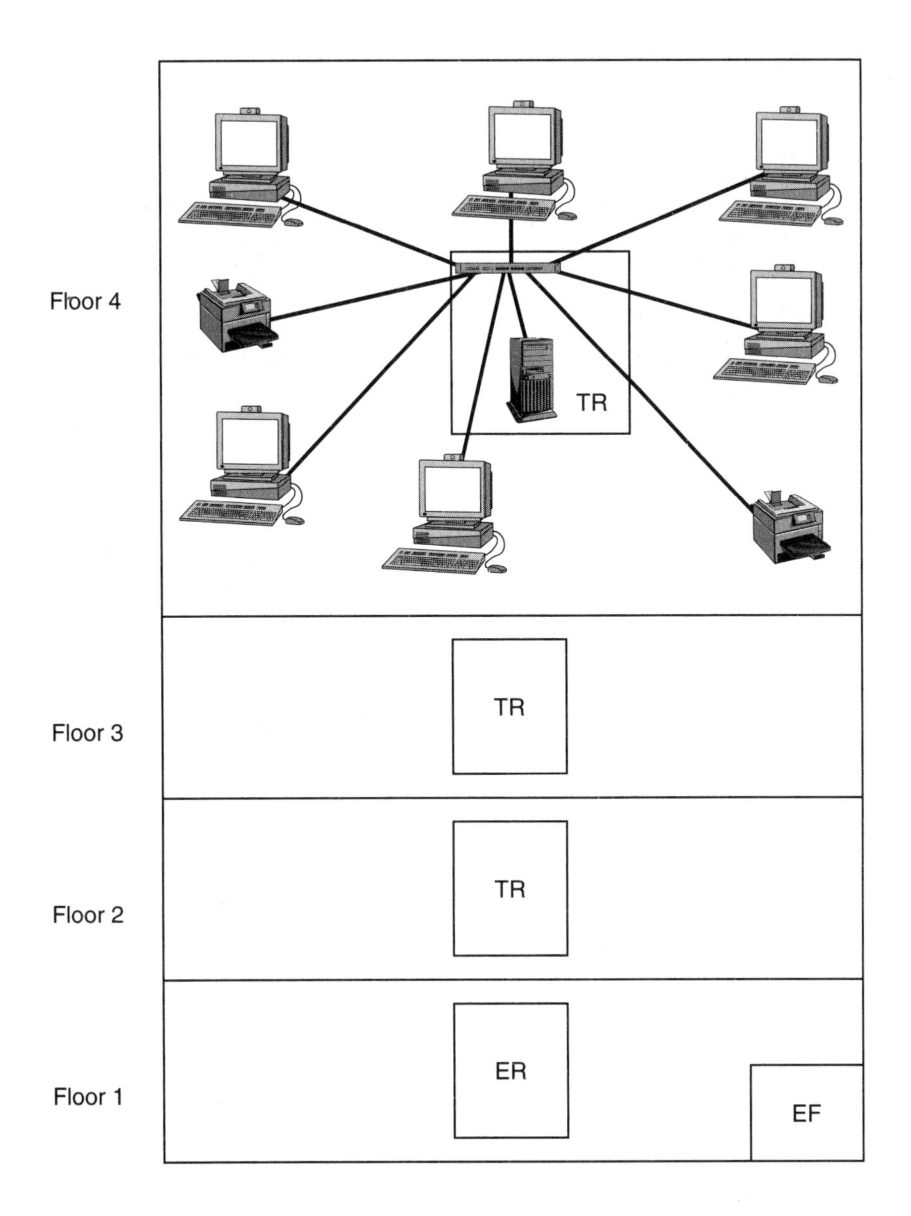

EF = Entrance facility
ER = Equipment room
TR = Telecommunications room

Figure C.4 Local area network architecture.

Remote Access

General

In this section, users are given the flexibility to connect to a departmental LAN from any location (e.g., home, hotel).

Level 1: communications paths

Telecommunications circuits are used to provide the communications paths for remote devices.

Level 2: devices

The basic hardware components of remote access are remote stations, modems, remote access servers, and remote access interface cards (see Figure C.5).

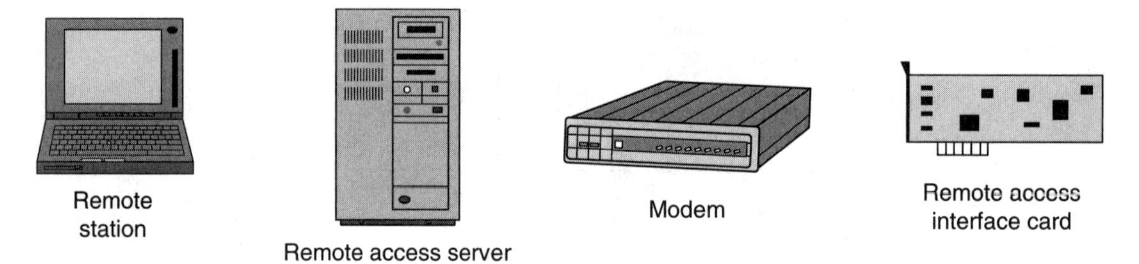

Figure C.5 Remote access devices.

Level 3: architectures

Both remote control and remote node architectures can be used to implement remote access services on an organizational network (see Figure C.6).

Level 4: administration and applications

When remote access services are implemented, network security must be expanded to include remote users.

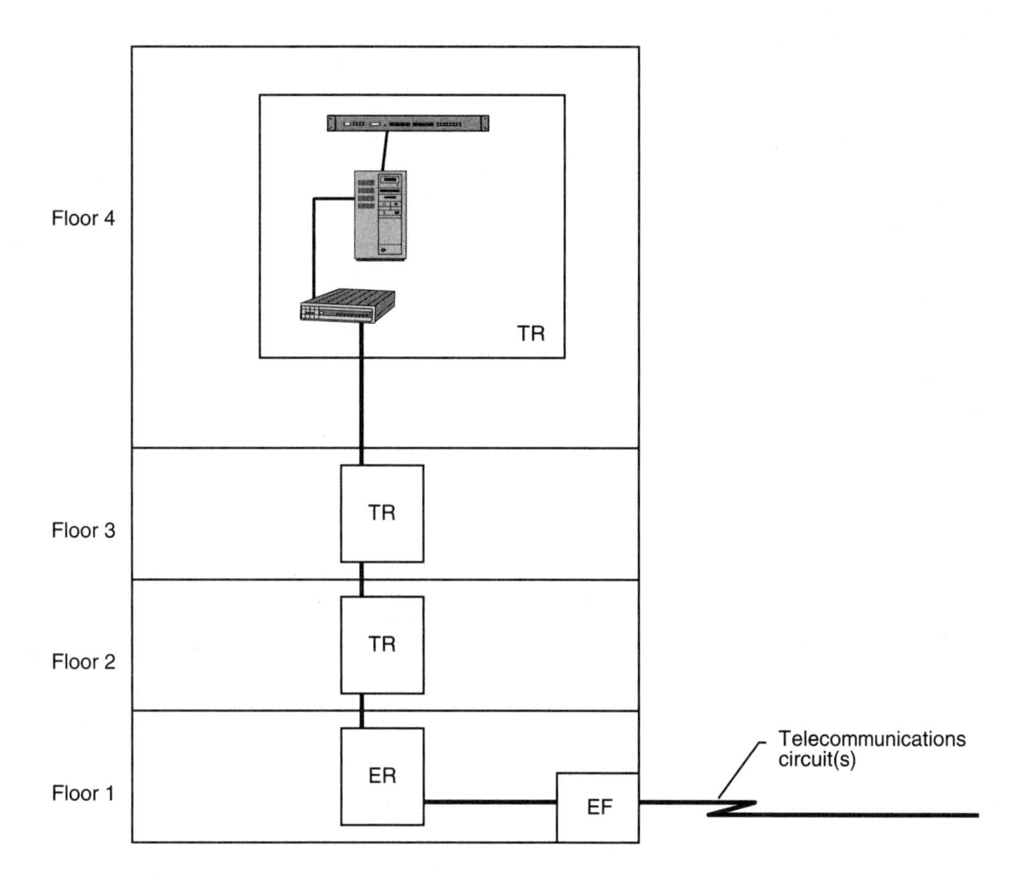

Figure C.6 Remote access architecture.

Building Internetwork

General

In this section, multiple departmental LANs within the building are linked using an internetwork.

Level 1: communications paths

A structured cabling system is typically used to provide the communications paths for a building internetwork.

Level 2: devices

The basic hardware component of a building internetwork is the network access device (e.g., switch [see Figure C.7]).

Switch

Figure C.7 Building internetwork device.

Level 3: architectures

If an internetwork is centralized, all LANs are connected to the internetwork at a common location (see Figure C.8).

Level 4: administration and applications

The expanded organizational network makes it necessary to expand the administrative services and user applications previously in place.

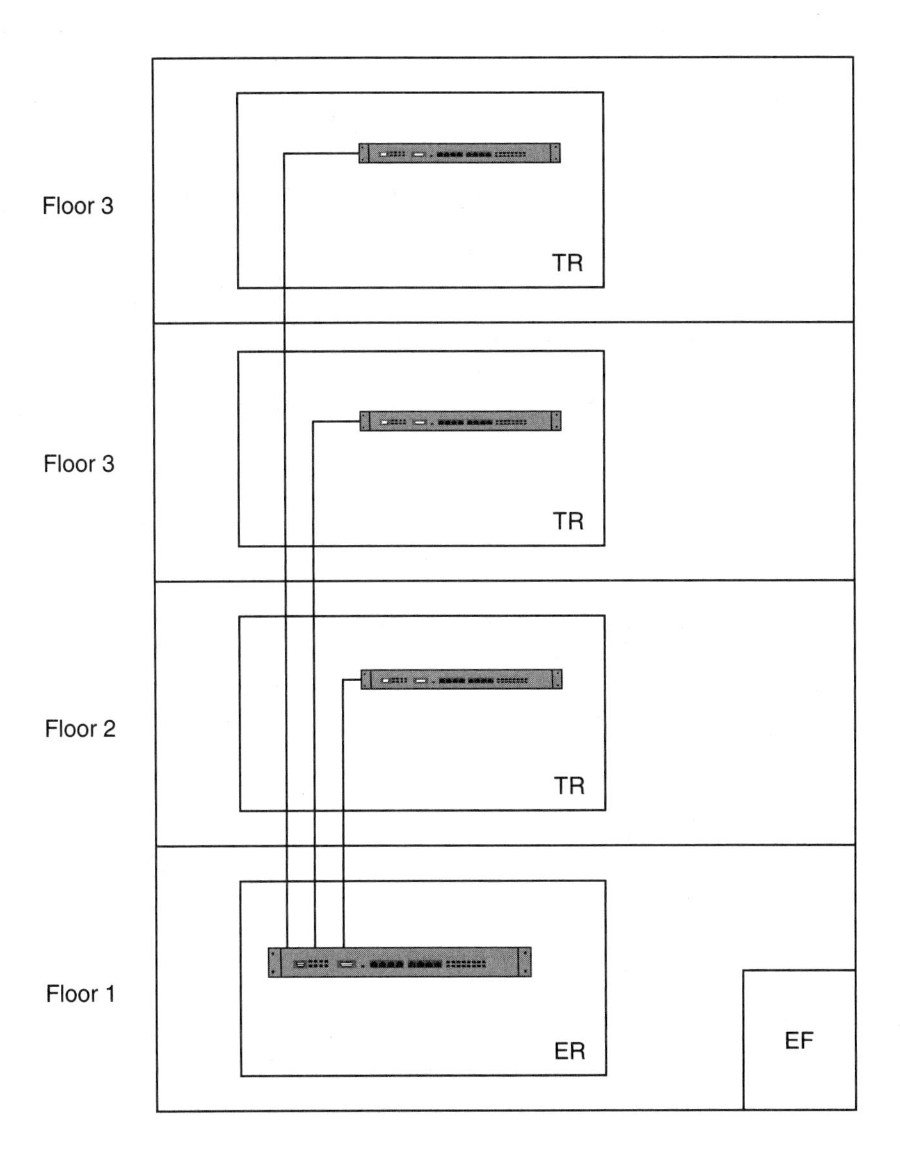

EF = Entrance facility
ER = Equipment room
TR = Telecommunications room

Figure C.8 Building internetwork architecture.

Campus Internetwork

General

In this section, LANs located in two adjacent buildings are linked to the existing internetwork.

Level 1: communications paths

A structured cabling system is typically used to provide the communications paths for a campus internetwork.

Level 2: devices

The basic hardware component of a campus internetwork is the network access device (e.g., switch [see Figure C.9]).

Switch

Figure C.9 Campus internetwork device.

Level 3: architectures

If network services are centralized on a large scale, additional network storage and access technologies can be used to provide a high level of performance (see Figure C.10).

Level 4: administration and applications

The expanded organizational network makes it necessary to expand the administrative services and user applications previously in place.

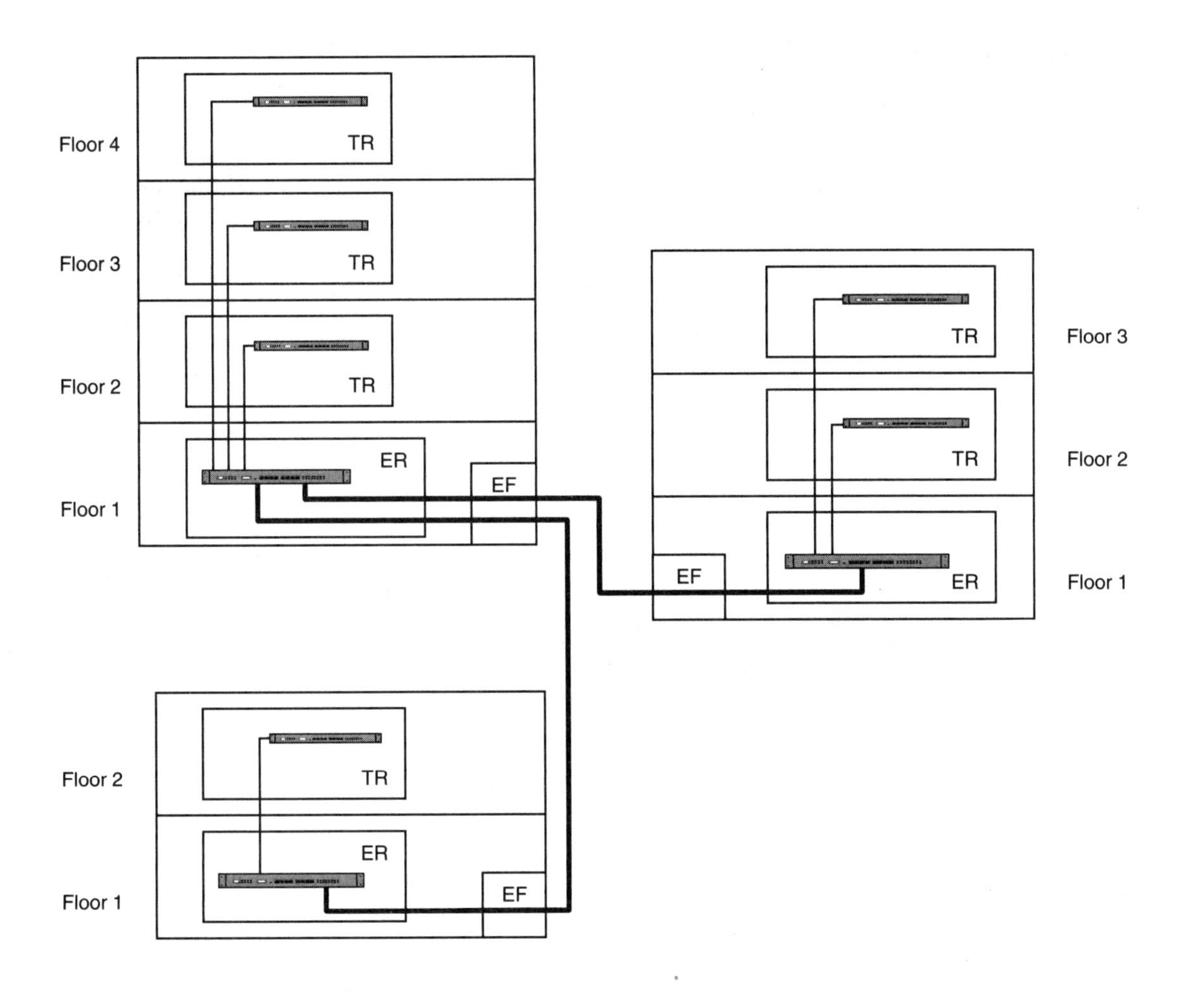

Figure C.10 Campus internetwork architecture.

Wide Area Network (WAN)

General

In this section, the organizational network is expanded to include the LANs located in a branch office in another city, as well as connections to the Internet.

Level 1: communications paths

Telecommunications circuits are used to provide the communications paths for a WAN.

Level 2: devices

The basic hardware component of an Internet-connected WAN is the network access device (e.g., router [see Figure C.11]).

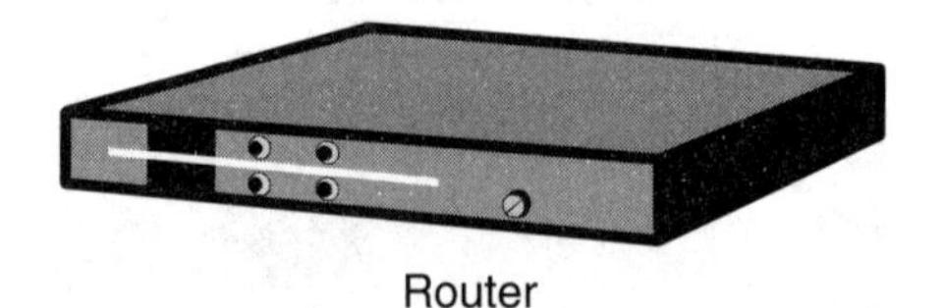

Figure C.11 Wide area network device.

Level 3: architectures

It is possible to optimize WAN performance by managing the flow of traffic over telecommunications circuits, using technologies described as quality of service (QoS) initiatives (see Figure C.12).

Level 4: administration and applications

When an organizational network is connected to the Internet, additional security is required to protect network resources.

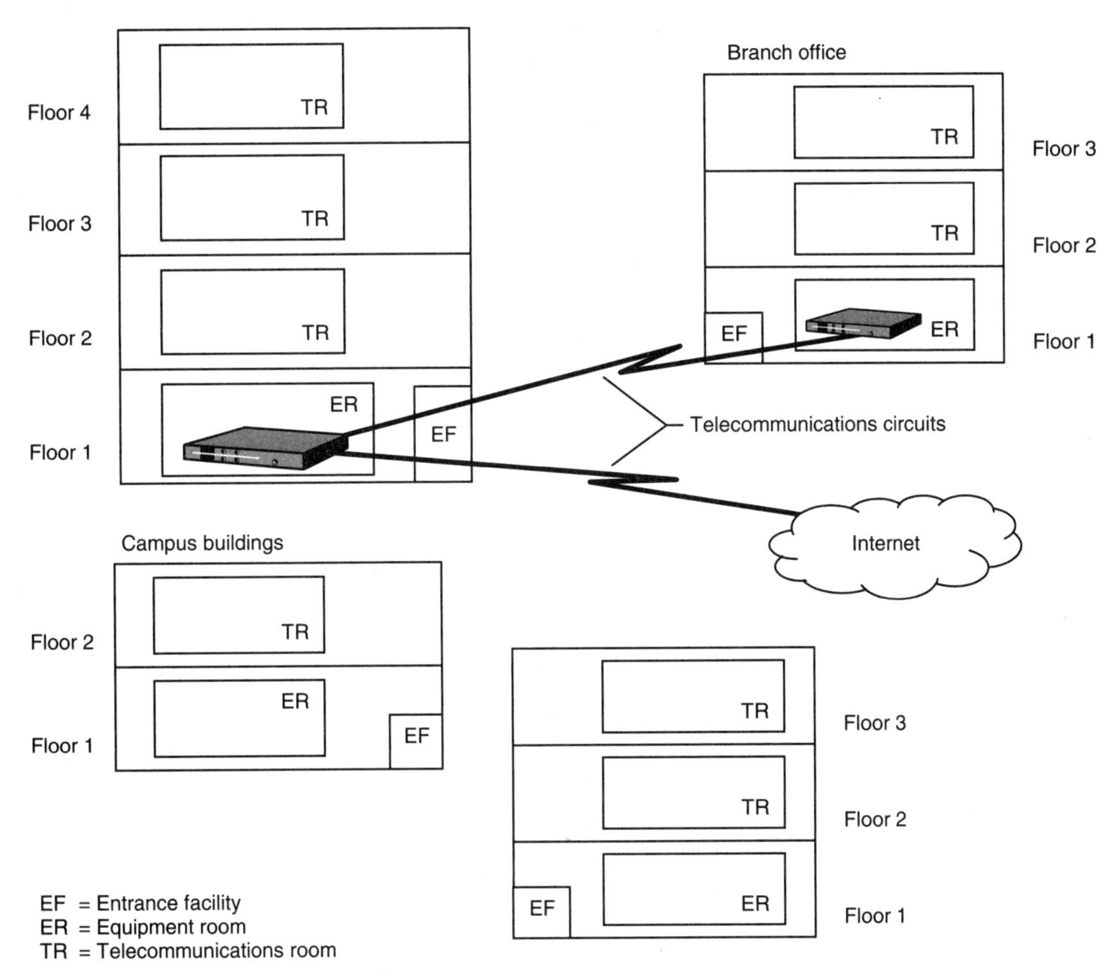

Figure C.12 Wide area network architecture.

Glossary

Glossary of common definitions, including frequently used symbols and acronyms.

Definitions

A

access The process of connecting to a network.

access control list (ACL) A mechanism used to control access to network services through assigned privileges for users and groups. Also referred to as an authorization profile.

access point (AP) The central or control point in a wireless cell that acts as a link for traffic to and from wireless devices in the cell. The access point also connects wireless devices to the wired portion of the network.

access provider (AP) A company, such as a telephone company, that provides a circuit path between a service provider and the client user. An access provider can also be the service provider. See service provider (SP).

address See device address and network address.

agent A mechanism to collect information about a device and make that information available to a network management system. Agents operate in managed devices on the network (e.g., hubs, switches, servers, and stations).

American National Standards Institute (ANSI) ANSI Federation is a private, nonprofit membership organization focused on meeting the standards and conformity assessment requirements of its diverse constituency. It provides a neutral forum for the development of consensus agreements on issues relevant to voluntary standardization. The United States representative to the International Organization for Standardization, and through the United States National Committee, to the International Electrotechnical Commission.

American wire gauge (AWG) A system used to specify wire size. The greater the wire diameter, the smaller the AWG value.

analog signal A signal in the form of a wave that uses continuous physical variables (e.g., voltage amplitude or frequency variations) to transmit information. Contrast with digital signal.

Application layer The Open Systems Interconnection model layer responsible for providing mechanisms that enable similar or dissimilar applications on different systems to use the services of a network to exchange information. Also referred to as Layer 7.

asymmetric digital subscriber line (ADSL) See xDSL.

asynchronous communication A form of signaling in which each data character is coded as a string of bits. The bits are separated by start-character and stop-character bits. See isochronous communication and synchronous communication.

asynchronous transfer mode (ATM) A high-speed switching transmission protocol that utilizes payload packages organized into cells to carry all types of traffic (e.g., voice, data, still image, or audio/video).

ATM Forum, The An international organization of manufacturers, service suppliers, and users of ATM.

attachment One or more files connected to and sent with an e-mail message.

attenuation A decrease in transmission signal strength between points, expressed as the ratio of output to input.

audit trail System records used to track access to network resources. Logs all access to specific devices and files.

authentication A security mechanism that confirms the identity of a user accessing the network.

authorization The process of establishing and enforcing the network activities that are permitted for a given user.

auto-negotiation (AUTONEG) A feature that determines link options and optimal settings for a given connection. When auto-negotiation is enabled, a network interface card can determine the capabilities of the device at the far end of the link and select the best mode of operation.

availability A measure of network response time or operating failure during a given period of time.

B

backbone cabling Cable and connecting hardware that provides interconnections between telecommunications rooms, equipment rooms, and entrance facilities.

backbone network See internetwork.

backup A copy of the data stored on a device.

backup path A secondary or alternate channel for signal flow. It is typically used when there has been a failure of the main (primary) path.

bandwidth A measure of the range of frequencies associated with a given signal or communications channel, typically expressed in hertz. It is used to denote the potential capacity of the medium, device, or system.

baseband signaling Transmission of an analog or digital signal at its original frequency. A method of signal transmission where the entire bandwidth of the medium is used to send a single signal.

baud A measure of signaling speed equal to the number of signal transitions per second, which may be equal to the data rate in bits per second.

bidirectional signaling Signaling that occurs in both directions. See dual-duplex signaling, full-duplex signaling, half-duplex signaling, and simplex signaling.

binary digit (bit) **1.** The smallest unit of information in digital systems. **2.** Zeros and ones used to represent data processed by digital devices.

biometrics Authentication techniques based on measurable physical characteristics of individuals (e.g., fingerprints).

bit per second (b/s) A unit of measure used to express the data transfer rate. Commonly used rates include kilobit per second (kb/s), megabit per second (Mb/s), and gigabit per second (Gb/s). Also referred to as bit rate.

bit rate Transmission of a binary signal measured in bits per second.

Bluetooth A radio-based wireless personal area network technology.

bottom-up design See physical design process.

branch access A form of resource access where connection to a local area network (LAN) is made from a LAN-connected station at another organizational site using a telecommunications link.

bridge An internetworking device used to connect separate local area networks or to link network segments. Data frames are either forwarded or discarded (filtered) by the bridge, depending on their destination address.

broadband signaling A transmission technique in which the bandwidth is divided into multiple channels. This allows the transmission of multiple signals to take place simultaneously.

broadcast A technique for sending data simultaneously to all devices attached to a network with a single transmission. See multicast and unicast.

broadcast domain The span of a network as represented by the devices capable of receiving a Network layer datagram addressed to all.

broadcast storm An uncontrolled series of broadcasts caused by unnecessary repetition of the original message.

buffer A temporary storage area in a networking device used to hold incoming data until it can be processed.

bus topology A linear configuration where all network devices are placed on a single length of cable. It requires one backbone cable to which all network devices are connected.

byte A data unit made up of eight bits, sometimes referred to as an octet.

C

cabling system A specific system of telecommunications cables, equipment/patch cords, connecting hardware, and other components that is supplied as a single entity.

caching **1.** A technique that provides faster access to the data on a hard disk by keeping a copy of the data in memory. **2.** Using a storage area on a hard disk to duplicate data found on a remote network.

Category A North American standard for classifying cabling.

cell **1.** A 53-byte data transfer unit used by asynchronous transfer mode networks. **2.** The fixed area in which a wireless device operates.

centralized cabling An optical fiber cabling configuration from the work area to a centralized cross-connect using pull-through cables, an interconnect, or splice in the telecommunications room.

channel A path between devices.

channel service unit (CSU) A customer premises device that connects the data terminal equipment to a digital line from the public network service provider.

checksum A value calculated from the content of a message. It is used by the receiving device to verify that the data has not been altered during its transfer from source to receiver.

circuit switching A communications method in which a dedicated communications path is established between two devices prior to message transfer.

cladding The transparent outer concentric glass layer that surrounds the optical fiber core and has a lower index of refraction than the core.

client A network device that requests services from a server.

client software Additions to a station's operating system that enable access to network resources.

cluster A collection of servers and associated storage devices interconnected using a dedicated, high-speed network. The collection appears as a single device to the network and all incoming requests are divided among the servers for quicker response. See load balancing.

coaxial cable An unbalanced cable consisting of a central metallic core surrounded by a layer of insulating material. This insulating (dielectric) material may be a solid material or air spaced. The entire assembly is covered with a metallic mesh or solid metallic sleeve and may be protected by an outer layer of nonconducting material (cable jacket).

collapsed backbone An internetwork contained in one device. Individual networks are connected to this central device and can then communicate with one another.

collision A normal event on an Ethernet network indicating that two or more devices have simultaneously accessed the communications channel.

collision detection The process initiated when two or more network devices on an Ethernet network attempt to send a message at the same time and their messages collide. A device stops transmitting when it detects a collision and only attempts to retransmit after waiting a random period of time.

collision domain A collection of network devices and segments connected by repeaters. When a device in a collision domain transmits a Data Link layer frame, all other devices in the same domain receive the transmission. Devices and segments separated by bridges, routers, or switches are said to be in different collision domains.

communications See telecommunications.

communications protocol See protocol.

compression A method of modifying data so that it requires less bandwidth to transmit or to store than the original format.

concentrator **1.** A centralized connecting device that houses multiple interconnected trunk coupling units in a Token-Ring Network environment. Also referred to as multistation access unit or controlled access unit. **2.** In a fiber distributed data interface (FDDI) environment, a device on the FDDI ring that allows for the connection of multiple devices to the ring.

congestion A state in which the volume of messages exceeds the capacity of a communications channel, resulting in transfer delays or failures.

content acceleration The process of loading data accessed frequently by a large number of users onto devices capable of transferring the data more rapidly than the existing servers can transfer.

content access The process of providing remote users the means to connect to a local area network to obtain specific information. See resource access and transactional access.

contention A network access method in which devices compete for use of the available communications channel.

contention domain See collision domain.

controlled access unit (CAU) See concentrator.

converter A type of repeater that changes the data signal from one transmission medium type to another (e.g., from copper to optical fiber).

core The central, light-carrying part of an optical fiber through which light pulses are transmitted.

crosstalk The unwanted reception of electromagnetic signals on a communications circuit from another circuit.

cut-through A switching method in which messages are forwarded as they are received, with no error checking. See modified cut-through and store-and-forward.

cyclic redundancy check (CRC) See checksum.

D

daisy-chaining The practice of connecting devices in series.

data integrity The assurance that a given data file has not been deleted, modified, duplicated, or forged without detection.

Data Link layer Open Systems Interconnection layer responsible for providing reliable data transfer in the form of frames on a local area network. Also referred to as Layer 2.

Data Link layer address See device address.

data protection Techniques used to guard the confidentiality, integrity, and availability of data.

data service unit (DSU) A customer premises device that frames and channelizes the user's data for transmission on the digital network.

data terminal equipment (DTE) **1.** The term used in the IEEE 802.3 standard to refer to a station (computer) or port that serves as the data transmission source, data transmission destination, or both, for the purpose of sending or receiving data over the network. **2.** A device producing data to be transmitted across an internetwork.

data transfer rate The rate at which information is transferred between network devices over a communications channel. Sometimes referred to as throughput or operating speed.

datagram A data unit created at the Network layer of the Open Systems Interconnection model. It contains the data and control information necessary to transfer a message from one network to another. Also referred to as a packet.

decibel (dB) A logarithmic unit for measuring the power or strength of a signal.

demultiplexing The process of reconstituting the individual channels from the composite signal.

device address An address to uniquely identify each device on a network. The address is coded in the physical hardware. See network address.

dialer Software used by remote stations to connect to the network over telecommunications circuits. Also called remote client software.

dielectric **1.** The nonconducting properties of an insulating material that resists the passage of electric current. The insulation surrounding a copper conductor is known as a dielectric. **2.** A material that is nonmetallic and nonconductive, often used to describe insulating materials.

digital certificate A security tool used to authenticate a message. It ensures the recipient that the message originated from a source whose identity has been verified by the issuer of the certificate.

digital key A security tool used to encrypt a message prior to transmission to keep its contents confidential. Also used by the recipient to restore encrypted messages into readable content.

digital signal Information in the form of a sequence of discrete pulses separated by intervals. Commonly, a binary signal with two values that are used to transmit the two states (0,1) used by digital computers. Contrast with analog signal.

digital signature A security tool used to authenticate a message. It ensures the recipient that the message was not modified after being transmitted by the sender.

digital subscriber line (DSL) See xDSL.

directional antenna An antenna in a wireless environment that sends signals in a specific direction. Contrast with omni-directional antenna.

directory A database of the resources available on a network. Typically, it contains records for devices, software applications, data files, users, and groups.

disk duplexing The simultaneous writing of all data to be stored onto two hard disks, where each hard disk is connected to a separate controller card. If either hard disk or controller card fails, the other hard disks and controllers continue to provide storage services.

disk mirroring The simultaneous writing of all data to be stored onto two hard disks, where both hard disks are connected to the same controller card. If one of the hard disks fails, the other continues to provide storage services.

dual-duplex signaling A form of bidirectional signaling in which data transfer can flow in both directions at the same time over a single communications channel. See full-duplex signaling, half-duplex signaling, and simplex signaling.

dual-ring topology A ring topology where each device or network has two connections to each adjacent device or network.

E

electromagnetic spectrum The full range of electromagnetic emissions, which includes all light and radio waves.

Electronic Industries Alliance (EIA) The alliance is organized along specific electronic product and market lines, and, as a standards association, develops and publishes industry guidelines.

emulation The technique of modifying a device with hardware or software to make it operate in the same manner as another device. See terminal emulation.

encoding See signal encoding.

encryption **1.** A security mechanism that transforms the readable content in a message into a seemingly random collection of characters, numbers, and/or symbols to provide confidentiality. **2.** A modification of a bit stream to make it appear random and control emissions.

end node A device attached to a network for originating or receiving information on that network. Examples of end nodes are user stations and servers.

Ethernet A local area network protocol using a logical bus structure and carrier sense multiple access with collision detection.

F

fabric An interconnection scheme that enables communications between any two connected devices or networks through a series of interlinked switches.

failback Restoration of normal operation from a failover.

failover The transfer of control from a primary to a backup system in a redundant environment.

Fast Ethernet A local area network protocol with a 10 times higher transmission rate (100 Mb/s) than Ethernet. See Ethernet.

fault tolerance The ability of a system to continue operations after the failure of one or more components or transmission paths.

fiber See optical fiber cable.

fiber distributed data interface (FDDI) A token-passing network based on dual optical fiber rings. The network operates up to 100 Mb/s.

fiber optics A communications system that uses optical fiber as its medium.

Fibre Channel A gigabit interconnect technology commonly associated with storage area network technologies.

filtering A process that examines all incoming data for specific characteristics (e.g., source address, destination address, or protocol) and determines whether to accept, forward, or discard that traffic based on the established criteria. See forwarding.

firewall One or more security mechanisms designed to prevent, detect, suppress, and/or contain unauthorized access to a network.

flooding The process used by switches/bridges and routers to direct a message to all outgoing ports, with the exception of the port or interface on which the traffic was received. Flooding is used typically when the destination address of the message is not recognized or for multicast and broadcast message distribution.

flow control A mechanism used to manage the frame exchange rate between full-duplex ports on a switch.

forwarding The transferring of a message to another network by an internetworking device. See filtering.

forwarding logic The set of rules used by a switch to process the messages received from local area network devices.

frame A data unit created at the Data Link layer of the Open Systems Interconnection model. It contains the data and control information necessary to transfer a message from one device to another on the same network.

frequency The measure of the number of cycles (waves) per second, expressed in hertz.

frequency band A range of frequencies within which a class of radio communications operates.

full-duplex signaling A bidirectional signaling method using two communications channels. See dual-duplex signaling, half-duplex signaling, and simplex signaling.

functional design process A network design process where the designer begins by examining the types of applications and services that are to be supported by the network. Also called top-down design. Contrast with physical design process.

G

gateway An internetworking service that is used to connect dissimilar applications running on different networks with different communications protocols. Gateways normally operate at one or more of the top four layers of the Open Systems Interconnection Reference Model.

Gigabit Ethernet A local area network protocol with a 100 times higher transmission rate (1000 Mb/s) than Ethernet. See Ethernet.

gigabit per second (Gb/s) A transmission rate denoting one billion bits per second.

H

half-duplex signaling A bidirectional signaling method in which data transfer can take place in either direction, but in only one direction at a time. See dual-duplex signaling, full-duplex signaling, and simplex signaling.

hardware address See device address.

header The initial part of a message, typically containing identifying information such as source and destination addresses.

heartbeat The signals issued periodically by a clustered device to indicate that it is functioning normally.

hertz (Hz) A unit of measure used to express the range of frequencies associated with a given signal or communications channel. This range is also referred to as bandwidth. A unit of frequency equal to one cycle per second. A commonly used rate is megahertz.

hierarchical star topology An extension of the star topology utilizing a central hub.

hierarchical topology A topology that links devices or network using a series of levels, similar to an organizational chart.

horizontal cabling The part of the cabling system that extends from (and includes) the work area telecommunications outlet/connector to the horizontal cross-connect (floor distributor) in the telecommunications room.

host **1.** In the Internet environment, the term used to describe any network-attached device that provides application-level services. For example, a server containing a Web site is considered to be a host, while a router is not. **2.** A generic term used to describe mainframe and minicomputers.

hub A network device that provides a centralized point for Ethernet local area network communications, media connections, and management activities of a physical star topology cabling system

I

inbound Connections to a network device from a source outside the network. Also referred to as incoming.

individual address See device address.

Infrared Data Association (IrDA) The organization responsible for producing the specifications for infrared-based connections over very short distances.

Institute of Electrical and Electronics Engineers, Inc.® (IEEE®) The organization responsible for the standardization of most local area network technologies.

integrated services digital network (ISDN) A fully digital communications facility designed to provide transparent end-to-end transmission of voice, data, audio/video, and still images across the public switched telephone network.

interface **1.** A shared boundary. A physical point of demarcation between two devices or systems where electrical signals, connectors, and timing are defined. **2.** The procedures, protocols, and codes that allow two devices to interact for the purpose of exchanging information.

intermediate network See internetwork.

International Electrotechnical Commission (IEC) The commission responsible for international electronics standards.

Internet Architecture Board (IAB) The technical advisory group for the Internet Society. Responsible for the overall development of the protocols and architecture associated with the Internet.

Internet Corporation for Assigned Names and Numbers (ICANN) The group that oversees Internet naming and addressing.

Internet Engineering Steering Group (IESG) The group that manages the process used to introduce or update Internet standards,

Internet Engineering Task Force (IETF) One of the organizations responsible for the overall development of the Internet and the standardization of internetworking technologies.

Internet protocol (IP) A Network layer protocol used on the Internet.

Internet Society (ISOC) The organization that oversees the overall development of the Internet.

internetwork The communications system connecting two or more networks.

internetwork operating system (IOS) The software used to control and coordinate internetwork devices.

intranet A collection of Internet-based technologies designed to provide content to users on an internal network. The content is viewed using a Web browser.

intrusion detection The process of tracking actual or attempted unauthorized access to a network or protected device.

IP address The Network layer address assigned to devices using the Internet protocol. Also referred to as an Internet address.

isochronous communication A signaling method where a set data transfer rate within a communications channel is guaranteed. See asynchronous communication and synchronous communication.

J

jitter An undesirable variation in the arrival times of a series of signals.

L

LAN address See device address.

LAN emulation (LANE) **1.** A set of specifications that permits different types of networks to communicate over an asynchronous transfer mode (ATM) network and exchange information transparently. **2.** The method used by ATM switching equipment to emulate the characteristics of shared-media local area networks.

latency **1.** The time it takes for a signal to pass through a device or network. **2.** The delay between the time a switch receives a message on an input port and forwards it to an output port.

Layer 1 See Physical layer.

Layer 2 See Data Link layer.

Layer 2 address See device address.

Layer 2 switching A technology that is functionally equivalent to bridging.

Layer 3 See Network layer.

Layer 3 switching A technology that integrates routing with switching, resulting in high data transfer rates.

Layer 4 See Transport layer.

Layer 5 See Session layer.

Layer 6 See Presentation layer.

Layer 7 See Application layer.

learning The process of associating port numbers with device addresses.

link aggregation See port aggregation.

load balancing A mechanism for distributing incoming requests among a collection of devices or circuits to reduce response times. A technology that complements server clustering.

lobby network A local area network, protected by firewalls, that is used exclusively to connect the servers providing public content on the Internet.

local area network (LAN) A geographically limited data communications system for a specific user group consisting of interconnected computers sharing applications, data, and peripheral devices such as printers and CD-ROM drives intended for the local transport of data, video, and voice.

local station A station that is directly connected to the local area network medium. Contrast with remote station.

logical topology The path taken by messages as they travel from one device to another on a network. Contrast with physical topology.

M

MAC address See device address.

managed hub A hub that can be monitored and controlled through network management software. See hub.

media interface connector (MIC) The fiber distributed data interface connector used to link devices to the network cabling.

medium The channel through which network devices communicate with each other.

medium access control (MAC) The set of procedures that enables a device to access a network (e.g., carrier sense multiple access with collision detection or token passing for local area networks).

megabit per second(Mb/s) A transmission rate denoting one million bits per second.

megahertz (MHz) A unit of frequency equal to one million cycles per second (hertz).

mesh topology A topology where each device or network is connected to all other devices or networks.

metric In routing, a value assigned to a path.

micron (µm) **1.** A unit of length equal to one millionth of a meter (0.000001 meter). **2.** Also, an abbreviation for micrometer.

microsegmentation The technique used to divide a network into multiple small networks to improve performance. Ultimately, each device can have its own dedicated local area network through an exclusive connection to a switch port.

mirroring A technique used to increase the fault tolerance of a system. A backup device is configured identically to the primary device and can replace the primary device if it fails. See disk mirroring.

mobile access A form of resource access where connection to a local area network is made from a remote station using a temporary telecommunications link.

modem An acronym for modulator/demodulator. A modulator-demodulator device that converts between analog signals and digital signals for transmission over communication networks.

modified cut-through A switching method in which messages are forwarded as they are received, with minimal error checking. See cut-through and store-and-forward.

modulation Any of several techniques used to combine the data to be transmitted with a carrier signal.

multicast A technique for sending data to a specific group of devices on a network using a single transmission. See broadcast and unicast.

multiplexing The merging of two or more, low-capacity communications channels into a common, high-capacity channel.

multiport repeater See hub.

multistation access unit (MAU) See concentrator.

N

nanometer (nm) **1.** A unit of length equal to one billionth of a meter (0.000000001 meter). **2.** The most common unit of measurement for optical fiber operating wavelengths.

network access device Equipment used to interconnect stations, servers, and shared peripherals devices on a local area network or internetwork.

network address An address used to uniquely identify each local area network connected to an internetwork. See device address.

network interface card (NIC) Circuitry in a device that provides the means to physically connect to the network.

Network layer Open Systems Interconnection layer responsible for transferring data in the form of datagrams from one network device to another on an internetwork. Also referred to as Layer 3.

network operating system (NOS) A collection of software programs designed to control and coordinate activities on a local area network.

newsgroup An Internet-based messaging environment in which all messages and replies are posted for shared public viewing.

newsreader Software used to read, compose, and post newsgroup messages.

nonblocking The ability of a device or network to successfully provide communications paths between as many devices as needed, at any time.

nonrepudiation services Network processes that provide proof that a message was sent from a specific source, thereby preventing that source from denying having sent the message.

O

octet A grouping of eight bits. Also referred to as a byte.

omni-directional antenna An antenna in a wireless environment that radiates signals in all directions equally. Contrast with directional antenna.

Open Systems Interconnection (OSI) Reference Model A seven-layer architecture developed by the International Organization for Standardization that is used as a foundation for the development of many standards for network communications.

optical fiber cable Cable made up of one or more strands of glass consisting of a central core and outer cladding (optical fibers), strength members, and an outer jacket.

organizational network An interconnected system of computers, peripheral devices, and software designed to enable the transfer of all types of messages between users and devices.

outbound Connections from a network to an external device. Also referred to as outgoing.

P

packet See datagram.

packet switching A data communications switching and transmission system in which an input data stream is broken into packets that are transferred between devices on different networks without first establishing a dedicated communications path between the devices.

peripheral device Equipment connected to and controlled by a station or a server. Examples include printers, CD-ROM drives, and modems.

permanent virtual circuit (PVC) A software-defined logical connection in a switched network in which users define logical connections and required bandwidths between end points while the switched network technology achieves the defined connections and manages the traffic. With this type of circuit, the line is always ready, which eliminates the delay associated with line setup and release.

physical address See device address.

physical design process A network design process where the network designer begins by assessing the site where the proposed network is to be implemented. Also called bottom-up design. Contrast with functional design process.

Physical layer An Open Systems Interconnection layer responsible for the transfer of bit streams over a specific medium. Also referred to as Layer 1.

physical topology The physical layout of a network as defined by its cabling architecture. Contrast with logical topology.

piconet A single Bluetooth wireless personal area network that can contain a maximum of eight active devices. See scatternet.

port A connection point on a network access device (e.g., a hub or switch).

port aggregation A mechanism that combines multiple network communications channels into a single large channel to improve data transfer efficiency. Also referred to as link aggregation and trunking.

port mirroring The process of copying the data passing through one or more ports of a switch to a network management port for monitoring purposes.

Presentation layer An Open Systems Interconnection layer responsible for various forms of message conversion, including compression and encryption. Also referred to as Layer 6.

private key encryption A security process in which information is encrypted with a key that both the sender of the information and the receiver possess. The parties involved are expected to agree on a key in a way that does not compromise the established security processes.

probe See agent.

propagation delay The time interval required for a signal to travel between network devices.

protocol The rules and procedures governing the formatting of messages and the timing of their exchange between devices on a network. See protocol stack.

protocol data unit (PDU) Information that is exchanged between peer entities on a network. It contains control information and may optionally contain data.

protocol stack A comprehensive set of specifications governing how devices exchange information over a network.

proxy A security mechanism that uses an intermediary device to represent one side of a connection to the other for a specific application. Messages and commands intended for the other side are inspected by this device before being forwarded to prevent unauthorized use of that application.

public key encryption A security process in which each user has two related keys. One is held privately and the other is distributed publicly. Anyone wishing to send a confidential message to the user encrypts it with the user's public key. When the message is received, the user decrypts it with the corresponding private key.

Q

quality-of-service (QoS) **1.** A commitment to performance, based on predefined service parameters. **2.** A measure of the level of service provided on a network.

queuing A technique that reduces transmission delays by classifying and sorting data prior to processing by the transmitting device.

R

real-time The recording, processing, storage, transmission, and viewing of an activity as it occurs. Contrast with store-and-forward.

reboot The process of shutting down and restarting a device or a group of devices on the network after configuration changes.

redundant array of independent disks (RAID) A technology that makes it possible to group together multiple hard disk drives and allow them to function as a single storage unit.

remote access A form of resource access where connection to a local area network is made from a remote station over a telecommunications circuit.

remote access interface card An interface card used to connect a remote access server to one or more telecommunications circuits. Also called a wide area interface card.

remote control A form of remote access where the remote station connects to a local station and controls its operation. Contrast with remote node.

remote node A form of remote access where the remote station accesses local area network resources through a remote access server, eliminating the need for a local station. Contrast with remote control.

remote station A station that is connected to the local area network (LAN) through a telecommunications circuit external to the LAN. Contrast with local station.

repeater See hub.

resource access The process of providing a remote user the means to connect to a local area network and obtain the same privileges as when connecting from local stations. See content access and transactional access.

ring topology A network topology in which nodes are connected in a point-to-point serial fashion in an unbroken circular configuration. Each node receives and retransmits the signal to the next node.

router An internetworking device, operating at the Network layer of the Open Systems Interconnection model, used to direct datagrams from one network to another.

S

scatternet A collection of two or more piconets located in a common area.

scripting The process of determining which configuration tasks can be performed unattended and then automating the process.

security policy The set of principles, rules, and practices that are used to implement security in an organization.

segment A portion of a network sharing a continuous length of cable.

segmentation The process of dividing a local area network into multiple independent segments to improve overall data transfer rates.

server A network device that combines hardware and software to provide and manage shared services and resources on the network.

service provider (SP) An entity that provides connection to a part of the Internet, or other services such as application programming interfaces. An SP, when reached by the user through an access provider (AP), becomes the AP to the service they provide. See access provider (AP).

Session layer An Open Systems Interconnection layer responsible for providing services used to organize, synchronize, and manage a given message exchange between network devices. Also referred to as Layer 5.

shielded twisted-pair (STP) cable Cable made up of multiple twisted copper pairs with an additional metallic shield covering each individual pair. The entire structure is covered with an overall shield or braid and an insulating sheath (cable jacket).

signal encoding The conversion of data into a form suitable for transmission over a medium.

simplex signaling A unidirectional signaling method in which data transfer can take place in only one direction, with no capabilities to change directions. See dual-duplex signaling, full-duplex signaling, and half-duplex signaling.

site survey A process used to identify the characteristics of an environment. Required for the design of a wireless network.

small computer system interface (SCSI) A specialized network intended to connect multiple storage and peripheral devices to computers over limited distances.

SONET node See synchronous optical network (SONET) node.

spanning tree algorithm(STA) An algorithm used by bridges to create a logical topology that connects all of the bridge-equipped networks to be interlinked. It ensures that no path loops exist on the internetwork.

spool A combination of hardware and software commonly used by print servers to redirect requests destined for a printer.

spread spectrum A radio transmission technology that distributes the transmitted signal over multiple frequencies within the assigned frequency band to increase the overall immunity of the signal to noise and prevent message interception.

stackable hub Multiple hubs that are in close proximity, capable of being connected to each other using a short length of specialized cable assembly, and functioning together as a single unit. See hub.

start bit A bit used in asynchronous communications to indicate the beginning of a character.

star topology A network topology in which services are distributed from a central point.

station A device used by an individual to access network services. See local station and remote station.

stop bit A bit used in asynchronous communications to indicate the end of a character.

storage area network (SAN) A specialized high-speed network used to interconnect storage devices.

store-and-forward **1.** The recording, processing, storage, transmission, and viewing of an activity after it has occurred. Contrast with real-time. **2.** A switching method in which messages are stored as they are received, fully checked for errors, and then forwarded. See cut-through and modified cut-through.

streaming A groups of technologies developed to enable the transfer of audio and video content over the Internet for immediate play at the receiving device.

striping A data storage technique in which the content of a given file to be stored is divided and placed on multiple hard disk drives for faster retrieval and improved fault tolerance.

subnet See subnetwork.

subnetting The process of dividing a single broadcast domain into two or more smaller broadcast domains by modifying Network layer address assignments. See subnetwork.

subnetwork A single broadcast domain in a network that is made up of multiple broadcast domains.

switch A multiport bridge on which each port represents a separate network communications channel. See bridge.

switch latency The amount of time it takes for an incoming message to be inspected, processed, and forwarded through a switch.

switched virtual circuit (SVC) A virtual circuit created on an as-needed basis. It is a temporary connection lasting only as long as the connected devices are communicating. When the data transfer is complete, the connections are terminated and the processing required to maintain the circuit is returned to the overall resource pool.

synchronous communication A form of signaling in which no start and stop bits are used. Each data character is coded as a string of bits and the sending and receiving devices are synchronized with each other, using a common clock. See asynchronous communication and isochronous communication.

synchronous optical network (SONET) A scalable transport technology designed to provide a uniform, consistent method of transferring data, typically using a fiber optic transmission infrastructure.

synchronous optical network (SONET) node The interface to a SONET network.

T

T-1 The fundamental digital transmission circuit, with a bandwidth of 1.544 Mb/s.

tagging The marking of a message for the purpose of specialized processing or handling on a local area network or internetwork.

telecommunications The transmission and reception of information of any nature by cable, radio, optical, or other electromagnetic systems.

telecommunications circuit Links that make it possible for an organizational network to grow beyond the physical boundaries of a building or a campus.

Telecommunications Industry Association (TIA) A standards association that publishes telecommunications standards and other documents.

terabyte (TB) A measure of disk storage capacity where one terabyte equals 1000 gigabytes.

terminal emulation The process that enables a personal computer to operate as a terminal for connecting to a mainframe or minicomputer.

token A signal sequence that is passed from device to device on a token ring or fiber distributed data interface network. The token ensures an orderly access to the shared medium.

token ring A network on which attached devices share a common cabling system for communications purposes without the possibility of a collision between transmissions. A device is only able to send a message when it is in possession of a special electronic sequence of bits called a token.

top-down design See functional design process.

topology The physical or communications path layout of a network or internetwork. See logical topology and physical topology.

traffic shaping A technique that directs data streams on the basis of their address, protocol, priority, or application content.

transactional access The process of providing remote users the means to connect to a local area network to conduct commercial activities. See content access and resource access.

transceiver **1.** A radio transmitter and receiver combined into a single unit. **2.** A device that acts as an interface between the network and the connected device.

translational bridge A bridge capable of converting frame formats from one type to another prior to forwarding messages (e.g., Ethernet to fiber distributed data interface).

Transport layer An Open Systems Interconnection layer responsible for providing a level of quality to the data transfer process. Also referred to as Layer 4.

trunking See port aggregation.

tunnel A term used to describe a virtual private network connection through the Internet.

twisted-pair cable A multi-conductor cable comprising two or more copper conductors twisted in a manner designed to cancel electrical interference.

U

unauthorized access In general, the use of any network resource without approval as established in the security policy.

unicast A technique for sending data to a single attached network device. A one-to-one mode of communication. See broadcast and multicast.

unidirectional signaling Signaling that occurs in one direction. Contrast with bidirectional signaling.

unshielded twisted-pair (UTP) cable Cable made up of one or more pairs of twisted copper conductors without additional metallic shielding. The entire assembly is covered with an insulating sheath (cable jacket).

V

virtual circuit A communications path through an internetwork that appears to be a dedicated circuit between two network devices.

virtual LAN (VLAN) A technique made possible by switching technologies that permits the logical grouping of any number of network devices into one or more subnetworks to improve traffic management and/or security.

virtual private network (VPN) A combination of hardware and software technologies designed to enable secure passage of organizational network traffic over the Internet. See tunnel.

W

Web Used as a noun, it is shorthand for the World Wide Web (WWW) services found on the Internet.

wide area interface card (WIC) See remote access interface card.

wide area network (WAN) Computer networks where devices are connected using telecommunications circuits, rather than a length of cable.

X

xDSL A family of digital technologies designed to provide high data transfer rates over existing (legacy) telecommunications circuits.

Z

zone In wireless networking, a zone consists of one access point (AP) and the group of devices associated with that AP. Also referred to as a cell.

Abbreviations, Acronyms, and Units of Measure

Numbers and symbols

μm	micron
μs	microsecond
3DES	triple DES

A

A/V	audio/video
AAL	ATM adaptation layer
ABR	available bit rate
ACL	access control list
ADSL	asymmetric digital subscriber line
AH	application header
AH	authentication header
AM	amplitude modulation
ANSI	American National Standards Institute
AP	access point
AP	access provider
ARL	adjusted ring length
ARPANET	Advanced Research Projects Agency Network
ATDM	asynchronous time division multiplexing
ATM	asynchronous transfer mode
ATU-R	ADSL transceiver unit remote
AUTONEG	auto-negotiation
AWG	American wire gauge

B

b/s	bits per second
BD	building distributor
BGP	border gateway protocol
B-ISDN	broadband integrated services digital network
bit	binary digit
BPDU	bridge protocol data unit

BRI	basic rate interface
BSS	basic service set
BSSID	basic service set identification
BUS	broadcast and unknown server

C

CA	certificate authority
CAN	cluster area network
CAU	controlled access unit
CBR	constant bit rate
CD	campus distributor
CD	compact disc
CD-ROM	compact disc-read only memory
CDDI	copper distributed data interface
CENELEC	Comité Européen de Normalisation Electrotechnique (European Committee for Electrotechnical Standardization)
CFI	canonical format indicator
CHAP	challenge handshake authentication protocol
CIR	committed information rate
CLP	cell loss priority
CP	consolidation point
CRC	cyclic redundancy check
CSMA/CA	carrier sense multiple access with collision avoidance
CSMA/CD	carrier sense multiple access with collision detection
CSU	channel service unit
CTR	classic token ring
CTS	clear to send

D

DAC	dual-attachment concentrator
DAP	directory access protocol
DAS	direct attached storage
DAS	dual-attachment station
dB	decibel
DCE	data circuit-terminating equipment

DCF	distributed coordination function
DD	drive distance
DDNS	dynamic domain name system
DES	data encryption standard
DHCP	dynamic host configuration protocol
DiffServ	differentiated services
DIX	Digital/Intel/Xerox
DMZ	demilitarized zone
DN	directory number
DNS	domain name system
DPAM	demand priority access method
DQDB	distributed queue dual bus
DS	digital signal
DSL	digital subscriber line
DSLAM	digital subscriber line access multiplexer
DSSS	direct sequence spread spectrum
DSU	data service unit
DTE	data terminal equipment
DTR	dedicated token ring

E

EF	entrance facility
EGP	exterior gateway protocol
EIA	Electronic Industries Alliance
ELAN	emulated LAN
e-mail	electronic mail
EMC	electromagnetic compatibility
EMI	electromagnetic interference
ER	equipment room
ESI	end system identifier
ESP	encapsulating security payload
ESS	extended service set
ETR	early token release
ETSI	European Telecommunications Standards Institute

F

FC-AL	Fibre Channel arbitrated loop
FCS	frame check sequence
FD	floor distributor
FDDI	fiber distributed data interface
FEP	fluorinated ethylene propylene
FHSS	frequency-hopping spread spectrum
FM	frequency modulation
FOIRL	fiber optic inter-repeater link
FR	Frame Relay
FRAD	Frame Relay access device
FT-1	fractional T-1
FT-3	fractional T-3
FTP	file transfer protocol
FTP	foil twisted-pair

G

GARP	generic attribute registration protocol
GB	gigabyte
Gb/s	gigabit per second
GBIC	gigabit interface converter
GFC	generic flow control
GMII	gigabit media independent interface
GTR	gigabit token ring
GVRP	GARP VLAN registration protocol

H

HBA	host bus adapter
HC	horizontal cross-connect
HDSL	high bit rate digital subscriber line
HEC	header error control
hex	hexadecimal
HomeRF	Home Radio Frequency
hostid	host identification

HR	high rate
HSTR	high-speed token ring
HTML	hypertext markup language
HTTP	hypertext transfer protocol
Hz	hertz

I

IAB	Internet Architecture Board
IANA	Internet Assigned Numbers Authority
IBM®	International Business Machines
IC	intermediate cross-connect
ICANN	Internet Corporation for Assigned Names and Numbers
IEC	International Electrotechnical Commission
IEEE®	Institute of Electrical and Electronics Engineers, Inc.®
IESG	Internet Engineering Steering Group
IETF	Internet Engineering Task Force
IHL	Internet header length
IMAP4	Internet message access protocol version 4
IOS	internetwork operating system
IP	Internet protocol
IPng	IP next generation
IPsec	Internet protocol security
IPv4	Internet protocol version 4
IPv6	Internet protocol version 6
IR	infrared
IrDA	Infrared Data Association
IRSG	Internet Research Steering Group
IRTF	Internet Research Task Force
ISDN	integrated services digital network
IS-LAN	integrated services LAN
ISM	industrial, scientific, and medical
ISO	International Organization for Standardization
ISOC	Internet Society
ISP	Internet service provider

ITU-T	International Telecommunication Union-Telecommunication
IW	inside wiring

K

kb	kilobit
kb/s	kilobit per second
km	kilometer

L

L2TP	Layer 2 tunneling protocol
LAM	lobe attachment module
LAN	local area network
LANE	LAN emulation
LAP-B	link access protocol-balanced
LAP-D	link access procedure-D channel
LCF	low-cost fiber
LCF-PMD	low-cost fiber-physical medium dependent
LDAP	lightweight directory access protocol
LEC	LAN emulation client
LECS	LAN emulation configuration server
LED	light-emitting diode
LES	LAN emulation server
LLC	logical link control
LMSC	Local and Metropolitan Area Network Standards Committee

M

MAC	medium access control
MAN	metropolitan area network
MAU or **MSAU**	multistation access unit
MAU	medium attachment unit
MB	megabyte
Mb/s	megabit per second
MC	main cross-connect
MCU	multipoint conference unit

MDI	medium dependent interface
MDI-X	medium dependent interface-crossover
MHz	megahertz
mi	mile
MIB	management information base
MIC	media interface connector
MIC	medium interface connector
MII	medium independent interface
MMF	multimode fiber
MMF-PMD	multimode fiber-physical medium dependent
MPLS	multiprotocol label switching
MPOA	multiprotocol over ATM
MUTOA	multi-user telecommunications outlet assembly
mux	multiplexer
mW	milliwatt

N

NAS	network-attached storage
NAT	network address translation
NAUN	nearest active upstream neighbor
netid	network identification
NH	network header
NIC	network interface card
N-ISDN	narrowband ISDN
nm	nanometer
NMS	network management system
NNI	network node interface
NNI	network-to-network interface
NNTP	network news transfer protocol
NOS	network operating system
NPDU	network protocol data unit
nrt-VBR	non-real-time variable bit rate
NSAP	network service access point
NSP	network service provider

NTA	network terminal adapter
NVT	network virtual terminal

O

OC	optical carrier
OFDM	orthogonal frequency division multiplexing
OSI	Open Systems Interconnection
OSPF	open shortest path first
OUI	organizationally unique identifier

P

PAD	packet assembler/disassembler
PAP	password authentication protocol
PAT	port address translation
PBX	private branch exchange
PC	path cost
PC	personal computer
PCF	point coordination function
PDA	personal digital assistant
PDU	protocol data unit
PE	polyethylene
PH	presentation header
PHY	physical layer device
PHY	physical layer protocol
PMA	physical medium attachment
PMD	physical layer medium dependent
PMD	physical medium dependent
POF	plastic optical fiber
POP3	post office protocol version 3
POS	personal operating space
PPP	point-to-point protocol
PPTP	point-to-point tunneling protocol
PRI	primary rate interface
PSDN	packet switched data network

PSTN	public switched telephone network
PTI	payload type indicator
PVC	permanent virtual circuit
PVC	polyvinyl chloride

Q

QoS	quality of service

R

RADIUS	remote authentication dial-in user service
RADSL	rate-adaptive digital subscriber line
RAID	redundant array of independent disks
RAN	regional area network
RAS	remote access server
RAS	remote access services
RGB	red, green, and blue
RF	radio frequency
RFC	request for comment
RFI	radio frequency interference
RI	routing information
RI/RO	ring in/ring out
RIP	routing information protocol
RMON	remote network monitoring
RPR	resilient packet ring
RSA	Rivest, Shamir, and Adleman
RSVP	resource reservation protocol
RTCP	real-time transport control protocol
RTP	real-time transport protocol
RTS	request to send
rt-VBR	real-time variable bit rate

S

SAC	single-attachment concentrator
SAN	storage area network

SAP	service access point
SAR	segmentation and reassembly
SAS	single-attachment station
SAT	source address table
SCS	structured cabling system
SCSI	small computer system interface
ScTP	screened twisted-pair
SDDI	shielded distributed data interface
SDH	synchronous digital hierarchy
SDSL	symmetric digital subscriber line
SEL	selector
SFS	start-of-frame sequence
SH	session header
SIG	special interest group
SILS	standard for interoperable LAN/MAN security
SLA	service level agreement
SMF	singlemode fiber
SMF-PMD	singlemode fiber-physical medium dependent
SMI	structure of management information
SMON	switched network monitoring
SMP	symmetric multiprocessing
SMS	selective multicast server
SMT	station management
SMTP	simple mail transfer protocol
SNMP	simple network management protocol
SONET	synchronous optical network
SP	service provider
SPE	synchronous payload envelope
SPVC	soft permanent virtual circuit
SRB	source routing bridge
SRT	source routing transparent
STA	spanning tree algorithm
STM	synchronous transport module
STP or **STP-A**	shielded twisted-pair

STPDDI	shielded twisted-pair distributed data interface
STS	synchronous transport signal
SVC	switched virtual circuit
SWAP	shared wireless access protocol

T

TA	terminal adapter
TACACS+	terminal access controller access control system plus
TAG	technical advisory group
TB	terabyte
TBI	ten-bit interface
TC	transmission convergence
TCI	tag control information
TCP	transmission control protocol
TCU	trunk coupling unit
TDM	time division multiplexing
TE	terminal equipment
TH	transport header
TIA	Telecommunications Industry Association
TOS	type of service
TP	transition point
TPDDI	twisted-pair distributed data interface
TPDU	transport protocol data unit
TPID	tag protocol identifier
TP-PMD	twisted-pair physical medium dependent
TR	telecommunications room
TRN	Token Ring Network
TTL	time to live
TXI	transmit immediate protocol

U

UBR	unspecified bit rate
UDP	user datagram protocol
UNI	user-to-network interface

U-NII	unlicensed national information infrastructure
UPS	uninterruptible power supply
UTP	unshielded twisted-pair
UTPDDI	unshielded twisted-pair distributed data interface

V

VAN	value-added network
VC	virtual channel
VCI	virtual channel identifier
VCSEL	vertical cavity surface emitting laser
VDSL	very high bit rate digital subscriber line
VID	VLAN identifier
VLAN	virtual local area network
VoIP	voice over Internet protocol
VP	virtual path
VPI	virtual path identifier
VPN	virtual private network

W

W3C	World Wide Web Consortium
WAN	wide area network
WG	working group
WIC	wide area interface card
Wi-Fi	wireless fidelity
WLAN	wireless local area network
WMAN	wireless metropolitan area network
WPAN™	wireless personal area network
WWDM	wide wave division multiplexing
WWW	World Wide Web

Bibliography

Ahuja, Vijay. *Network & Internet Security*. San Diego, CA: Academic Press, 1996.

American National Standards Institute/Institute of Electrical and Electronics Engineers, Inc. ® (IEEE®). ANSI/IEEE 802.11 – 1999 (ISO/IEC 8802-11 – 1999). *Information Technology—Telecommunications and Information Exchange Between Systems—Local and Metropolitan Area Networks—Specific Requirements—Part 11: Wireless LAN MAC and Physical Layer Specifications*. New York, NY: Institute of Electrical and Electronics Engineers, Inc., 1999.

———. ANSI/IEEE 802.12 – 1998 (ISO/IEC 8802-12 – 1998). *Information Technology—Telecommunications and Information Exchange Between Systems—Local and Metropolitan Area Networks—Specific Requirements—Part 12: Demand Priority Access Method (DPAM), Physical Layer and Repeater Specification for 100 Mb/s Operation*. New York, NY: Institute of Electrical and Electronics Engineers, Inc., 1998.

———. ANSI/IEEE 802.1B – 1995 (ISO/IEC 15802-2 – 1995). *Information Technology—Telecommunications and Information Exchange Between Systems—Local and Metropolitan Area Networks—Common Specifications—Part 2: LAN/MAN Management*. New York, NY: Institute of Electrical and Electronics Engineers, Inc., 1995.

———. ANSI/IEEE 802.1D – 1998 (ISO/IEC 15802-3 – 1998). *Information Technology—Telecommunications and Information Exchange Between Systems—Local and Metropolitan Area Networks—Common Specifications—MAC Bridges*. New York, NY: Institute of Electrical and Electronics Engineers, Inc., 1998.

———. ANSI/IEEE 802.1G – 1998 (ISO/IEC 15802-5 – 1998). *Information Technology—Telecommunications and Information Exchange Between Systems—Local and Metropolitan Area Networks—Common Specifications—Part 5: Remote MAC Bridging*. New York, NY: Institute of Electrical and Electronics Engineers, Inc., 1998.

———. ANSI/IEEE 802.2 – 1998 (ISO/IEC 8802-2 – 1998). *Information Technology—Telecommunications and Information Exchange Between Systems—Local and Metropolitan Area Networks—Specific Requirements—Part 2: Logical Link Control*. New York, NY: Institute of Electrical and Electronics Engineers, Inc., 1998.

———. ANSI/IEEE 802.3 – 2000 (ISO/IEC 8802-3 – 2000). *Information Technology—Telecommunications and Information Exchange Between Systems—Local and Metropolitan Area Networks—Specific Requirements—Part 3: CSMA/CD Access Method and Physical Layer Specifications*. New York, NY: Institute of Electrical and Electronics Engineers, Inc., 2000.

———. ANSI/IEEE 802.4 – 1990 (ISO/IEC 8802-4 – 1990). *Information Processing Systems—Local Area Networks—Part 4: Token-Passing Bus Access Method and Physical Layer Specifications*. New York, NY: Institute of Electrical and Electronics Engineers, Inc., 1990.

———. ANSI/IEEE 802.5 – 1998 (ISO/IEC 8802-5 – 1998). *Information Technology—Telecommunications and Information Exchange Between Systems—Local and Metropolitan Area Networks—Specific Requirements—Part 5: Token Ring Access Method and Physical Layer Specifications*. New York, NY: Institute of Electrical and Electronics Engineers, Inc., 1998.

———. ANSI/IEEE 802.6 – 1994 (ISO/IEC 8802-6 – 1994). *Information Technology—Telecommunications and Information Exchange Between Systems—Local and Metropolitan Area Networks—Specific Requirements—Part 6: Distributed Queue Dual Bus (DQDB) Access Method and Physical Layer Specifications*. New York, NY: Institute of Electrical and Electronics Engineers, Inc., 1994.

American National Standards Institute/Telecommunications Industry Association/Electronics Industries Association. ANSI/TIA/EIA-568-B.1. *Commercial Building Telecommunications Cabling Standard, Part 1: General Requirements*. Arlington, VA: Telecommunications Industry Association/Electronic Industries Alliance, 2001.

———. ANSI/TIA/EIA-568-B.2. *Commercial Building Telecommunications Cabling Standard, Part 2: Balanced Twisted-pair Cabling Components*. Arlington, VA: Telecommunications Industry Association/Electronic Industries Alliance, 2001.

———. ANSI/TIA/EIA-568-B.3. *Optical Fiber Cabling Components Standard*. Arlington, VA: Telecommunications Industry Association/Electronic Industries Alliance, 2000.

———. ANSI/TIA/EIA-569-A. *Commercial Building Standard for Telecommunications Pathways and Spaces*. Arlington, VA: Telecommunications Industry Association/ Electronic Industries Alliance, 1998.

———. ANSI/TIA/EIA-569-A-1. *Addendum 1 – Surface Raceways*. Arlington, VA: Telecommunications Industry Association/Electronic Industries Alliance, 2000.

———. ANSI/TIA/EIA-569-A-2. *Addendum 2 – Furniture Pathways and Spaces*. Arlington, VA: Telecommunications Industry Association/Electronic Industries Alliance, 2000.

———. ANSI/TIA/EIA-569-A-3. *Addendum 3 – Access Floors*. Arlington, VA: Telecommunications Industry Association/Electronic Industries Alliance, 2000.

———. ANSI/TIA/EIA-569-A-4. *Addendum 4 – Poke-Thru Fittings*. Arlington, VA: Telecommunications Industry Association/Electronic Industries Alliance, 2000.

———. ANSI/TIA/EIA-569-A-5. *Addendum 5 – Commercial Building Standard for Telecommunications Pathways and Spaces*—SP 4722-A-1. Arlington, VA: Telecommunications Industry Association/Electronic Industries Alliance, draft.

———. ANSI/TIA/EIA-569-A-6. *Addendum 6 – Commercial Building Standard for Telecommunications Pathways and Spaces—Multi-Tenant Pathways and Spaces*—SP 3-2950-AD6. Arlington, VA: Telecommunications Industry Association/Electronic Industries Alliance, draft.

———. ANSI/TIA/EIA-570-A. *Residential Telecommunications Cabling Standard*. Arlington, VA: Telecommunications Industry Association/Electronic Industries Alliance, 1999.

———. ANSI/TIA/EIA-606. *Administration Standard for the Telecommunications Infrastructure of Commercial Buildings*. Arlington, VA: Telecommunications Industry Association/Electronic Industries Alliance, 1993.

———. ANSI/TIA/EIA-607. *Commercial Building Grounding and Bonding Requirements for Telecommunications*. Arlington, VA: Telecommunications Industry Association/Electronic Industries Alliance, 1994.

———. ANSI/TIA/EIA-758. *Customer-Owned Outside Plant Telecommunications Cabling Standard*. Arlington, VA: Telecommunications Industry Association/Electronic Industries Alliance, 1999.

Berkowitz, Howard C. *Designing Addressing Architectures for Routing and Switching*. New York, NY: Macmillan Technical Publishing, 1999.

Blacharski, Dan. *Maximum Bandwidth*. Indianapolis, IN: Que Corporation, 1997.

Black, Darryl P. *Building Switched Networks: Multilayer Switching, QoS, IP Multicast, Network Policy, and Service Level Agreements*. Reading, MA: Addison-Wesley, 1999.

———. *Managing Switched Local Area Networks: A Practical Guide*. Reading, MA: Addison-Wesley, 1998.

Black, Uyless D. *Frame Relay Networks*. New York, NY: McGraw-Hill, 1998.

Brown, Steven. *Implementing Virtual Private Networks*. New York, NY: McGraw-Hill, 1999.

Bryce, James Y. *Using ISDN*. Indianapolis, IN: Que Corporation, 1995.

Clayton, Jade. *McGraw-Hill Illustrated Telecom Dictionary*. New York, NY: McGraw-Hill, 1998.

Dhawan, Chander. *Remote Access Networks: PSTN, ISDN, ADSL, Internet, and Wireless*. New York, NY: McGraw-Hill, 1998.

Dodd, Annabel Z. *The Essential Guide to Telecommunications*. Englewood Cliffs, NJ: Prentice-Hall, 2000.

Downes, Kevin, Merilee Ford, H. Kim Lew, Steve Spanier, and Tim Stevenson. *Internetworking Technologies Handbook*. Indianapolis, IN: Cisco Press, 1998.

Enck, John and Dan W. Blacharski. *Managing Multivendor Networks*. Indianapolis, IN: Que Corporation, 1997.

Flanagan, William A. *T-1 Networking: How to Buy, Install, and Use T-1 From Desktop to DS-3*. New York, NY: Telecom Books, 1997.

Goncalves, Marcus and Kitty Niles. *IP Multicasting: Concepts and Applications*. New York, NY: McGraw-Hill, 1999.

Goralski, Walter. *ADSL and DSL Technologies*. New York, NY: McGraw-Hill, 1998.

———. *Frame Relay for High-Speed Networks*. New York, NY: John Wiley & Sons, 1999.

Greer, Tyson. *Understanding Intranets*. Redmond, WA: Microsoft Press, 1998.

Habraken, Joe. *Practical Cisco Routers*. Indianapolis, IN: Que Corporation, 1999.

Held, Gilbert. *LAN Performance: Issues and Answers*. New York, NY: John Wiley & Sons, 1996.

———. *Virtual LANs: Construction, Implementation, and Management*. New York, NY: John Wiley & Sons, 1997.

Hopkins, Gerald L. *ISDN Literacy Book*. Reading, MA: Addison-Wesley, 1995.

Huitema, Christian. *IPv6: The New Internet Protocol*. Englewood Cliffs, NJ: Prentice-Hall, 1996.

Institute of Electrical and Electronics Engineers, Inc.® (IEEE®). IEEE 802.10 – 1998. *IEEE Standards for Local and Metropolitan Area Networks: Interoperable LAN/MAN Security*. New York, NY: Institute of Electrical and Electronics Engineers, Inc., 1998.

———. IEEE 802.11. *IEEE Standard for Information Technology—Telecommunications and Information Exchange Between Systems—Local and Metropolitan Area Networks—Specific Requirements—Part 11: Wireless LAN MAC and Physical Layer Specifications*. New York, NY: Institute of Electrical and Electronics Engineers, Inc., 1997.

———. IEEE 802.15. *Standard for Telecommunications and Information Exchange Between Systems—LAN/MAN—Specific Requirements—Part 15: Wireless MAC and Physical Layer Specifications for Wireless Personal Area Networks*. New York, NY: Institute of Electrical and Electronics Engineers, Inc., draft.

———. IEEE 802.1D. *IEEE Standard for Information Technology—Telecommunications and Information Exchange Between Systems—Local and Metropolitan Area Networks—Common Specifications—MAC Bridges*. New York, NY: Institute of Electrical and Electronics Engineers, Inc., 1986.

———. IEEE 802.1G. *IEEE Standard for Information Technology—Telecommunications and Information Exchange Between Systems—Local and Metropolitan Area Networks—Common Specifications—Part 5: Remote MAC Bridging*. New York, NY: Institute of Electrical and Electronics Engineers, Inc., 1998.

———. IEEE 802.1Q. *IEEE Standards for Local and Metropolitan Area Networks—Virtual Bridged Local Area Networks*. New York, NY: Institute of Electrical and Electronics Engineers, Inc., 1998.

———. IEEE 802.1v. *IEEE Standards for Local and Metropolitan Area Networks: Virtual Bridged Local Area Networks- Amendment 2: VLAN Classification by Protocol and Port*. New York, NY: Institute of Electrical and Electronics Engineers, Inc., 2001.

———. IEEE 802.3. *Part 3: CSMA/CD Access Method and Physical Layer Specifications, Information Technology—Telecommunications and Information Exchange Between Systems—Local and Metropolitan Area Networks—Specific Requirements*. New York, NY: Institute of Electrical and Electronics Engineers, Inc., 1985.

———. IEEE 802.5. *Standard for Information Technology—Telecommunications and Information Exchange Between Systems—Local and Metropolitan Area Networks—Part 5: Token Ring Access Method and Physical Layer Specifications*. New York, NY: Institute of Electrical and Electronics Engineers, Inc., 1989.

———. IEEE 802.7. *IEEE Recommended Practices for Broadband LANs*. New York, NY: Institute of Electrical and Electronics Engineers, Inc., 1989.

———. IEEE 802.8. *IEEE Recommended Practice for Fiber Optic Local and Metropolitan Area Networks*. New York, NY: Institute of Electrical and Electronics Engineers, Inc., 1998.

———. IEEE 802.9. *Information Technology—Telecommunications and Information Exchange Between Systems—Local and Metropolitan Area Networks—Specific Requirements—Part 9: Integrated Services (IS)-LAN Interface at the MAC and Physical Layers*. New York, NY: Institute of Electrical and Electronics Engineers, Inc., 1996.

International Organization for Standardization/International Electrotechnical Commission. ISO/IEC International Standard 11801:2000 – *Information Technology—Generic Cabling for Customer Premises*. Geneva, Switzerland: International Organization for Standardization/ International Electrotechnical Commission, 2000.

Kaeo, Merike. *Designing Network Security*. Indianapolis, IN: Cisco Press, 1999.

Kanter, Joel P. *Understanding Thin-Client/Server Computing*. Redmond, WA: Microsoft Press, 1998.

Knapp, James. *Nortel Networks: The Complete Reference*. Berkeley, CA: Osborne/McGraw-Hill, 2000.

Kosiur, Dave. *Building and Managing Virtual Private Networks*. New York, NY: John Wiley & Sons, 1998.

Loshin, Peter. *Extranet Design and Implementation*. Alameda, CA: SYBEX, 1997.

Lu, Cary. *The Race for Bandwidth: Understanding Data Transmission*. Redmond, WA: Microsoft Press, 1998.

Marcus, J. Scott. *Designing Wide Area Networks and Internetworks*. Reading, MA: Addison-Wesley, 1999.

Martin, James, Kathleen Kavanagh Chapman, and Joe Leben. *Enterprise Networking: Strategies and Transport Protocols*. Englewood Cliffs, NJ: Prentice-Hall, 1996.

Martin, Michael J. *Understanding the Network: A Practical Guide to Internetworking*. Indianapolis, IN: New Riders Publishing, 2000.

Maufer, Thomas A. *IP Fundamentals: What Everyone Needs to Know About Addressing and Routing*. Englewood Cliffs, NJ: Prentice-Hall, 1999.

Miller, Mark A. *Managing Internetworks with SNMP*. Foster City, CA: M&T Books, 1999.

Minoli, Daniel. *Internet & Intranet Engineering*. New York, NY: McGraw-Hill, 1997.

Minoli, Daniel and Emma Minoli. *Web Commerce Technology Handbook*. New York, NY: McGraw-Hill, 1998.

Muller, Nathan J. *Desktop Encyclopedia of Voice and Data Networking*. New York, NY: McGraw-Hill, 2000.

Nemzow, Martin. *Web Video Complete*. New York, NY: McGraw-Hill, 1998.

Newton, Harry. *Newton's Telecom Dictionary*, 15th ed. Billerica, MA: CMP Books, 1999.

Norton, Peter and Dave Kearns. *Peter Norton's Complete Guide to Networking*. Indianapolis, IN: Sams Publishing, 1999.

Parnell, Tere. *LAN Times Guide to Wide Area Networks*. Berkeley, CA: Osborne/ McGraw-Hill, 1997.

Pfaffenberger, Bryan. *Webster's New World Dictionary of Computer Terms*. Indianapolis, IN: Que Corporation, 1997.

Quinn-Andry, Terri and Kitty Haller. *Designing Campus Networks*. Indianapolis, IN: Cisco Press, 1998.

Roberts, Dave. *Internet Protocols Handbook*. Poway, CA: Coriolis Group Books, 1996.

Roese, John J. *Switched LANs: Implementation, Operation, Maintenance*. New York, NY: McGraw-Hill, 1998.

Rybaczyk, Peter. *Novell's Internet Plumbing Handbook*. Foster City, CA: Novell Press, 1998.

Salamone, Salvatore. *LAN Times Guide to Managing Remote Connectivity*. Berkeley, CA: Osborne/McGraw-Hill, 1997.

Saunders, Stephen. *McGraw-Hill High-Speed LANs Handbook*. New York, NY: McGraw-Hill, 1996.

Scott, Charlie, Paul Wolfe, and Mike Erwin. *Virtual Private Networks*. Sebastopol, CA: O'Reilly & Associates, Inc., 1998.

Shafer, Kevin. *Novell's Dictionary of Networking*. Foster City, CA: Novell Press, 1997.

Sheldon, Tom. *Encyclopedia of Networking, Electronic Edition*. Berkeley, CA: Osborne/ McGraw-Hill, 1998.

Shnier, Mitchell. *Dictionary of PC Hardware and Data Communications Terms*. Sebastopol, CA: O'Reilly & Associates, Inc., 1996.

Spohn, Darren L. *Data Network Design*. New York, NY: McGraw-Hill, 1997.

Sportack, Mark A. *Windows NT Clustering Blueprints*. Indianapolis, IN: Sams Publishing, 1997.

Taylor, Ed. *Network Architecture Design Handbook*. New York, NY: McGraw-Hill, 1998.

Thomas II. Thomas M. *Thomas' Concise Telecom & Networking Dictionary*. New York, NY: McGraw-Hill, 2000.

Telecommunications Industry Association/Electronic Industries Alliance. TIA/EIA-758-1, *Addendum No. 1 to ANSI/TIA/EIA-758*. Arlington, VA: Telecommunications Industry Association/Electronic Industries Alliance, 1999.

Ward, Andrew F. *Connecting to the Internet: A Practical Guide About LAN-Internet Connectivity*. Reading, MA: Addison-Wesley, 1999.

Zacker, Craig and Paul Doyle. *Upgrading and Repairing Networks*. Indianapolis, IN: Que Corporation, 1996.

Zwicky, Elizabeth D., Simon Cooper, and D. Brent Chapman. *Building Internet Firewalls*. Sebastopol, CA: O'Reilly & Associates, Inc., 2000.

Index

NOTE: Boldface numbers indicate illustrations.

Technical Publications

Since 1974, BICSI has been developing a library of technical publications, several of which have become industry standards. These reference books span the subjects of network design, outside plant design, cabling installation, and more. Many of the books double as study guides for BICSI courses and exams. BICSI members receive a significant discount on manuals. *Note: Most manuals available on CD-ROM at manual price.*

- BICSI's *Telecommunications Distribution Methods Manual* (*TDMM*), now the accepted guideline of the industry, is a valuable reference tool for those who design the telecommunications infrastructure. The *TDMM* also serves as a detailed study guide for those preparing to take the Registered Communications Distribution Designer (RCDD®) exam. US$179 BICSI member; US$329 nonmember. ISBN 1-928886-04-3 (manual); ISBN 1-928886-05-1 (CD).

- The *Network Design Reference Manual* (*NDRM*) describes all aspects of networking—LANs, local internetworks (backbone/campus), and wide area internetworks and associated telecommunications links. Content is the same as that found in *Network Design Basics* and *Networking Technologies*, published by McGraw-Hill. US$179 BICSI member; US$329 nonmember (3-ring binder version). ISBN 1-928886-10-8 (manual); ISBN 1-928886-11-6 (CD).

- Ideal for those with previous outside plant design experience, our *Customer-Owned Outside Plant (CO-OSP) Design Manual* offers an overview of outside plant design, including pathways and spaces, CO-OSP cabling infrastructure, and more. US$99 BICSI member; US$179 nonmember. ISBN 1-928886-06-X (manual); ISBN 1-928886-07-8 (CD).

- In the *Telecommunications Cabling Installation Manual* (*TCIM*), you'll find step-by-step procedures for installing telecommunications cable and useful information included in the BICSI Installer and Technician exams. Also available as a hard cover, 3-ring binder. US$99 BICSI member; US$99 nonmember (3-ring binder version). ISBN 1-928886-08-6 (manual); ISBN 1-928886-09-4 (CD).

- The newest BICSI publication, the *Residential Network Cabling Manual* (*RNCM*) provides detailed how-to information in residential voice, data, and video distribution design, installation, and systems integration. US$49 BICSI member; US$49 nonmember. ISBN 0-07-138211-9 (manual).

- BICSI's *Introduction to Commercial Voice/Data Cabling Systems* Video and Workbook provides a visual tour of voice/data cabling for the modern commercial building, as well as the spaces and systems that comprise its infrastructure. US$249 BICSI member; US$349 nonmember.

Introduction to Commercial Voice/ Data Cabling Systems Video and Workbook

- Learn design guidelines for the integration of emerging applications into existing LANs in the *LAN and Internetworking Applications Guide*. US$29 BICSI member; US$29 nonmember. ISBN 1-928886-00-0.

- The *BICSI Telecommunications Dictionary* is a compilation of glossaries from all of BICSI's publications. Acronyms, abbreviations, symbols, and international telecommunications standards can all be found in this convenient book. US$19 BICSI member; US$19 nonmember. ISBN 1-928886-03-5.

- BICSI also offers the *On-the-Job Training Booklet*, which provides a performance checklist of key cabling installation tasks—perfect for candidates studying for registration as a BICSI Installer or Technician, or for the supervisor looking to evaluate employee performance. US$9 BICSI member; US$9 nonmember.

- And finally, BICSI developed the *Telecommunications Quick Reference Guide for Code Officials: Summary and Excerpts from the NEC® 2002*, which outlines portions of the National Fire Protection Association (NFPA 70), *National Electrical Code® (NEC), 2002 Edition*. Free to both BICSI members and nonmembers.

New Member Instant Reference Library Deal

Save over US$250! For a limited time only, BICSI is offering a New Member Instant Reference Library Deal, allowing new members to purchase all of the above publications (minus the *RNCM*) with their CD-ROM counterpart at a significant savings—20% off of the discounted member price! New members who purchase the entire library within 30 days of becoming a BICSI member pay only US$995 plus shipping!

Check out BICSI's publications in more detail. Visit the BICSI Web site (www.bicsi.org) for additional information on any of BICSI's reference books. There you'll find sample chapters of all BICSI manuals, a complete price listing, and online ordering. You may also call BICSI's Customer Service Department at 813-979-1991 or 800-242-7405 (USA/Canada toll free) to request a publications brochure (also available on the BICSI Web site). An information request form may be found on the following page.

Membership

To remain competitive in our changing environment, you need to stay abreast of telecommunications issues, standards, and technology—locally and around the globe. Fortunately, BICSI is here to keep you informed and knowledgeable in all aspects of the telecommunications profession. With more than 22,000 members, BICSI offers members substantial discounts on our quality technical publications, design courses, and conference fees. In fact, the cost of membership can more than pay for itself with the purchase of just one BICSI manual.

We recognize that you may have questions about BICSI, so we encourage you to contact BICSI for a BICSI Information Packet. BICSI membership is your key to a successful career in telecommunications, and we want you to appreciate all of the member benefits.

BICSI Information Packet Available

To request a BICSI Information Packet, complete the form below and fax it to 813-971-4311. You may also request a packet by calling 813-979-1991 or 800-242-7405 (USA/Canada toll free) or e-mailing bicsi@bicsi.org.

Yes, I want to find out more about BICSI. Please send me a BICSI Information Packet:

name

address

city state zip

phone fax

e-mail

For more information or to request a BICSI Information Packet, contact BICSI today!

8610 Hidden River Parkway, Tampa, FL 33637-1000 USA
813-979-1991 or 800-242-7405 (USA/Canada toll free)
fax: 813-971-4311
e-mail: bicsi@bicsi.org
Web site: www.bicsi.org